Researching Human Geography

Keith Hoggart
Loretta Lees
and
Anna Davies

Department of Geography, King's College London

A member of the Hodder Headline Group
LONDON
Co-published in the United States of America by
Oxford University Press Inc., New York

First published in Great Britain in 2002 by
Arnold, a member of the Hodder Headline Group,
338 Euston Road, London NW1 3BH

http://www.arnoldpublishers.com

Co-published in the United States of America by
Oxford University Press Inc.,
198 Madison Avenue, New York, NY 10016

British Library Cataloguing in Publication Data
A catalogue record for this book is available from the British Library

Library of Congress Cataloging-in-Publication Data
A catalog record for this book is available from the Library of Congress

ISBN 0 340 67674 4 (hb)
ISBN 0 340 67675 2 (pb)

1 2 3 4 5 6 7 8 9 10

Production Editor: Anke Ueberberg
Production Controller: Iain McWilliams
Cover Design: Terry Griffiths

Typeset in 10/14pt Gill Light by Phoenix Photosetting, Chatham, Kent
Printed and bound in Great Britain by MPG Books Ltd, Bodmin, Cornwall

Contents

Preface

A primary reason we agreed to thump fingers on plastic keys to produce this book was the sense that students and researchers engaging with the issues of dissertation/thesis/research project preparation and implementation would benefit from a text that sought to link issues of research method with epistemological*[1] understandings. Even in writing this first sentence we are desperately aware that it is being interpreted in different ways. The word methodology* is not one that attaches itself to universal interpretation, but what word does? Here we find ourselves in strong agreement with Barnes and Duncan (1992) in acknowledging that we do not 'own' what we write. Readers impose their own interpretations on texts. These might conform or be at variance with the messages the writer intended. What readers believe is most conducive to making 'sense' of a text inevitably owes much to the cultural baggage they carry with them at the point they engage with it. For those who are new to epistemological considerations, this might seem like a strange opening for a book on human geography research. Yet it is appropriate to start with recognition of the complexity, uncertainties and at times contested nature of approaches to research. Given this recognition, one might ask what 'right' these academics have to write this volume. On this score no special claims are made. However, direct engagement with many of the issues and methods described in this volume provides a basis for our contribution. Add to this, we each bring a different perspective to the issues addressed. In part this arises from the primary emphasis in our initial training and research expertise, which inevitably impacts on perspectives, despite re-evaluation and critical self-reflection. With university careers starting at different times, we embrace initiation periods into the academic world that run from the tail end of quantification with early infusions of phenomenology and the onslaught of structuralist critiques, to postmodern and post-structural interpretations that are more sensitive to the epistemological underpinning of research methods. The insights we gained from bringing dissimilar research perspectives to this book were enhanced by different practical research experience. Anna Davies has extensive focus group and interviewing experience, Keith Hoggart has most commonly

1 An explanation or definition of concepts identified with an asterisk is given in the Glossary. Some of these concepts are contentious. The Glossary provides a guide to the uninitiated rather than a detailed examination of themes or disputes.

utilized questionnaires, in-depth interviews and official statistics, and Loretta Lees has based much of her work on interviewing, ethnography and interpretative methods. Offering a further basis for questioning interpretations and understandings, the main themes in our research have been grounded in different subfields. Anna Davies works in the environmental field, Keith Hoggart works mostly on rural and political issues, while Loretta Lees predominantly undertakes research in urban-cultural geography. All of us work largely in advanced capitalist economy settings, and these settings provide the focus for most of our comments, even if we draw in material from other contexts to illuminate themes and provide prompts for readers about lack of universality in particular trends. We stress that many of the ethical and power-relational issues of research in advanced capitalist economies are felt with heightened intensity in other settings; however, the treatment that is needed for such locations merits specialist writing.

In this book we draw on our dissimilar backgrounds, expertise and experience to draw out the varied ways in which research can be conducted, alongside recognition that methods of data collection and analysis require different interpretation depending on their underlying epistemological framework. Put simply, what information you extract from (say) in-depth interviews is likely to be quite different if your analysis is grounded in positivism compared with post-structuralism. In this regard it is appropriate to spell out for the reader the boundaries within which we operated in this book. First of all we should make clear what we see as the target audience. Here we have sought to write for those who are engaged in student dissertations/theses, whether at undergraduate or graduate level, but the book should also provide ideas and prompts for those at the start of their research careers. In addressing these audiences we have been driven by two further concerns. The first is that an understanding of research methods needs to be cognizant of the implications of their application. To put this more directly, no application of research methods takes place without it being grounded in an underlying (albeit often implicit) philosophical base. What conclusions can be derived from a research project thereby depend not simply on what data are collected and analysed but also on our understanding of what knowledge is (epistemology*) and what is 'reality' (ontology*). In this regard this book seeks to draw out the implications of research methods for two quite different perspectives on society – between those who believe there is a real world and those who believe the world is not fixed in any real sense.

Readers might think the philosophical underpinnings of world views are more complex than this simple division. You will find no argument from this quarter. However, we have not written this text as a contribution to epistemological debate. It is a book about research methods. But it is also a text that has been written partly because of a sense that too many methods texts are just about methods, shearing their application from underlying philosophical stances. Our intention is not to become embroiled in the intricacies of minute differentiation. Rather we highlight key differences, so as to reveal the dissimilar ways research methods can be used. In doing so we have sought to be critical – to pinpoint weaknesses in all approaches – in order to emphasize the merits of utilizing a plurality of methods. Added to which, the reader will find we approach epistemological ideas through a number of central themes, rather than becoming embroiled in fine-tuned distinctions. Thus, in Chapter 1 we have placed positivism and critical realism under the same heading, on account of their adherence to a similar acceptance of the existence of a real world (even if they approach its identification in different ways). We have not felt it was necessary to include sections on other 'isms' that fall within this broad framework. So when we asked ourselves whether we should include a section on structuralism, given that post-structuralism is considered, we concluded the basic tenets of structuralist ideas were sufficiently covered in the context of what we are seeking to achieve without a special section on it. Inevitably compromises have been made, given considerations of length, coherence and maintaining reader interest. It is likely that others would place their emphasis differently. This is the nature of book writing. We are not seeking to capture the world, just inform readers in helpful ways (for some at least).

Having this as a guiding principle has also impacted on the manner in which we have written the text. For those who struggle through the obscure (and for them often obfuscatory) prose of many contributions on epistemology, we hope we have been able to provide a text that is intelligible. By this we mean we have sought to write this text in as jargon-free a manner as possible. By sticking to the main themes, rather than delving into minute detail, we have also sought to pursue lines of explanation that are more easily intelligible for those who find most comfort in empirical research (rather than epistemological refinement). In this we are well aware that we will not be pleasing some readers. We make no apologies, for we see too large a gulf between the epistemological understanding of many researchers and their empirical enactment. Even if clearer lines of connection between the two result in a streamlining of argument, this is justified if it enhances empirical researchers' appreciation of broader issues of research project formulation and implementation.

More specifically, we are seeking to encourage investigators to recognize that what some see as a wide gulf, one that distinguishes those who are seeking insight on a 'real world' from those who acknowledge no such entity, is not as great as often imagined. It is certainly fuzzy at the edges. If those who approach their work from such dissimilar perspectives have a stronger grasp of the similarities (and dissimilarities) in each others' data collection and usage, this could provide more appreciative understanding of viewing issues from different perspectives. At present we see the gulf between these two 'factions' as too great. On the one side, issues of representation are too easily ignored, while on the other dilution of concern for materiality diminishes research efforts.

As a final insight on our intentions, we should lay bare the scope of this text. The words 'research methods' encapsulate activities that fall under a broad church. If you think about what materials come under this heading, you will soon appreciate that any attempt at comprehensive coverage would lead to a book many, many times the length of this one. As with issues of epistemological refinement, what we have sought to do in this book is draw the reader's attention to texts they can go to for the specifics of particular methods. For instance, we do not provide any discussion on the techniques of statistical analysis, since abundant volumes exist on this topic. What we have focused on is the nature of data used in research projects. This means we draw on issues related to how data are collected. It also means, since these are considerations that should be considered from the outset of a project, that we have deliberations on how data might be analysed and on writing up research results. In treating these issues, we illustrate principles and problematics of approaches. You will not find detailed commentary on technical issues. For this we will direct you to other work that provides the detail they deserve. The questions we wish to pose centre on setting up a project and thinking about its implementation. These questions relate to relative advantages, and to potentialities for complementarity in data forms, not to fine-scale technicalities of handling the data once obtained. In our view, the methods literature in geography is too heavily biased towards technicalities. Too often analysts appear to associate single methods with particular epistemological visions. Insufficient weight is given to how different (multiple) methods can add lustre to a research design. In providing a text we trust readers will find readable, we hope to encourage more thinking along these lines.

1
Method and methodology in human geography

To know the history of science is to recognize the mortality of any claim to universal truth. Every past vision of scientific truth, every model of natural phenomena, has proved in time to be more limited than its adherents claimed. The survival of productive difference in science requires that we put all claims for intellectual hegemony in their proper place – that we understand that such claims are, by their very nature, political rather than scientific.

(Fox Keller 1985: 178–9)

The primary aim of this chapter is to examine claims for renewed attention to questions of method in human geography (e.g. Jones *et al.* 1997). In doing so, a fundamental consideration will be the identification of links between methodology* and method* (words with an asterisk are described in the Glossary). The distinction between these is well recognized in some geographical circles, where researchers acknowledge the consequences of different epistemologies* for research practices, but in others issues of methodology are seen in the narrowly technical terms of data collection and data analysis. In the terminology used in this book, data collection and analysis are issues of method. Methodology is a more encompassing concept that embraces issues of method but has deeper roots in the bedrock of specific views on the nature of 'reality' (namely, in ontology*) and the grounds for knowledge (namely, in epistemology*). One of the distinctive features of this book is the manner in which it draws links between issues of method and epistemology. The geographical literature is already replete with contributions that focus on theoretical and epistemological issues (D. W. Harvey 1969, 1989; R. J. Johnston 1986; Cloke *et al.* 1991; J. P. Jones *et al.* 1993; G. Rose 1993; Gregory 1994; Peet 1998), without exploring in depth their implications for research methods. In similar vein, a large number of methods texts present techniques of data collection or analysis with little regard for their philosophical underpinnings (Ebdon 1985; Sheskin 1985) or present the two as somewhat dislocated (Flowerdew and Martin 1997; G. Robinson 1998; Kitchin and Tate 2000). We believe that epistemology and method are closely and complexly intertwined.

While this book focuses more on method than on epistemology, its starting point and its continuing reference point are necessarily epistemology. However, as we state in the Preface, we do not offer a comprehensive overview of all epistemological concerns.

There are many good books that focus explicitly on philosophy and theory. These issues have become increasingly important – and contentious – as geographers have come to grips with the sceptical, anti-foundational* currents of postmodernism* (Box 1.1). Unlike books about social theory, our focus is on the relationship between epistemology and research method. In so doing we seek to bring together researchers who rarely cross each others' paths. To proponents of more traditional scientific methodologies we hope to demonstrate the importance of thinking explicitly about epistemology and considering the implications of various postmodern currents for their research (Box 1.2). For those human geographers who have embraced postmodernism and pursue various anti-foundational methodologies, we hope to demonstrate the utility of traditional social scientific methods and approaches. Basically we would like the two factions to be cognizant of each others' work and to engage in sensible, open conversation.

Box 1.1 Postmodernism: a reaction against the Enlightenment Project

Modern enlightenment claim	Postmodern response
There is an independent world of objects that exists independent of the way it is represented to be (ontology).	The phenomenon of reality depends on how it is represented to be.
The rational human subject is the universal foundation for knowledge (Descartes: I think therefore I am).	The human subject is decentred, driven by all sorts of unconscious and irrational desires.
True knowledge corresponds to how the world actually is (epistemology: correspondence theory to truth).	Truth is whatever we agree to call it, there is no archimedian point from which to observe the world that is independent of it.
Observation is theoretically neutral and has no effect on the thing being observed.	Knowledge and power are mutually constituted (Foucault, etc.).
Writing is an unproblematic representation of knowledge.	Representation is *not* transparent; it inevitably influences, inflects, distorts the thing it represents.

Source: David Demeritt, designed for class in Philosophy and Epistemology, King's College London.

A primary reason why human geographers have become so concerned with epistemology in recent years is doubt over the acceptability of previous research practices. Lying at the heart of such doubts was/is a so-called crisis of representation. The reasons behind this crisis of confidence in the foundations of knowledge and the truth of representation are various but long-standing. Yet only in recent years has the social

Box 1.2 Three strands of postmodernism

1. Some emphasize new and non-universalist *models of human subjectivity* and rationality suggested by postmodernism (Pile 1991). These challenge both the neo-classical economic ideas of *homo* economicus (popular especially in economics, behavioural psychology and human ecology) and the long-standing liberal tradition of social contract theories of rights, justice and limited government. The postmodern attack on the essential universality of human nature presents problems for Marxist and feminist appeals to the common and undifferentiated interests of the working class and women. For this reason, postmodernism has been received uneasily by Marxists and some feminists.
2. Others see postmodernism as an essentially *epistemological project* attacking notions of Truth and universal knowledge. In this guise postmodernism is a challenge to so-called meta-narratives and claims to foundational knowledge (in the sense of True for all times and all places). This is perhaps the most widely accepted notion of postmodernism, wrapped up with the work of post-structuralist philosophers and deconstructionists. From this direction comes the widely proclaimed 'crisis of representation' in the social sciences. If the social sciences can no longer be said to be about representing the truth about society, then what is their purpose?
3. Others emphasize *the radical phenomenological* and social constructivist** implications of postmodernism in suggesting that our knowledge of nature, and, in some sense, the nature of reality itself (and of nature), are culturally relative. In this way, postmodernism has become embroiled in the so-called Science Wars and the debates over the social construction of science and nature. This argument is perhaps most closely associated with Baudrillard (1983). For a review of these debates, see Demeritt (1998).

Source: David Demeritt, designed for class in Philosophy and Epistemology, King's College London

science literature seen such regular and angst-ridden declarations of uncertainty about the epistemic status of its knowledge and the truth of its representations of the world. For some this uncertainty results from changes in society. Shifts in emphasis from production to consumption, the much touted decline of Fordism* and the rise of post-Fordism*, the contradictions of globalization alongside localization, and the growth of unstable, multidimensional and multifaceted social identities, have combined to challenge the foundations and principles of established interpretative perspectives. For others the weakened base of contemporary research arises less from empirical* factors than from the epistemological (and indeed political) frailties of prevailing analytical modes. It is the latter of these that is the focus of this chapter.

Fundamental to claims about a crisis of representation is the notion that there is no self-evident and universally accessible reality against which the 'truth' of research conclusions can be checked. One (but by no means the only) important channel through which such claims have emerged is feminism (I. M. Young 1990). The feminist critique has helped undermine the authority of 'conventional' social science by challenging the notion of value-free research. Feminists have demonstrated that supposedly value-free research regularly embodies gender specific values and that there is a white, Western, male, middle-class bias to most social theory, social policy and social research (Fox Keller 1985; Lather 1991; G. Rose 1993; Women and Geography Study Group 1997). Many feminists have gone a step further and claimed that objective and unbiased knowledge is impossible because our access to reality is never unmediated. Our experience of the world, they claim, depends on language and on preconceived theories and practices. This means that knowledge must always be situated and partial, in the double sense of incomplete and biased (Haraway 1991). As we will see, feminism is not the only source of these radical critiques. There are other intellectual sources as well.

These critiques are important because they challenge the philosophical foundations of scientific approaches to the investigation of human societies. As we will discuss, there is no single scientific approach. There are important philosophical differences between the empiricism* of the long-standing fieldwork tradition of human geography and the positivism* that came to prominence in geography during the 'quantitative revolution' of the late 1950s and 1960s (Billinge *et al.* 1984). What is more, the scientific underpinnings of positivism sit uncomfortably alongside the philosophical presumptions of Marxist-inspired critical realism (Sayer 1984). Despite their philosophical and political differences, the dominance of empiricism-positivism-realism in geography ensured that the discipline was wedded to the idea of itself as a science and that scientific methodologies dominated post-war human geography (R. J. Johnston 1991). This is despite the fact that the basic tenets of different scientific approaches so differ that many criticisms of positivism are as appropriately made from a Marxist or realist standpoint as they are from 'non-scientific' perspectives like hermeneutics* or postmodernism*. What this demonstrates is that there is a basic divide not simply between scientific and non-scientific approaches, but also amongst scientific approaches, and amongst non-scientific approaches.

Scientific methodologies

Positivism and the scientific method

As Hammersley and Atkinson (1995: 3) note: 'Today, the term "positivism" has become little more than a term of abuse among social scientists.' Perhaps it stretches the analogy somewhat, but the tone of this abuse is such that, for some researchers, to proclaim an adherence to positivism carries overtones similar to a member of the general public openly declaring that he or she is racist or sexist. Such extreme reactions are often born of a polemical refusal to acknowledge distinctions among different scientific epistemologies. As reviews of the scientific method make clear, what we find under this heading is a rather disparate array of philosophical ideas (J. Hughes 1980). This applies even if we focus on similar scientific traditions. For example, even amongst those who hold that methodologies from the natural sciences can be transferred to the investigation of human phenomena, there are divergent understandings of what constitutes *the* scientific method (D. W. Harvey 1969; Amedeo and Golledge 1975). Positivism itself is not a unitary approach (J. Hughes 1980).

Positivism has a long and complex history stretching back to the dawn of the Enlightenment Project in the eighteenth century. At this time, philosophers like Comte (1903) saw positivism as a methodology for applying reason to distinguish scientific truth from religious dogmatism and superstition (Kolakowski 1972). Influential philosophers, like Popper (1965), shared this faith in reason, but argued that claims about absolute truth cannot be verified. Just because all previously observed swans were white does not provide a logical basis for assuming *all* swans, even those not yet observed, are white.

Popper responded to this problem by reformulating positivism into logical positivism (sometimes also called critical rationalism). First, he distinguished analytical statements, *a priori* propositions, like mathematical proofs, whose truth was guaranteed logically, from synthetic statements that made reference to the external world based on empirical observation. Second, he argued that, although such synthetic statements could not be verified, they could be falsified. Thus he argued that science should be based upon falsification of hypotheses rather than their verification.

Despite these important differences between positivist verificationism and logical positivist falsificationism, they share certain basic tenets with each other and with other scientific methodologies. Positivism is founded on five philosophical beliefs. The first two of these are shared in common with empiricism.

1. Scientific observations are grounded in a direct, immediate and empirically accessible experience of the world.
2. Statements about those empirical observations can be made, and their truth evaluated, independently of any theoretical conclusions that might be constructed about them.

Where positivist geographers responsible for the so-called quantitative revolution of the 1950s and 1960s differed from the long-established tradition of empirical regional geography was in their commitment to the nomothetic ideal of building universal scientific laws from empirically observed particulars. Capturing the mood of the times, Bunge's *Theoretical geography* proclaimed that 'The basic approach to geography is to assume that geography is a strict science' (Bunge 1966: x), wherein: 'Regional geography classifies locations and theoretical geography predicts them' (Bunge 1966: 199).

Alongside D. W. Harvey's *Explanation in geography* (1969), Bunge provided a philosophical manifesto for the quantitative revolution in geography.[1]

Their pursuit of general explanatory theories of spatial science committed positivist geographers to several further beliefs.

3. Scientific observations had to be repeatable, and this universality was guaranteed by a unitary scientific method.
4. Science would advance through the construction of formal theories, which, if empirically verified, would assume the status of universal laws.
5. Those scientific laws would pertain only to matters of necessary fact and would be properly distinguished from normative questions of value.

Within this positivist framework, the classical formulation of the scientific method proceeds in seven analytical steps (Haines-Young and Petch 1986; see also Fig. 1.1). The starting place for these steps is the 'discovery' of scientifically derived propositions (steps 1–3), wherein it is assumed that investigators do not have theoretically derived prior expectations. The process through which data are analysed to generate theoretical propositions is called *induction**.

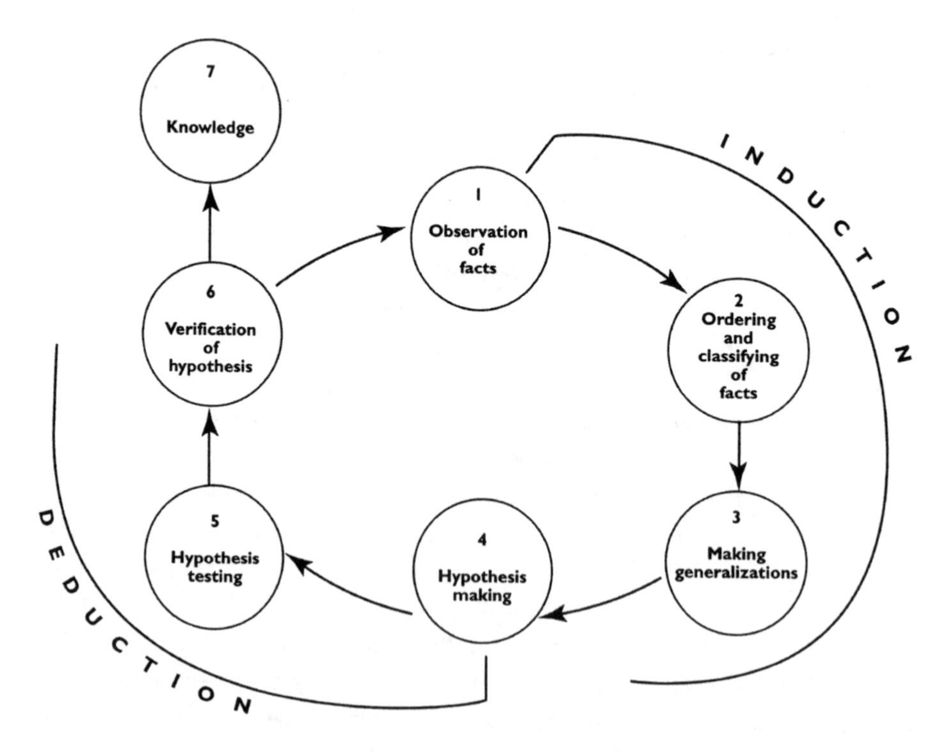

Figure 1.1 *Seven steps of the scientific method*

1 Just as the 'conversion' of these researchers to critical social theory epitomized the fall from dominance of positivist epistemologies (Bunge 1971; D. W. Harvey 1973).

The seven analytical steps of the scientific method

1 Observation of facts

Like empiricists, positivists believe that science begins with direct and immediate empirical observation of the world. For them observation is the necessary starting point for research, because it lays the groundwork both for formulating hypotheses and for testing them. Both critical realists and anti-foundationalist critics of positivism argue that this starting point is problematic because it leaves unexamined the manner in which the researcher arrives at a research problem (Gregory 1978; Sayer 1984). These critics maintain that, far from being self-evident, as sometimes claimed (J. U. Marshall 1985), empirical observation depends on theoretically determined observational categories, concepts and preconceptions that help the researcher identify 'facts' and distinguish them from 'noise' and other irrelevant information.

In contrast to naive inductionist versions of positivism, more sophisticated positivists, such as Popper (1965: 36), acknowledge that

> *at no stage of scientific development do we begin without something in the nature of a theory, such as a hypothesis, or a prejudice, or a problem ... which in some way guides our observations, and helps us to select from the innumerable objects of observation those which may be of interest.*

So positivism does not necessarily imply starting with a 'blank sheet' and making observations 'blind' (without focus is perhaps better). Instead, what is clear is that, far from being a linear seven-step process, the positivist model of research is iterative and involves tacking back and forth *hermeneutically* (see below). Analysts begin their observations and data collection based on having identified 'holes' in understanding to be clarified through future research, so science progresses though a process of bold *conjectures and refutations* (Popper 1963).

2 Ordering and classifying of facts

Having made observations, the positivist search for explanatory laws calls for efforts to classify similar events or occurrences, in order to establish the basis for hypotheses that might explain them (see step 3). We will discuss in later chapters some of the different techniques for classifying data, such as content analysis and computer-based qualitative data analysis software such as Atlas.ti. For the moment our concern is with the philosophical presumptions underwriting the recording and classification of observed facts. First, there is an assumption about the mimetic fidelity of language as a medium of representation. Positivists, like empiricists, regard our observational language as a direct and transparent representation of the phenomena it records. This understanding of language has been challenged by a variety of hermeneutic and postmodern theories.

These theories have been a driving force behind the crisis of representation we alluded to earlier and the recent turn away from scientific methodologies and towards more interpretative ones.

Second, there is an assumption that the classifications researchers devise correspond to real differences in observed phenomena, or, to use Plato's gruesome metaphor, valid classifications 'cut nature at its joints' (J. R. Brown 1994: 125). As John Marshall (1985: 117) explained: 'scientists constantly sift and sort their observation of results, arranging them into classes and hoping to find a pattern of recurring relationships among measurements of different kinds. At length a regularity emerges ...'. This understanding of the process of classification is by no means unchallenged. Critical realists pay much more attention to the practice of devising conceptual categories, as an essential, if also problematic, part of the research process. In contrast to Marshall's understanding of classification as something that the facts seem to do for themselves, critical realists insist that classification is a theoretical practice and depends in good part on the purpose of the exercise. The key factor in classification is to make 'significant' distinctions among groups of events, occurrences or processes, in order to link groupings with the uneven incidence of 'causal' forces. In idealized terms, as the identification of causal forces has not taken place at this stage, a multitude of classifications is (hypothetically) feasible. A critical question, therefore, concerns the 'accuracy' of different classifications. Here we confront divergent epistemological stances, for the emphasis on identifying empirical regularities in positivism clashes with other understandings of causality. The positivist position is clear – in order to identify causation we need to identify similar results. By contrast, critical realists hold that empirical classification is flawed because it deals only with surface appearances, rather than real causation (L. Harvey 1990: 19) and therefore runs the risk of ascribing causality to 'chaotic concepts' that are little more than contexts in which particular behaviour forms commonly arise (Sayer 1984).

An obvious geographical example is the false association of certain values and lifestyles with residence in the city or countryside (Pahl 1966). However, the fact analysts have incorrectly credited the chaotic concepts 'rural' and 'urban' with explanatory force is not necessarily an indictment of positivist methodology. Indeed, the research undertaken to refute the explanatory force of these concepts was undertaken within a positivist frame of reference. Thus, this anecdote might be cited as an example of the success of a positivist methodology in advancing hypotheses and testing them against the observed facts in accordance with the seven-step positivist procedures.

3 Making generalizations

Fundamental to positivism is the search for explanatory laws to explain and predict the behaviour of the world. In order to reach that step, researchers must first classify data according to regularities in observed behaviour. Dey (1993) compares the formulation of such generalizations to the process of putting mortar between the building blocks that a classification system lays down. As such, the process of filling in the mortar rests on comparing different ways of classifying observations so as to evaluate whether the resulting structure of generalizations about cause and effect can be linked in a logically convincing explanation of their association. The difficulty is that in the real world there are inevitable uncertainties. No single generalization will explain perfectly all the relationships between events or processes.

For example, it might seem obvious on examining a data set that men behave differently from women, but then Indians might behave differently from Pakistanis, and, while most Pakistanis in a data set are Muslim, the Indian population might embody a diversity of religions, further confounding any generalization you might make about the behaviour of the category 'Indian' or even 'Indian men'. Even ignoring complications such as age, caste or class, the search for generalization linking cause to effect requires that a variety of potential causes of behaviour differences be evaluated. This involves splitting categories and reconstituting them. As an illustration, the behaviour of women of different nationality might be examined, to establish whether nationality still discriminates once gender differences are excluded from consideration (that is, men are ignored in this element of the analysis). The process is very like completing a jigsaw puzzle, except, unlike a jigsaw puzzle, which has only one possible solution, in social research the potential number of 'pieces' – observed facts and classifications of them – is limitless and thus, so too, are the potential ways in which they can be combined.

The procedures outlined in steps 1–3 above are rightly referred to as *naive inductionism* if we do not add Popper's caveat (1965) that observation requires some preconceived ideas about what it is that is important to observe. In Dey's words (1993: 15): 'Data are not "out there" waiting collection, like so many rubbish bags on the pavement.' If we do not have a preconceived idea of what might be theoretically important, then the number of potential variables we have to observe, collect and classify just in case they might prove critical is limitless. Without prior knowledge there is no reason to discount the idea that human behaviour is decided by hair colour, eye colour, toe-nail length, ear shape, day of week, hour of birth, hour of closest sibling's birth, and you can go on *ad infinitum*. Peter Mann (1985: 20) provides an apt description, in considering what needs to be examined to understand a church service:

> *The totality of the situation would have to include not only physical description of every person present and the clothes they wore, but also every movement they made and word they spoke during the service. Obviously, then, observation of a genuine total situation is neither feasible nor is it likely to be much use to anyone.*

This does not mean flashes of inspiration never lead to original understandings. Rather it suggests that even seemingly inspirational insights are often the result of incremental changes in the knowledge base. For the vast majority of analysts approaching a research problem inductively, the search for generalization is conditioned by previous scientific knowledge. In posing questions, in defining aspects of human behaviour as 'data', and in moving to interpret the behaviour patterns they identify, investigators are influenced by evidence and conceptualizations from previous studies in allied areas (see e.g. A. M. Williams *et al.* 1989; Woods 1997). This might seem obvious but its implications are profound, for it means that the observation and interpretation of data are not independent of pre-existing understandings of how the world is organized, as naive inductionism suggests (see the discussion of hermeneutics below).

Hence, induction tends to be built on an accretion of knowledge within a particular frame of reference. This frame of reference can change over time, when its basic tenets are found wanting and more satisfactory accounts are put forward (as with 'overthrowing' the belief that volcanoes erupt when the gods are angry). When this happens, the search for generalization takes place within a new frame of reference (Kuhn 1970). It was not so long ago that those searching for cures for diseases would have been dumbfounded by the idea that viruses exist. Quite simply, an element that is today taken to be critical to our understanding of disease was not recognized in the past. For positivism this is not a problem. Scientific progress dictates that as our knowledge increases we gain more insight on the 'real' variables that distinguish behaviour patterns. Getting it wrong at one point in time does not invalidate positivism as a methodology. Indeed, for Popper it was proof of success in driving science forward.

4 Hypothesis making

We test our generalizations by formulating hypotheses and deducing the logical consequences that follow from them. Classically, hypotheses take the form of 'If X, then Y' statements that can be used to test against the empirical evidence. The implication, or in many cases the fear in students' minds, is that a hypothesis is a highly formal statement, often embodying 'grand notions' of causality and requiring rigid, formalized structures of measurement and evaluation such as those used in inferential statistical tests. Such an idea is nonsense. If, just before you turn a street corner, you suggest to yourself that 'you bet' you will meet your sister in the next street, you have generated a hypothesis (and one that is easy to 'test' by turning the corner). Now it is obvious that hypotheses can be generated about almost anything. The anticipation that if you turn a street corner you will meet your sister could be based on prior knowledge about the timing and route of her journeys home. If so, your hypothesis about meeting her is based on a generalization from previously observed patterns of behaviour (induction). To move from this stage to the setting of a hypothesis, and then on to testing that hypothesis, shifts the research strategy from one of induction to deduction*, which constitutes steps 4–7 in the scientific method.

This is an appropriate point at which to shift towards consideration of hypothesis testing, but we wish to retain the reader's attention on hypothesis formulation for a short while longer to emphasize the formulation process and to reiterate the role of previous work in generating hypotheses. Too commonly students seem to think hypotheses (or more generally research problems) should be 'original'. As we have sought to stress, this idea is based on a misperception. There are very few Einsteins in the history of the world, and the chance of real innovation in research is arguably much less likely in social science than in the natural sciences. An obvious indication of this is the fact that the writings of Sigmund Freud, Karl Marx, Adam Smith and Max Weber are still providing core underpinnings for much contemporary social science theory. The key idea to grasp is that the evaluation of existing ideas constitutes the major part of social science research, even if investigators are commonly attracted to a problem because they feel insufficient attention has been paid to a potential cause for behaviour differences or to a specific type of activity. In the research literature it is extremely rare for there to be no literature on which to derive some potential insight on that question (see e.g. Box 1.3). Rather it is common for the existing literature to offer general but not specific ideas on the question

Box 1.3 Secondary hypotheses introduce divergent research literatures

In Buller and Hoggart's work (1994) on British home buying in rural France, the start for the project was complaints in the French press about foreigners buying French land, alongside comments by local government officials in Brittany that British people were 'invading' the region. The starting assumption of the investigators was that British home-buyers were likely to be second home-owners, so the literature was trawled to establish previous findings on reasons for second home purchase. Not knowing if this was the case, a questionnaire was constructed that enabled first and second home-buyers to respond. As the interviews were completed and mail questionnaires returned, it was found that many had shifted their main place of residence to France, which meant that readings on causes of international migration came into play. As this literature is dominated by material on work-related movement, the expectations (hypotheses) derived from this literature were found to be unhelpful in explaining home-buyer actions. One reason was that around a quarter were retired (so the literature on retirement migration became relevant). Another was that, amongst those who moved their main home, many moved primarily for lifestyle reasons, with a notable preparedness to live on less bountiful and more insecure income sources. As the bulk of movement was into rural areas, this led to consideration of the counter-urbanization literature, so as to establish whether this literature helped explain home-buyer behaviour. While none of these literatures was that satisfactory, the key point is that literatures did exist that offered ideas on why people might buy homes and relocate to France. These literatures provided different hypotheses on reasons for home purchase. That these literatures posed general, then specific, questions for the study leads on to a second point to stress. This is that exploring a hypothesis is not a single event within a study.

at hand. This might mean that a precise hypothesis does not provide a starting point, or central guide, for an investigation. Yet during an investigation, as the 'story' unfolds, it is common for secondary hypotheses to be raised. These secondary hypotheses can be 'nested' within the primary research question (or even primary hypothesis).[2]

5 Hypothesis testing

Testing a hypothesis within a positivist framework involves making a deductive prediction about what will happen (the 'then') based on certain preconditions (the 'if'), then collecting data to see if the prediction is correct. One consideration here is the issue of circular logic in hypothesis testing. Particularly in statistical analyses it is important that a hypothesis be tested against data that are not used in formulating the initial hypothesis. If the goal of positivist science is generalized prediction, then the only way to test the generality of hypothetical predictions is in comparable but different situations from those used to dream up the hypothesis. This requires either collecting new data or re-examining old data in the light of hypothesized relationships.

Take the determination of house prices as an example. Models of residential location in cities suggest that land values increase as we move towards the CBD, so we would expect house prices to be higher in the central city than in the suburbs (let us exclude potential sectoral differences for illustrative purposes). This is a hypothesis about the relationship between price and location. Its general applicability can be tested by collecting data about those variables and seeing if they confirm or confound expectations. But the data used in such a test must be different from those used initially to specify the variables of our hypothetical regression model.[3] Otherwise, we have no way of knowing how generally applicable our model is. These relationships soon get more complicated because house prices are also determined by the size of a property, with larger houses likely to cost more. They can also be determined by the age of a property, as older properties require more repair work. These three considerations – location, size and age – are all related in certain ways. In general terms, the age of urban property declines as you move farther from the CBD, just as the size of residential properties has its own geography. Approaching evaluation in terms of these three factors, following the scientific method, analysts would seek to control for the effect of

2 Purists might state we have not expressed Box 1.3 in hypothesis terms. The question 'Does the literature on elderly migration in Britain help us understand elderly British home-buying decisions in France?' could be expressed as 'We hypothesize that the literature on elderly migration in Britain helps us understand elderly British home-buying decisions in France'.

3 For those who are unfamiliar with the technique, multiple regression identifies the strength of (linear) relationships between variables (see O'Brien 1992; Walford 1995). The requirements of the method are that a dependent variable is specified (variation in which is to be 'explained'), and a series of explanatory variables are specified. In simple terms, for there are many pitfalls that can mean results are statistically invalid, the technique assesses the relationship between each explanatory variable and the dependent variable, as if the effect of the other explanatory variables was 'controlled'. This indicates how a change in one explanatory variable is expected to impact on the dependent variable, assuming values in the other explanatory variables do not change.

Box 1.4 'Controlling' for multiple effects when evaluating causality

Continuing with the work of Buller and Hoggart (1994) from Box 1.3, to provide a simple illustration of the process, let us start with the hypothesis that retired British home-buyers will be able to stay longer in their French home than will those who are not retired. From the 405 households for whom information is available in this data set, 252 households had all of their adult members of pre-retirement age, while 95 had all members of retirement age. These are the two groups we will focus on here. On examining the data, we find agreement with the hypothesis, as the mean average period of time spent at a French home in the previous 12 months was 38.5 weeks for retired households and 21.6 weeks for pre-retirement households. Yet this simple relationship is potentially confusing, in that it does not take account of the role of the home in people's lives. It is to be expected that those who have moved their main home to France will spend more time there than those whose residence in France is a second home. If we want to see what effect retirement has on home usage, it is appropriate to *control* for different home roles. This point is brought home when we contrast these roles for retired and pre-retirement households, for 70.5 per cent of retired households indicated that their French home was their main place of residence, as against 37.0 per cent for pre-retirement households. To *control* for home usage, we first examine how many weeks were spent at a main home in France by retired and pre-retirement householders. Here the figures are 42.7 weeks for those of pre-retirement age and 47.2 weeks for those who are retired. Making the same computations for second home-owners, the figures are 8.5 and 15.5 weeks. In both cases, analysis of variance tests show that these differences are 0.05 statistically significant. From this it can be concluded that, irrespective of the residential purpose of their home, retired householders stay longer in France than those who are in work. Undertaking this analysis strengthens the conclusion, yet the introduction of a control for home role is only one of the potential factors that could cause dissimilar lengths of stay in France. Without giving further numerical examples, we can note that the location of employment could be an important consideration for pre-retirement householders. This is because some who lived permanently in France still gained their employment through UK-based offices (which had to be visited periodically) or from undertaking work involving a lot of travelling. If comparison is made between retired people and pre-retired householders who had (temporary, part-time or full-time work) in France, then the difference between length of residence in France for retired and pre-retirement householders disappears (for those whose main residence was in France at least). This points to the conclusion that, for those who move their main residence to France, being retired or not is not the critical factor that determines how many weeks of the year are spent in that country.

For a broader commentary on this kind of issue, which requires nothing more complex than the cross-tabulation of social survey responses, see Hellevik (1984).

each variable, while the relationship between house prices and each variable would be explored separately. Put simply, the aim is to identify whether the age of a property has an effect on its price once the impact of CBD access and property size have been 'neutralized' (i.e. assumed to have no effect). A simple illustrative example of the procedure is presented in Box 1.4.

This procedure is often used but is subject to significant criticisms. Most important in this regard is the way variables interact to give new meaning. As one example, the survival of older, larger properties close to a CBD could well provide a social cache to dwellings that cannot be captured by (even the combined effects of) access, age and size (see e.g. Firey 1947). Moreover, holding the influence of one variable 'constant' while the relationship between others is explored is regarded by some as untenable, because of the non-random distribution of variable values (Lieberson 1985). As an extreme example, we would not get far in exploring the importance of CBD access on the prices of bungalows in UK cities, since these are built in such restricted (predominantly suburban) parts of metropolitan areas. If analysts keep looking for finer and finer comparisons in data to convince themselves that relationships between two variables are 'genuine' (rather than artefacts of relationships with 'other' intervening factors), the prospects of grasping the wider theoretical picture are much diminished (Tilly 1984). Of course, the theoretical framework in which such evaluations take place, along with the understanding of causality* employed, bear on this outcome. Both are considered later in the text. For now, let us assume that the hypothesis has been tested and it is agreed that a causal relationship has been identified. In order to build towards a theory, this conclusion is in need of verification.

6 Verification of hypothesis

For positivists, the process of accepting or rejecting a hypothesis is straightforward. Starting with a hypothesis in the form of an 'if, then' statement, scientists test hypotheses by comparing what is observed (the 'then') against their prior predictions from theory (the 'if'). Empirical observation is assumed to be essentially self-evident and theoretically neutral. Therefore, the results of a test provide an independent basis on which to decide whether the hypothesis is true or not.

Critics have suggested a number of problems with this understanding of hypothesis evaluation. Popper (1963) argued that the logical problem of induction meant that there could be no grounds for verifying, absolutely, synthetic statements about the world. For Popper, hypotheses could only ever be accepted provisionally and therefore science must proceed through the falsification, rather than verification, of hypotheses. Other critics have offered more radical criticisms that undermined Popper's belief in empirical falsification as a methodology. According to the Duhem–Quine thesis, some philosophers of science argue that the outcome of any test is logically underdetermined by empirical results because the criteria for 'acceptable' support for a hypothesis and for what counts as empirical observation are conventionally determined and theoretically dependent (A. F. Chambers 1982). For instance, is a statistical study producing an R^2 value of 0.20 an adequate basis for rejecting a hypothesis? In a statistical sense this shows that 20 per cent of variation (that is, of different values) is (statistically) accounted for by variation in the values of other variables in the study. Is this a good level of explanation or not? While some statistical studies appear to accept this level of 'explanation', others do not (J. S. Armstrong 1967; Gould 1970). For the historian of science Thomas Kuhn

(1970) the basis for such divergence depends not so much on the nature of evidence as on evaluative norms within the academic community about what counts as an adequate test. These paradigms* draw heavily on prevailing views on acceptable theories and laws. Kuhn's critique of the theoretical independence of empirical observation and the value-neutrality of hypothesis evaluation has been very influential. This link is a major reason critical realists reject positivism as a methodology. Since value-neutrality is impossible, they emphasize the importance of internal reflection on and critique of theoretical concepts, categories and assessment criteria in the research process.

Supporters of positivism have responded vigorously in defence of the objectivity and rationality of hypothesis evaluation. The philosopher Imre Lakatos famously declared that Kuhn's ideas of paradigm shifts reduced science to 'mob psychology' (Lakatos and Musgrave 1974: 178). Fearing that without some universal criteria scientific truth would become a matter of raw power, Lakatos (1978: 169) sought to reformulate Popper's notion of falsification to 'give us guidance as to when the acceptance of a scientific theory is rational or not'. He noted that if a rigid – one count and you are out – interpretation of falsification is adhered to, then most explanatory ideas would be discarded. Instead he advocated a 'sophisticated falsificationism' that recognizes that we cannot expect theory to be 100 per cent accurate; some contradictory findings are inevitable. Thus the decision about whether to accept or reject a hypothesis cannot be based on a single test. It must be based on a whole 'research programme', based on a coherent theoretical core guiding the development of research as well a 'protective belt' of auxiliary hypotheses that provide a way to accommodate new, contradictory findings without rejecting the theoretical core. For Lakatos research programmes could either be progressive or degenerative depending on whether they led to the successive discovery of new phenomena.

7 *Knowledge (in the form of theories, laws and explanation)*

These contested terms of debate also feature in different understandings of scientific laws. For positivists, a law is a universally true statement about a condition of constant conjunction: 'If X, then Y.' It is the generality of laws that distinguishes them from empirical statements referring to some specific time or place. Given certain preconditions (the 'if'), a law states that some event or phenomenon must necessarily follow (the 'then'). Repeated testing provides a rational basis for believing that a hypothesis represents a law and for using it to predict behaviour. Actual usage, however, does not always live up to the rigid logic of the seven-step scientific methodology, as David Harvey (1969: 105) explained:

> *A scientific law may be interpreted as a generalization which is empirically universally true, and one which is also an integral part of a theoretical system in which we have supreme confidence. Such a rigid interpretation would probably mean that scientific laws would be non-existent in all of the sciences. Scientists therefore relax their criteria to some degree in their practical application of the term.*

For critics of positivism and the scientific method Harvey's qualification raises two key issues. The first concerns the universality or repeatability of research results. As we will see below, proponents of interpretative approaches to human geography insist that, unlike the natural sciences, the human sciences are concerned with the intersubjective understanding(s) of other people, who, unlike inert matter, respond consciously to the ways we understand them. What is important, therefore, is not explaining the universal physical properties of human behaviour but what those actions *mean* in context. Geertz (1973/1975: 6–7) tells a story about understanding three boys, one winking, another twitching involuntarily, and a third doing likewise, possibly in parody of the other two:

> *[The] movements are, as movements, identical; from an I-am-a-camera, 'phenomenalistic' observation of them alone, one could not tell which was twitch and which was wink, or indeed whether both or either was twitch or wink. Yet the difference, however unphotographable, between a twitch and wink is vast; as anyone unfortunate enough to have had the first taken for the second knows … the point is that between … the 'thin description' of what the rehearser (parodist, winker, twitcher …) is doing ('rapidly contracting his right eyelids') and the 'thick description' of what he is doing ('practicing a burlesque of a friend faking a wink to deceive an innocent into thinking a conspiracy is in motion') lies the object of ethnography: a stratified hierarchy of meaningful structures in terms of which twitches, winks, fake-winks, parodies, rehearsals of parodies are produced, perceived, and interpreted …*

Critical realists draw a somewhat different conclusion from the realization that humans are conscious subjects. For them, the difference between the natural and the social sciences is not based on an absolute ontological distinction between the 'meaningless' matter studied by natural scientists and the conscious human subjects studied by social scientists; after all, animals can also respond consciously depending on how we approach them, as anyone whose otherwise friendly dog sometimes growls at the sight of a stranger will understand. Instead, critical realists believe the key difference is between open and closed systems and the ability of different sciences to devise either empirically or (more typically) experimentally closed systems. The laws of positivism are based on the idea of constant conjunction: the consequence of a law (the 'then') necessarily follows if and only if its specified preconditions (the 'if') are met. These necessary conditions can only be met in a

4 Indeed, controlled conditions can be difficult to achieve even in the laboratory. What is usually at stake in scientific controversies are the constancy of experimental conditions and the experimental competence of the scientists to achieve them. For example, in the debate over the existence of the gravity waves predicted by relativity theory, negative experimental results were not decisive, because physicists could not decide if they were the correct result or the methods were flawed. Thus, the sociologists of science Collins and Pinch (1993: 3) conclude: 'Experiments tell you nothing unless they are competently done, but in controversial science no one can agree on a criterion of competence. Thus in controversies … scientists disagree not only about results, but also about the quality of each other's work.'

closed system, like a laboratory, where close control ensures that all factors except those of experimental interest are held constant.[4] When a system is open and other conditions are allowed to vary, it becomes impossible to tell whether the outcome of a prediction is due to the success (or failure) of the prediction or to the operation of intervening and contingent factors.

Social systems are open systems and subject to constant change. Such change severely limits the ability of the social sciences to provide laws to predict behaviour based on past experience. For instance as technology develops, will the specified conditions under which past theories operated continue to hold so that they may predict future behaviour? Compare traditional industrial location theories with their emphasis on distance to markets and raw materials and the locational possibilities afforded by the Internet, jet travel and supertankers. Precisely this understanding lies behind Glaser and Strauss's emphasis (1967) on encouraging *grounded theory*. Their claim is that the scientific method provides an inappropriate methodology for evaluating social theory. In good part this arises because of social change itself. If we wait for studies to seek confirmation of tests of a hypothesis, then our wait will be fruitless. By the time the re-tests have been undertaken, society will have moved on. The conclusions reached will be different, not because the original theory was incorrect, but because human behaviour has changed. As such, grounded theory rejects the deductive approach to theory development, emphasizing the need to use induction to generate theories of short duration and limited (geographical) validity.

The second issue raised by Harvey's comment about the scientific practice of 'relax[ing] their criteria' is the question of value-neutrality. Echoing Kuhn (1970), it could be argued that in practice what is commonly regarded as good explanation owes much to the values of the scientific community (Longino 1990: 4). Positivists reply that the progression from hypothesis to general explanatory principle is not based on the faith or will of the investigator, but on repeated and rational empirical testing (or, after Popper, repeated failure to falsify). Critics dispute this idealized image of science both theoretically (Feyerabend 1975) and empirically through actual studies of practicing scientists (Collins 1985). They charge that the insistence of positivists on maintaining value neutrality and a detached and objective distance both from the people they study and from the effects of their being studied has important political effects. Offering a taster of the basis for antagonism toward this research, Mercer (1984: 164) notes how

> *by far the most important purpose of empirical positivistic social science is the perpetuation of the status quo through the continual underscoring of the 'rightness' and 'inevitability' of existing institutions, modes of production and ways of living. At one level, then, the most dominant (positivist – empiricist) form of social science practice can be viewed as a kind of propaganda device for legitimating existing policies.*

Put simply, by emphasizing that explanation comes from existing behaviour, positivism legitimizes what is happening now, irrespective of whether present actions are determined by the grossest inequalities and discriminations. By taking what is for granted, rather than interrogating it, positivist approaches do not uncover the 'real' structural determinants of human behaviour. This critical impulse not just to study society but to change it was a starting place for the emergence of a realist epistemology.

(Critical) Realism

As we have suggested already, realism[5] differs from positivism both politically and philosophically. Whereas the positivist search for universal scientific knowledge has often had a liberal political orientation, particularly in the nineteenth century when positivist science was a bulwark against the received authority of the Church, realism has often been associated with the radical politics of Marxism* (hence 'critical' realism). As Sayer (1984: 39) explained: 'in order to understand and explain social phenomena, we cannot avoid evaluating and criticizing societies' own self understanding'. This political critique of the supposed value-neutrality of positivism is connected to a philosophical critique of the empiricist foundations of positivism.

Realists reject the positivist ideal of value-neutral or theory-free observation and the ontological distinctions between abstract, mental theorizing and concrete, empirical observation. Realists insist that empirical observation is conceptually saturated:

> *What a layperson and a biologist claim to be able to see under a microscope will differ considerably, not just in the sense that they see the same shapes but interpret them differently … but because they have learned to see or 'discriminate' different patterns in the first place. The distinction between the observable and the unobservable is therefore not simply a function of the physical receptivity of our sense organs [as empiricists would have it]: it is also strongly influenced by the extent to which we take for granted and hence forget the concepts involved in perception. (Sayer 1984: 53)*

Realists respond to the inevitable theory ladenness of observation by putting a premium on the theoretical work of conceptualization. While this explicit emphasis on theorizing may raise concerns about bias, particularly given the radical political purposes to which critical realist research is often put, critical realists insist that, since observation necessarily depends on theoretical prejudices, there can be no alternative to the ongoing critical interrogation of the conceptualizations by which we make sense of the empirical world.

5 The word 'realism' also refers to a style of literary representation, first developed by nineteenth-century novelists, that used the naturalistic conventions of omniscient, third-person narration to conceal the artifice of the author and create an illusion of lifelike immediacy. In ethnographic* writings, the conventions of realism involve a documentary style of presentation, focused on the minute, mundane details of everyday life, to create the effect for the reader of being there in person (van Maanen 1988: 51).

This methodological emphasis on conceptualization is related to a second, more ontological criticism of positivism. Positivism is based on empiricism and the ontological belief that the only things that (can be said to) exist are those that are immediately accessible to the senses (Doyal and Harris 1986). Strictly speaking, empiricists do not believe in unobservable entities or structures, like say class or quarks, because they are not directly observable. Realists charge that this positivist understanding of reality is superficial and thus so too is its conception of scientific explanation based on laws of constant conjunction and empirical regularity. The problem with positivist laws of constant conjunction is that they tend to commit the fallacy of affirming the consequence. Positivists assume that if they follow their seven steps to affirm the hypothesis 'If X, then Y', this shows X causes Y. But, as Sayer (1984: 110) counters, 'what causes an event has nothing to do with the number of times that it has been observed to occur'. Instead realists like Sayer subscribe to a 'depth ontology', encompassing the ontologically necessary structures of causation lying beneath the surface appearance of things. From this perspective, realists regard laws as statements of causal necessity rather than of universal empirical regularity.

Realist research seeks to disclose ontologically necessary structures of causation. It does so through a combination of extensive research, designed to determine how widespread a phenomenon is, and intensive research, designed to identify through conceptual abstraction underlying causes for why and how it happens. To illustrate these distinctions, it is appropriate to direct the reader to a specific example (Box 1.5). Extensive research tends to involve quantitative methods and intensive research qualitative methods, but these are (to use a realist terminology) contingent as opposed to necessary relations. The intensive research of conceptualizing underlying causal structures does not necessarily involve qualitative methods, nor does the use of qualitative techniques necessarily imply a commitment to intensive research.

Though critical realism has had its greatest influence on the social sciences, its concern for disclosing the ontologically necessary structures of causation is not unique to them. Indeed, Roy Bhaskar (1978), one of the leading philosophers of realism, has proposed a *Realist theory of science* that would unify the social and natural sciences around the methodological principles of realism. Recently, physical geographers have engaged in philosophical discussion of realism (Richards 1990), but in many ways this represents a convergence of independent research interests, rather than direct influence. Long before realist philosophy became popular in geography, the geomorphologists Brunsden and Thornes (1979) were calling for the need for physical geographers to penetrate 'behind' the external appearances of phenomena and identify the essential mechanisms of causation.

Aside from these philosophical differences, what does the practice of critical realist research involve? How is it different from positivism? One way to answer these questions is to consider some specific examples (Box 1.5). Yet, while examples tell us something about what realist researchers believe, they are often not helpful in explaining how to do

Box 1.5 Realism and contingency in the farm structures of three US regions

In the United States, three different agricultural systems have dominated much of the last 100 years (Pfeffer 1983). California has some extremely large farms, with many farm labourers, most with short-term contracts. On the Great Plains, family-owned farms dominate, with few farm labourers. In parts of the Old South, the agricultural system has been characterized by share-cropping, with tenant farmers often reliant on landowner credit and/or paying land rent as a share of their farm output. In positivist terms, these agricultural systems do not possess the uniformity that would suggest a shared cause (D. W. Harvey 1969).

This understanding is rejected by realists, who, like Pfeffer (1983), identify common explanations behind these very different farm systems. The reason dissimilar systems are said to have emerged is not because what caused them is different but because the circumstances in which they originated induced the same cause to have a different outcome. This is known as a contingency* effect. In the case of US farm systems, the distinctive form of the California system arose because of the (illegal) retention of large Spanish landholdings after the region joined the Union. These were farmed as large operations, with profits being maximized through the use of cheap imported labour. Over time this has come from a variety of source regions, with China, Japan, the 1920s Dust Bowl states and, more recently, Mexican *braceros* providing a supply of poorly paid workers, whose employment rights were regularly broken, owing to their weak bargaining position as immigrants or poverty-stricken, landless itinerants. The contrast with the US South could not be more stark. Here, another readily accessible workforce was available, which likewise occupied an insecure labour market position. However, a key difference was that, after the Civil War, former slaves were not inclined to return to work on plantations, and the plantation economy was threatened. With white workers unwilling to take on plantation work, and immigrants drawn to better-paid city work or to farm land in the West, without their former slaves plantation-owners would be unable to make an economic return on their land. Their solution was to entice black workers to produce commodities for them by offering land for rent. As former slaves had insufficient resources to pay for the land, animals and equipment needed to establish a farm, landowners agreed to take a share of their crop in payment. With landowners having best access to supplies and markets, an exploitative labour relation soon developed, which saw large profits accruing to landowners. The same option was not available on the Plains. There were still large landowners, as railroad companies and a few others had acquired very large land tracts. However, with no large local workforce, and given the wages they offered, landowners were unable to lure workers, who preferred higher rates of city pay or, in the case of former Union troops in the Civil War, could obtain their own land for free. To maximize profit on the land they held, the best solution was not to farm but to sell the land in small plots as family farms.

Landowners in three different regions started with very large land tracts. This common beginning was associated with different outcomes, yet large landowner motives were the same in each case – profit maximization. Stated more broadly, the driving force for landowner actions are the structural demands of capitalist production relations. These are 'structural' in the sense that they impose on participants (those who wish to be engaged in such activities) conditions for staying in operation. Failure to comply with these 'rules' commonly results in 'removal' from the stage, with others willing to take vacated places; seizing on possibilities for profit that become apparent. As these opportunities cannot be determined *a priori*, the causal impetus behind structural forces is dynamic and unpredictable (see e.g. N. Smith 1984).

realist research, as Yeung (1997) explains in a review. Yeung suggests that realism is a philosophy more than a methodology. Despite empirical studies proclaiming a realist logic (Allen and McDowell 1989; Massey and Wield 1992; A. C. Pratt 1994), Yeung (1997: 56) finds existing work 'opaque' on method. The problem is that a satisfactory link between philosophy and method has not been established. Indeed, as 'reality' is taken to be beyond direct human consciousness, it is conceptualized as an abstraction. Yet critical realists have provided little by way of guidance on how such abstractions are to be theorized or tested (A. C. Pratt 1995). This brings us back to the falsification ideas of Popper (1965). Until a methodological approach is developed that provides guidance on how to identify, and when to accept or reject, theoretical propositions, the realist approach will not offer much of a practical guide to doing research.[6]

That stated, the underlying ideas of realism are close to the hearts of many geographers. Moreover, the stratified conception of an ontologically deep reality wherein the 'facts' we observe are just surface impressions, caused by underlying processes of ontological necessity, has already scored highly as an explanatory framework in other scientific endeavours. Molecules, DNA, proteins and viruses all qualify as an underlying reality with real power to explain (Norris 1995). Discovering these previously unimagined entities offered an explanation for processes that had not previously been understood. This is the attraction of critical realism. Especially for those for whom positivism is anathema, realism offers an interpretative vehicle that does not assume 'reality' is unique to each individual. Moreover, in so far as it seeks to understand the links between investigated behaviour and broader social structures, realism focuses on assessing whether observed social processes are mediated by often unquestioned structures within which people make behaviour decisions. Despite a lack of guidance on how to conceptualize these structural conditions, realism provides scope for the combination of a variety of research methods, from in-depth ethnographic enquiry of small groups (Willis 1977) to statistical insights on the context in which behaviour decisions are made (Crenson 1971). In both cases, the hallmark is to go beyond surface impressions (L. Harvey 1990).

6 As Newby (1986) points out, the geological metaphor proposed by Massey (e.g. 1995), whereby social change is conditioned by the social environment that existed prior to the change process beginning (much as an existing geological stratum impacts on new landform processes, by offering differential 'resistance' to the new 'imposition'), is only a metaphor. It is a descriptive statement, not an explanation of why places change in particular ways.

Despite these political and philosophical differences, positivism and realism share a number of points in common.

1. Both claim to provide a general philosophical basis for research that applies equally to the physical and the social sciences.
2. Both believe in a reality that is ontologically independent of the consciousness of human action.
3. Notwithstanding the realist critique of empiricism, both emphasize the role of empirical observation in providing a foundation for distinguishing true from false belief about that ontologically independent reality.

Each of these points is contested by interpretative and anti-foundational approaches to geographical research that in different ways seek to go beyond scientific representation.

Hermeneutics and interpretative human geography

In contrast to empiricists, positivists and critical realists, many others insist human geography should 'not be an experimental science in search of law but an interpretative one in search of meaning' (Geertz 1973/1975: 5). This concern with the interpretation of meaning is sometimes called the cultural (because meanings are cultural) or linguistic (because they are communicated linguistically) turn. By whatever name, this concern with the interpretation of meaning has been hugely influential not just in human geography but across the social sciences. Methodologically this cultural turn has involved an embrace of various qualitative methods, such as discourse analysis and semiotics (Chapter 4), in-depth interviewing (Chapter 6), and ethnography (Chapter 7). Politically it has been associated with the development of cultural, feminist and post-colonial geographies that seek to give 'voice' to the different understandings and perspectives of marginalized 'others' (Philo 1992). Conceptually, the cultural turn has been fuelled by an ontological understanding of the world as meaningful and therefore textlike, in the sense that its meanings must always be interpreted. Inspired by such metaphors of cultural production as text, theatre and icon (Demeritt 1994), cultural geographers have turned away from the scientific conception of cultures, landscapes and other objects of geographical knowledge as things to be explained. It looks at them instead as symbols whose meaning must be interpreted (Daniels and Cosgrove 1988; I. Cook *et al.* 2000).

This ontological understanding of the world is heavily indebted to hermeneutics*. Hermeneutics is the study of interpretation and meaning. In contrast to the liberalism associated with positivism and the political radicalism associated with realism, hermeneutics has often had a conservative political orientation, both because its historical origins lie in the Romantic reaction to the Enlightenment and because it tends to privilege the interpretative authority of experienced experts over novices. Hermeneutics first rose to prominence as a philosophy for interpretation in the late eighteenth and early nineteenth centuries among biblical scholars struggling to clarify the word of God as expressed in ancient manuscripts. Its principles were soon generalized, particularly in the German-speaking world, beyond texts to the understanding of art (Panofsky 1962), history (Dilthey 1988) and sociology (M. Weber 1949), even to *The interpretation of dreams* (Freud 1976). All of these culturally meaningful objects can be understood hermenutically as texts to be read and interpreted (Ricoeur 1971).

The hermeneutic model of interpretation recognizes that, in trying to understand the meaning of something, we bring to it a whole set of preconceptions. These preconceptions provide a context in which we make sense of and interpret the meaning of the text. Interpretations are formed by tacking back and forth between our evolving contextual preconceptions of the text and the text itself, and between individual parts of the text and the text as a whole. This iterative movement between text and interpretative context is called the hermeneutic circle (see Fig. 1.2).

Figure 1.2 *The model of the double hermeneutic. Source: David Demeritt, designed for class in Philosophy and Epistemology, King's College London*

Hermeneutic interpretation demands of the researcher explicit reflection on one's biases and preconceptions as an integral part of the research process. As Duncan and Ley (1993: 4) explain:

> *Rather than setting up a model of a universal, value-neutral researcher whose task is to proceed in such a manner that s/he is converted into a cipher, this approach recognizes that interpretation is a dialogue between one's data – other places and other people – and the researcher who is embedded within a particular intellectual and institutional context. It is precisely the interpersonal – and intercultural – nature of the hermeneutic method which poses a challenge to mimesis, since a 'perfect copy' of the world clearly is not possible if the interpreter is present in that textual copy.*

This practice of reflecting upon and questioning the assumptions and statements of one's own research is termed reflexivity* (Bassett 1995). To be clear, reflexivity does not simply involve asking whether the research could have been improved upon or whether someone else might have come to a different conclusion. It embodies questioning research issues posed by the investigator, the interpretative biases of the researcher, and the way in which the identity of the researcher impacts on the research process. One researcher who has shown sensitivity to these possibilities is John Western. In discussing his research on Cape Town under the apartheid regime, he observed that: 'An investigator with a different personality make-up, a different set of basic beliefs, and particularly (I sometimes think) a female investigator, would have come up with an alternative *Outcast Cape Town*' (Western 1986: 27). Although the importance of such candour has long been recognized (Myrdal 1969), it has not always been practised. For instance, Baxter and Eyles's review (1997) of 31 qualitative studies in human geography found only 10 studies that indicated how informants were selected, virtually no discussion of the principles on which evidence was assessed and presented, and only five studies that mentioned how they verified findings.

Hermeneutic methodologies are by no means alone in demanding such reflexivity. In their own ways, both positivism and critical realism also put great store in the self-awareness and critical reflexivity of the researcher. Indeed, a number of philosophers have argued that the natural sciences are also hermeneutic, in so far as they involve scientists in an iterative move between their contextualizing prejudgements about what the evidence means and the evidence itself (Rouse 1987). What then distinguishes hermeneutic human geography from other methodologies?

Epistemology of creative interpretation

In the hermeneutic tradition, meanings clarified through the hermeneutic circle of interpretation are not found, they are made. As Geertz (1973/1975: 15) explains, interpretations 'are thus "fictions"; fictions in the sense that they are "something made", "something fashioned" – the original meaning *fictio* – not that they are false, unfactual, or merely "as if" thought experiments' with no reference to the material world to which they refer and represent. But from a hermeneutic perspective the active involvement of the researcher in the fashioning of interpretation, far from invalidating the resulting representations, is the precondition for true understanding, which is imagined as the outcome of an ongoing dialogue and engagement between the researcher and the meaningful objects he or she studies. Thus hermeneustics offers a very different perspective on truth from empiricism, positivism and critical realism, philosophies that all distinguished sharply between fact and fiction. From a hermeneutic perspective, these distinctions are more problematic. Indeed, as Haraway (1989: 3–4) notes fact and fiction both share the same Latin root, *facere*, to make:

> *It seems natural, even morally obligatory, to oppose fact and fiction; but their similarities run deep in Western culture and language. Facts can be imagined as original, irreducible nodes from which a reliable understanding of the world can be constructed. Facts ought to be discovered, not made or constructed. But the etymology of facts refers us to human action. … A fact is a thing done, a neuter past participle in our Roman parent language. … Fiction can be imagined as a derivative, fabricated version of the world … or as an escape through fantasy into a better world than 'that which actually happened'. … We hear the root of fiction in poetry and we believe, in our Romantic moments, that …fiction can be true … fiction seems to be an inner truth which gives birth to our actual lives. This, too, is a very privileged route to knowledge in western cultures. …*
>
> *Fiction's kinship to facts is close, but they are not identical twins. Facts are opposed to opinion, to prejudice, but not to fiction. Both fiction and fact are rooted in an epistemology that appeals to experience. However, there is an important difference; the word fiction is an active form, referring to a present act of fashioning, while fact is a descendant of the past participle, a word form which masks the generative deed or performance. A fact seems done, unchangeable, fit only to be recorded; fiction seems always inventive, open to other possibilities. … But in this opening lies the threat of merely feigning, of not telling the true form of things.*

A number of further implications follow from this hermeneutic approach to understanding and truth.

Anti-scientistic humanism

As a methodology hermeneutics stands opposed to claims for a scientific geography unified, as both positivists and realists would have it, by a universal method. Instead, proponents of hermeneutics make a strong separation between the methodologies appropriate to the human and natural sciences. This distinction is based on the ontological differences between their respective subject matter and the different methods appropriate to understanding them. Human geographers study the conscious and meaningful human actions of people who change in response to their own understandings of how they are understood. As a result, social scientists are involved not in a single hermeneutic between their own preconceptions and the unchanging thing-itself they are studying, but in a more interactive and dialogical, double hermeneutic with other conscious subjects (Fig. 1.2). Understanding in this context must necessarily be mutual and intersubjective. It requires a hermeneutic methodology very different from the positivism applicable to the natural sciences.[7]

By engaging with the human subjects they study on such a subject-to-subject basis, hermeneutic social scientists are involved in a very different relationship from the subject-to-object relationship of positivism, as Plummer (1983: 77–8) explains:

> *Many sociologists [or geographers] start with a view of the person as an active, creative, world builder but before they have completed their theoretical endeavours they have enchained, dehumanised, rendered passive and lost that same person. The subject has become the object, the person has become the statistic, the creative has become the constrained, the human being has become an abstraction.*

Thus hermeneutics is based on humanism and its commitment to some essential unity of the human species. This essential unity is what makes intersubjective dialogue, translation and interpretation possible. Although proponents of hermeneutics recognize social and cultural difference, most believe that the hermeneutic methodology of intersubjective dialogue provides a way to bridge cultural, historical and other differences, and make mutual understanding possible.

This ideal and its humanistic foundations have been challenged on a number of fronts. Structuralists, often pursuing a critical realist methodology, have criticized its model of human agency for ignoring the structures limiting free choice and determining human action. Likewise psychoanalytic critics have questioned the humanistic idea of a conscious human subject (Pile 1991), insisting that many actions are not consciously symbolic, and therefore amenable to hermeneutic interpretation of implicit meaning, but habitual or driven by unconscious drives and desires. These psychoanalytic arguments have been influential in moves within geography to go beyond the prevailing concern in the discipline with the representation of meaning and engage with so-called non-representational issues. Post-structural critics argue that hermeneutics does not represent the 'other' so much as construct it discursively. Post-colonial critics have sought to take this insight a step further by deconstructing the ethical and political effects of anthropological studies of the 'other' (Clifford 1988). Their work has had an important influence on the practice of ethnography (Chapter 7).

---

7 For critical realists both the natural and the social sciences, properly understood, use the same hermeneutic methods to conceptualize abstract underlying structures of causal necessity.

Cultural relativism

Proponents of hermeneutics have usually seen their role as fostering mutual understanding rather than judging or seeking to change research subjects. This is quite different from critical realists and advocates of so-called Action Research (Chapter 7) for whom the purpose of research is to effect change in the lives of the people they study. Though realists and action researchers may well employ hermeneutic *methods* of interpretation, they do not accept the non-judgemental ethic of non-intervention and cultural relativism that is the foundation for hermeneutics as a *methodology*. As a methodological principle, the hermeneutic ethic of cultural relativism has an important practical-instrumental dimension, in so far as tolerance of others is usually necessary to get them to engage in dialogue with you. It is difficult to understand the meaning and cultural significance of Voodoo for Haitian immigrants if from the outset you dismiss their beliefs as superstitious nonsense with no basis in reality (K. M. Brown 1991). Cross-cultural understanding is possible only if you accept other viewpoints on their own terms and refrain from judging them.

This methodological principle is sometimes called the 'symmetry principle' (Bloor 1976: 4) because it demands that investigators describe beliefs generally held to be true in the same symmetrical terms as those generally held to be false. Such agnosticism about truth is often mistaken for epistemological relativism. Although proponents of hermeneutics may refuse to judge the truth of their informants' beliefs, they do apply certain standards for evaluating their own work. Hermeneutic interpretations should be judged by the faithfulness and fidelity with which they represent their informants' viewpoints (Duncan and Ley 1993). The difficulty, which we address when we turn to post-structuralism and deconstruction, is that an interpreter's representation does not simply re-present an informant's views; it also constructs them textually. Plummer (1983: 123) discusses this difficulty in the context of oral history. Although designed to let the subject describe his or her own life, the resulting text is always shaped somewhat by the researcher: 'The very process of allowing the subject to ramble on about his or her own life will confirm the tenets of interactionist theory about the rambling and negotiable aspects of life.' Researchers necessarily make decisions on what aspects of the informant's 'story' to exclude, as well as deciding when to end the interview text.

For many proponents of hermeneutics cultural relativism is not simply a methodological principle for research, but an ethical end in itself. Cultural relativism has had the important effect of unsettling Western ethnocentrism among geographers and anthropologists studying other cultures. In the late nineteenth century, it became an increasingly important principle for professional social scientists studying other cultures to distinguish themselves from 'mere' travellers whose descriptions of so-called primitive peoples were written from a judgemental position of assumed cultural and evolutionary superiority. As such, professional geographers and anthropologists tried to avoid judging research subjects. In this way, they sought to identify themselves as scientists committed to the kind of value-neutrality long heralded by positivism (Stocking 1987). But ironically, as Clifford (1988) has noted, the anthropological commitment to the liberal ethic of tolerance for the 'other' is, itself, a kind of ethnocentric value that was imposed tacitly by anthropologists on their research subjects. Other critics of hermeneutics complain that in a contemporary context its cultural relativism towards others and refusal to judge their beliefs is an abdication of moral and epistemological responsibility (cf. Norris 1995). Consider a research project into female circumcision. Although illegal in the UK, this practice is still relatively common among immigrants from East Africa. It is common practice to afford anonymity to your research informants, but what if they are engaged in the illegal activity of circumcising young girls? Should you report them to the police in order to protect the girl? How far should liberal tolerance for others be extended?

Linguistic constructionism

Linguistic constructionism suggests the difficulty of ever stepping outside the received linguistic categories through which we make sense of the world. It challenges the self-evidence of sensual perception that was the foundation for empiricism and positivism. While critical realism acknowledges that our perception of reality depends upon our linguistic and conceptual constructions of it, they insist that we can test the accuracy of our representations of the world by comparing them with how the world actually is. By contrast, the linguistic constructionism of hermeneutics is more sceptical that it is possible to sit outside language.

Hermeneutic approaches emphasize the importance of language, not just in defining us as human, but also in constructing what we take for 'reality'. In this regard, Berger and Luckmann's *The social construction of reality* (1967) was hugely influential for the sociology of knowledge, and later geography. They argued that the 'taken-for-granted' worlds we live in are not natural 'objective' phenomena, but depend on the relations a person has with a place and the people met there. As such, 'social reality' is socially constructed intersubjectively through social practices and linguistic categories that the researcher can recover using a hermeneutic methodology. Berger and Luckmann's initial formulation has perhaps come to be seen as rather limited and unreflexive, but this volume did put social constructionism* on the research agenda (Box 1.6)

A great deal of ink has been spilled over this term and so it is important to be clear about just what is being constructed (Demeritt 2001c). What Berger and Luckmann (1967: 169) and others who have followed in their wake refer to is the construction of *social* reality. This does not imply any ontological commitments about the construction of gravity and what might be called 'brute reality'. From this perspective social problems can be understood 'as products of particular constructions of social reality, rather than necessarily of actual physical conditions' (Spector and Kitsuse 1987: 38). This distinction between 'social reality' and 'actual physical conditions' is important because it signals the cultural relativist's agnosticism about both the existence of social problems under investigation and the truth of any claims made by informants about them. Rather than judging whether the problem was 'real', as positivist or critical realist methodologies

Box 1.6 A typology of social constructivisms

	Commonsense realism	Social object constructivism	Social institutional constructivism	Artefactual constructivism	Neo-Kantian constructivism
Chief tenets	Observational statements refer directly to a pre-existing, independent, and, in this sense, objective reality.	Taken-for-granted beliefs about reality, e.g. gender, constitute a social reality no less 'real' in its causal effects than reality itself.	Science is a social construction in the sense that its institutions and the social contexts of its discoveries are socially conditioned and constructed.	The reality of the objects of scientific knowledge is the contingent outcome of social negotiation among heterogeneous human and non-human actors.	The objects of scientific thought are given their reality by human actors alone.
Key proponents	Gross and Levitt (1994)	Berger and Luckmann (1967)	Merton (1938/1970)	Latour (1987); Haraway (1992)	Woolgar (1988); Collins and Pinch (1993)
Ontology	Nature/society, subject/object, mind/matter are ontologically distinct realms.	Socially constructed reality is distinct from objective facts given by nature, e.g. sex.	Objective reality is distinct and independent from beliefs about it.	No absolute ontological distinction exists between representation and reality, nature and society.	Nature is whatever society makes of it.
Epistemology	Truth value is determined by correspondence between representation and reality.	Scientific truth is explained by nature; socially constructed belief is the cause of scientific falsehood.	Ignorance and socially constructed bias explain belief in scientific falsehood.	Ultimate truth is undecidable.	Truth is what the powerful believe it to be.

Source: Demeritt (1998).

would demand, a hermeneutic approach would 'attempt to account for the emergence, organization, and maintenance of claims making activity' – the linguistic categories and rhetorical practices – through which social problems are linguistically identified and constructed *as problems* (Burningham and Cooper 1999: 304).

While realists recognize that we construe our world through our language and concepts, and thereby 'construct' a 'social reality', they are uncomfortable with the ontological idealism that hermeneutics sometimes seems to imply about reality. Critical realists insist on upholding 'the difference between the acts of material construction and the acts of construing, interpreting, categorizing or naming' (Sayer 1997: 468). Just because you may believe that by flapping your arms you will fly like a bird does not necessarily make you do so. However, such arguments against the ontological idealism of linguistic constructionism work more readily with such 'death-and-furniture-type' examples than with social phenomena like social class whose ontological existence is what Sayer (1984: 30) calls 'concept dependent'. Careful thinkers, like Sayer (1984), recognize these distinctions. But in the furore over constructionism these distinctions have sometimes been lost, and the linguistic construction of hermeneutics is at times identified with some kind of all purpose relativism. For instance, the physicist Alan Sokal (1996: 63–4) complains:

> *Intellectually, the problem with such doctrines [as constructionism] is that they are false (when not simply meaningless). There is a real world; its properties are not merely social constructions: facts and evidence do matter. What sane person would contend otherwise? And yet, much contemporary academic theorizing consists precisely of attempts to blur these obvious truths. ... Politically, I'm angered because most (though not all) of this silliness is emanating from the self-proclaimed Left ... For most of the past two centuries the Left has been identified with science and against obscurantism ... The recent turn of many 'progressive' or 'leftist' academic humanists and social scientists toward one or another form of epistemic relativism betrays this worthy heritage and undermines the already fragile prospects for progressive social critique. Theorizing about the 'social construction of reality' won't help us find a cure for AIDS or devise strategies for preventing global warming.*

For Sokal and others on the traditional left, the political problem with social construction talk is that it leaves no epistemologically secure foundation from which to speak 'the truth'. The fear is that the powerful stand to gain the most from such anti-foundationalism. But it is by no means clear that hermeneutics implies such hostility to truth claims.

Does hermeneutics provide a methodology for uncovering true meanings?

This question has long been debated among proponents of hermeneutic approaches to social science. Historically there have been two broad schools of thought. The first, associated with the scientific claims of Freud and Weber, insists that hermeneutics offers a methodology for achieving metaphysically objective, universal and true knowledge of human societies. Though different in kind, such hermeneutic knowledge is no less universally true or epistemologically secure than that of physics. Proponents of scientific history insist that through painstaking archival research with primary sources (Chapter 4) the historian can come closer to converging on a true, objective understanding of the past and be able, as the late-nineteenth-century German historian Leopold von Ranke famously put it, 'to say how it was' (quoted in Novick 1988: 26). Although the resulting histories would be constructed 'fictions', they are not judged as such:

> *The stories we write ... are judged not just as narratives, but as nonfictions. We construct them knowing that scholars will evaluate their accuracy, and ... judge the fairness and truth of what we say. Because our readers have the skill to know what is not in a text as well as what is in it, we cannot afford to be arbitrary in deciding whether a fact does or does not belong in our stories. Someone among our readers ... will eventually inform us of our failings ... Criticism ... keep[s] us honest by forcing us to confront contradictory evidence and counternarratives. We tell stories with each other and against each other in order to speak to each other. (Cronon 1992: 1373–4)*

The iterative procedures of hermeneutic interpretation – scholarly contextualization and interpretation of histories and the primary sources they re-present – provides historians with a secure epistemological basis for distinguishing good histories from poor ones, thick descriptions from thin ones.

While this first understanding of hermeneutics sees its procedures as a route towards scientific truth, it also understands the process as ongoing and never finished. As Geertz (1973/1975: 29) explains: 'Anthropology, or at least interpretative anthropology, is a science whose progression is marked less by a perfection of consensus than by a refinement of debate. What gets better is the precision with which we vex each other.' Geertz's image of endless scholarly discussion and debate suggests common ground with the vision of truth – as what is accepted through unforced agreement – advanced by the pragmatist philosopher Richard Rorty. For Rorty (1991: 23, 22), ' "knowledge", like "truth", is simply a compliment paid to the beliefs we think so well justified that, for the moment, further justification is not needed'. The only distinction pragmatists like Rorty recognize 'between knowledge and opinion ... is simply the distinction between topics on which agreement is relatively easy to get and topics on which agreement is hard to get'. Positivists often express great discomfort with this anti-foundationalism. But, as Demeritt (2001b) notes, there are important similarities between pragmatism and the later work of Popper, for whom the epistemological authority of science rested on a social process of organized scepticism and a vision of knowledge as conditional – the best we can do for the moment.

A second, more radical reading of hermeneutics is associated with the deconstructive post-structuralism of Jacques Derrida (for an accessible introduction, see Sarup 1993). According to Derrida, hermeneutics is founded on the assumption that there is some fixed and original meaning to recover through its procedures of interpretation. Derrida connects this assumption to a long tradition of Western metaphysics, which conceives the world in dualistic terms, prioritizing speech as a truer and original expression of thought over writing. This binary opposition of speech to writing is hierarchical and analogous to other hierarchical dualisms, such as nature/culture, presence/absence, signified/signifier, meaning/text, essence/appearance, original/copy, mind/body, spirit/matter, essence/form and so on. In each case the privileged first term of the binary provides an unquestioned foundation for truth and meaning. Think of the Cartesian *cogito*, the idea that the ground for truth and knowledge is centred in the individual's mind ('I think therefore I am'), which is the basis for empiricist and positivist epistemologies. Through a method of careful reading (Chapter 4), Derrida argues that these binary oppositions can be deconstructed logically to show how definition of the privileged first term actually depends upon the existence of the second, and thus that the first term cannot provide a logical foundation for truth.

The philosophical claims of post-structuralism are complicated, but, to get a sense of how deconstruction works and how it challenges the idea of a fixed, original meaning, imagine being in a foreign country and hearing a word whose meaning you do not understand. You might consult a foreign-language dictionary to learn the meaning of the word. The fact that the original meaning of a spoken word is specified in writing already points to the instability of the hierarchical opposition of speech to writing that Derrida deconstructed. But it is possible to push his argument about the instability of meaning farther. If you open the dictionary you will find more than one definition of the word, as the original meaning of the word depends on context. To understand differences between the various definitions the dictionary associates with the single word you are checking, it is common to have to look up these other definitions. Any definition of a word calls up other words and associations that complicate the first definition you find. As the meaning of the first word is endlessly deferred in a never-ending cascade of other possible supplemental meanings, the original meaning you searched for is never fully present. It is not that signs (signifiers) are meaningless; rather they mean too much. Thus there can be no final end, or ground, that guarantees the ultimate meaning of language.

This more radical take on hermeneutics has often been accused of relativism. It is commonly criticized in association with the more nihilistic strands of postmodernism. But such critiques are often simplistic, for postmodernism means a variety things. Indeed postmodernism is not a very helpful term. As Dear (1986) points out, the umbrella of postmodernism can refer to the period or epoch (Ley 1987), to a style, such as postmodern architecture (Jencks 1991), or to beliefs, theory or epistemology (Cloke *et al.* 1991). It is the latter that we have tended to focus on, for this body of ideas rejects the totalizing (abstracting and generalizing) ambitions of social science.

For geographers, this second reading of hermeneutics complicates the turn towards qualitative methods as the best way to interpret human societies. In effect, this hermeneutical position sees the way to understand a text (or any other human creation) as putting it into an interpretative context that makes it meaningful (to you). Thus, through the successive contextualization of texts, hermeneutics has sought to recapture the meaning of a text as intended by its author (Lees 2001). But, after Derrida, hermeneutics cannot be seen to provide a methodology for decoding that original intention, for what basis can there be for distinguishing good readings from bad ones?

Geographers have been understandably discomforted by this realization. As a result many have tried to limit the radicalism of post-structuralism. Typical here would be Duncan and Duncan (1988), who try to ground and limit the potential range of meanings of texts by invoking social factors and power. The basis for such critiques is usefully articulated by Hoy (1985/1990: 44):

> *[Derrida] … focuses not on the central ideas or arguments, but on marginal metaphors and other rhetorical devices that most interpreters gloss over. As opposed to interpreters who purport to enable us to read the text, Derrida would make us unable to read it. Instead of assuming the text succeeds in establishing its message, Derrida's strategy is to get us to see that it does not work. Instead he does not reconstruct the text's meaning, but instead deconstructs it.*

Such debate need not detract from the insight Derrida's method of deconstruction can yield (Doel 1999). As a methodology it has proven important in helping to deconstruct received assumptions about the world and its ultimate meaning.

Beyond representation

More recently, concern has again been voiced at the limits of representation encapsulated in the hermeneutic tradition. In particular, critics charge that the interpretative tradition seems to have paid scant attention to action, practice and performativity. Thrift (1996: 33) has voiced concern that 'academic accounts have not only downgraded the importance of practical activity by trying to represent it as representations … but may also have understated its power'. Basically he argues that researchers have ignored the importance of how people *do* things. Take, for example, research into urban sustainability. There has been a lot of research into what people think about global warming, traffic congestion, and so on, and how it is represented as related to urban sustainability, but little research into what people actually do, how they *practice* urban sustainability. In moving away from the hermeneutic tradition, Thrift (2000a: 274) seeks to broaden analysis of space(s) by emphasizing everyday practices in/of those spaces: 'These are unreflective, lived, culturally specific, bodily reactions to events which cannot be explained by causal theories (accurate representations) or by hermeneutical means (interpretations).'

Thrift (2000b) seeks to move geography in the direction of 'non-representational theory'.

His idea of non-representational theory takes on board insights from a variety of theoretical traditions, from actor-network theory to performance studies, Judith Butler's queer theory and various post-structuralisms. The roots of 'non-representational theory' can be found in the French reaction to structuralism, in the Lacanian psychoanalytical tradition (Sarup 1988) and the French feminist triumvirate, Cixous, Irigaray and Kristeva (see e.g. Shurmer-Smith 2000). Fundamental to such post-structuralist interpretations, as we have said, was the epistemological challenge it posed for Enlightenment thought about language and representation being transparent vehicles for the communication of prior, original meaning and intention. What psychoanalysis added was a reformulation of ideas on human subjectivity. These ideas were taken on board in geography through the 'psychic turn' (Pile 1993; Pile and Thrift 1995). As stated earlier, the 'psychic turn' draws our attention to how people think and act and that some of this is unconscious. The unconscious that Pile (1996) identifies is not simply the opposite of our conscious but rather a parallel process that has its own language. It is *not* accessible to ourselves or to researchers. This means that we cannot fully interpret human actions as we cannot understand the unconscious dimension of these actions. The process of understanding never attains the goal of knowability, knowing is always 'contradictory, multiple, and always becoming' (G. Rose 2000: 654). It is at this point that performativity (J. Butler 1990; 1993) through which identities are made and remade becomes important; for performance is 'a living demonstration of skills we have but cannot ever articulate fully in the linguistic domain' (Thrift 1996: 34).

This move beyond representation demands different methods or different ways of using methods from the hermeneutic tradition. It demands methods that are concerned with documenting and doing. Thrift (2000b) argues that researchers pursuing such a project must be *observant participants* rather than *participant observers*. The documentation of actions, of performances, is important. Methods such as interviews become important not for the interpretations of their text but for the practices of their speech. Lees (2001) argues that an ethnographic approach also enables researchers to move away from an emphasis on representation and interpretation towards theories of practice. She demonstrates the utility of the ethnographic approach by 'reading' the new Vancouver Public Library. She argues that, if we are to take seriously the suggestion that (architectural) geography must address itself to something beyond the symbolic to questions of use, process and social practice, important methodological implications follow:

> *Traditionally, architectural geography has been practised by putting architectural symbols into their social (and especially historical) contexts to tease out their meaning. But if we are to concern ourselves with the inhabitation of architectural space as much as its signification, then we must engage practically and actively with the situated and everyday practices through which built environments are used. In this regard, ethnography provides one way to explore how built environments produce and are produced by the social practices performed within them. (Lees 2001: 56)*

Lees goes on to argue that, although ethnography is informative, by itself it is no more sufficient than (say) a political-semiotics approach. She advocates a triangulation of different approaches. This is an important point for the interpretative turn, especially as moves beyond representation are far from mutually exclusive terrains. Thus adopting an ethnographic approach to understanding architecture does not mean abandoning questions about the meaning of built environments. Rather it means approaching them differently, as an active and engaged process of understanding, rather than as a product to be read off retrospectively from its social and historical context. As Richard Bernstein (1983: 126) noted: 'meaning is not self-contained – simply "there" to be discovered; meaning comes to realization only in and through the "happening" of understanding.' What Bernstein calls the 'happening' of understanding is something performed by investigators engaging actively with the world around them and in the process changing them both. Lees explains that Thrift's work is useful here in answering the question: what is place? He argues that cultural geographers have looked at place as 'animated by culture', as if place exists before it is lived in. Thrift (1997a: 196–7) sees place simply as 'there as a part of us', 'something that we constantly produce, with others as we go along'. Elsewhere, Thrift (1997b: 139) uses dance to illustrate non-representational theory, quoting Isadora Duncan: 'If I could tell you what it meant, there would be no point in dancing it.'

Significantly, both the interpretative tradition and moves away from it have more visibly opened the eyes of geographers to the ethics and politics of their research. Although consideration of ethics in the interpretative turn sometimes came at the expense of political activity, the emergence of a critical geography movement has sought to remedy this. For example, critical geographers argue that researchers have a responsibility to communicate their research to wider audiences (see Shaw and Matthews 1998), to make it more accessible to non-academics, even if this is not as straightforward as it seems (see Lees 1999).

The politics of research

We want the reader to recognize the different philosophical positions that underpin research in human geography. In stating this we are aware that for some readers this might be a first encounter with ideas of this kind; or perhaps more accurately with the ideas being expressed overtly rather than being implicit in publications. The reader should have picked up a sense of diverging views within a single strain of thought. If you follow specific references, you will soon identify that we have provided a much abbreviated version of the complexity that exists. This is not a chapter that seeks to draw out the fine nuances and precise dimensions of intricate debate. But the chapter is fundamental for understanding the use to which data collection and analysis can be put. As we show in the chapters that follow, the same method can be used to explore an issue but with different implications for what is sought in the data obtained. Good research requires understanding how method is embedded in methodology in real-life applications. As John Hughes (1980: 13) reminds us: 'every research tool or procedure is inextricably embedded in commitments to particular versions of the world.'

What we wish to emphasize in this last section of the chapter is the manner in which 'particular versions of the world' are communal visions. This arises because 'a claim to knowledge has social dimensions. Our claims and justifications work, if they do, because of collectively held conceptions about the world and how we relate to it' (J. Hughes 1980: 8). Whatever method is used to gather research information, that method is not self-validating. The use of research tools requires a philosophical justification that guides interpretations of what knowledge is (ontology) and how it is derived (epistemology). These understandings are not static. The physical sciences have been restructured on various occasions, as with the deliberate use of science in struggles against the Catholic Church and the feudal state (Harding 1991). The implication is that political and social interests are not 'add-ons' to research endeavours but central to it (Katz 1994). A second implication relates to understanding what is meant by quality research:

> *'better' research does not mean the production of 'better' definitive data through improved techniques. It implies a new framework of political will to confront inequalities in the research process and in wider society, and to be politically committed to contributing to social change. The need for rigour is no less important . . . (Truman and Humphries 1994: 1)*

Viewed in this manner, the search for quality research should not be restricted to considering questions of method or methodology, but should take into account the socio-political consequences of the research process (Mercer 1984). A third implication is that prevailing power relationships penetrate research frameworks. Longino (1990) provides two insights on this. First, by drawing attention to the manner in which 'normal science' is sexist (Eichler 1988) – the implication being that the best methods in the world will not prevent researchers reaching sexist conclusions unless we change this paradigm*. Secondly, in reminding us that Darwin's *On the origin of species by means of natural selection* (1861) challenged orthodoxies of the time, Longino brings out how such challenges can be manipulated by the powerful. So, when the initial outrage over the 'origin of species' died down, its main ideas were embraced within social theory as a legitimization of social inequality. As Mercer (1984) stridently charges, contemporary economic, political and social conditions place pressure on academic institutions to undertake particular kinds of enquiry. Such outside pressures are evident in the association of modernization impulses in the 1950s and 1960s with the so-called quantitative revolution in geography, just as it was with growing attention to gender, ethnicity and sexuality in the 1980s and 1990s (McDowell 1992b). In this link between academic research and broader societal trends, market forces have a strong role to play (D. Rose 1990), as is evident through the publishing industry (Barnett and Low 1996), with government inputs transparent in setting national research priorities (Dalyell 1983; Demeritt 2000).

The sociology of disciplinary 'progress' cannot be taken out of this equation. One only has to look at university geography in the USA to recognize the serious consequences of a discipline failing to maintain a premier identity for itself.[8] Similarly, the embracing of neo-classical economic theories during the quantitative revolution was undoubtedly assisted by the perceived higher status of economics (Barnes and Duncan 1992). Quantification itself, of course, also has a certain cache. Governments in particular are drawn to the 'facts' quantitative approaches impart (for example, from social surveys), often treating them in an unproblematic manner (Finch 1986; Cantley 1992). In so far as its leaders seek to promote a discipline's standing, such considerations are important. Yet, in reality, much of the pressure towards disciplinary change comes from individual academics seeking personal advancement. In so far as publishing and grant winning are important for career promotion, researchers are likely to be conditioned by priorities in the publishing and grant-awarding worlds.

Added to which, research in human geography is a communitarian production process (Ragin 1994). Human geographers judge one another's work and seek to convince others of arguments. In doing so, they have been socialized to adhere to a set of standards governing the collection, analysis and presentation of research results (Cantley 1992). This applies no matter what the research approach. Thus, we need to be careful about too ready an acceptance of postmodern claims that 'others' are given voice, with the implication that the researcher has a subdued role:

> *I distrust those who admit to 'letting the data' tell them what to do and think, because what they leave to let the data do presents only a marginal difference from structuring it as self-consciously as one can. Both the follow your nose and the rational, pre-planned field styles rely upon a socialization to the same canonical features. (D. Rose 1990: 34)*

An important feature of this socialization is its broad base. Despite the fact that a relatively small group of so-called master weavers appear to exist in geography, whose work is quoted widely by others and who are commonly seen as 'leaders' in particular research fields (Bodman 1991, 1992), the communitarian (rather than 'elite') basis of disciplinary socialization is apparent in the breadth of involvement in refereeing manuscripts for academic journals (Boots 1996). 'The whole aim of this socialization is at one level to duplicate the achievements of the discipline, and at another level to contribute uniquely to the development of knowledge' (D. Rose 1990: 14). Of course, as Kuhn (1970) so famously articulated, the socialization process also limits vision; for the designation of 'acceptable' practice limits perception of alternatives, with the potential of the disciplinary cold shoulder awaiting those whose work does not conform:

8 Examining the more prestigious universities in the USA, geography departments have been closed in Chicago, Michigan, Northwestern and Yale, as well as in notable second-string institutions. Flirtations with geography at Harvard and Princeton barely got off the ground, while other key departments, such as Berkeley, have come under threat of closure in the past. Having failed to establish a key presence in high schools, geography has a problem projecting a dynamic disciplinary image, especially as many geographers find themselves in allied fields such as geology or urban studies (R. J. Johnston 2000).

> *Life chances within a career are associated closely as a result of publishing books in genres that resemble, or at least address in a confirming way, the literature cited in the ones that preceded them. If you write a nonconforming text, then the rewards of the discipline may be withheld because the book does not read as a legitimate contribution to knowledge. (D. Rose 1990: 14)*

If such 'chilling' processes exist, a corollary is that other 'structured' biases penetrate the review practices that lead to research output. One area that has seen adverse comment in this light is nepotism and sexism in processes of peer review, whether for grants or publications (e.g. Weneras and Wold 1997). But let us retain a focus on Kuhn for the moment.

First, acknowledge that Kuhn's ideas do not suggest that disciplines do not change. An image of slow change is consistent with Kuhn's ideas on periods of normal science, in which a particular paradigm dominates researcher visions. The essential point of Kuhn's work is that over time inadequacies in prevailing paradigms grow in researchers' minds, with potentially rapid change in world views resulting as new perspectives seem to circumvent existing shortcomings, or specify that they are irrelevant. Longino (1990: 27) captures the flavour rather well: 'One accepts or rejects a theory not because of rational deliberation about the evidential support of a theory but as one acquires or loses (religious) faith. To change one's theory (paradigm) involves changing one's world view and hence one's world' (see D. W. Harvey 2000, in which he sometimes seems painfully to hang tooth and nail to a Marxist geography). As John Marshall (1985) points out, the rejection of a paradigm involves an accumulation of evidence (or philosophical argument) rather than a single refutation.

There is no scale along which an accumulation of 'negative' evidence or argument can be assessed, other than that of personal evaluation. Of course, such assessments are embedded in a social context, with broader views in the discipline playing a key conditioning role. While it might seem somewhat antithetical for supposedly questioning academics, a key element in this is that: 'The authority criterion is common: people will accept a theory because of *who* proposed it not because of the evidence for it' (Vaus 1991: 21). One example might be the rush with which many human geographers grasped Castells's ideas (1977) on urban social movements in defining the nature of the city (for a more recent critique of the 'authority criterion' in human geography, see Barnett 1998).[9] But going beyond this, there is also the character of a discipline. David Harvey (1993: 5) offers pointed insight on geography in this regard:

9 Castells (1983) himself was shortly to reject key ideas in his earlier work.

the incorporation of space into existing social theory, of whatever sort, always seemed to disrupt its power. The innumerable contingencies, specificities and 'othernesses' which geographers encountered could be (and often were) regarded by geographers as fundamentally undermined (dare I say deconstructing) of all forms of social scientific metatheory.

For us the 'messiness' that a geographical dimension imposes on social life goes a long way towards explaining why divergent philosophies continue to persist in geography. Rather than matching the Kuhnian model (1970) of one paradigm overthrowing another, within geography positivism, realism and postmodernism/post-structuralism exist side by side. For sure the popularity of each is uneven, but the continued vibrancy of each is not in question. The journal *Geographical Analysis* offers one clear sign of the ongoing popularity of positivism. Reading this beside *Ecumene* might be akin to chewing chalk with cheese, but they both represent current arms of enquiry within human geography. These arms might not talk to one another,[10] but in some measure they use the same empirical materials and can employ the same data-gathering methods (e.g. Demeritt 1994 compares the subdisciplines of environmental history and 'new' cultural geography). As the chapters that follow unfold, we explore different understandings of what information resulting from data-collection methods are able to tell us. In this regard, it is worth carrying forward the message that: 'It is not the manner of data collection but the approach to evidence that is important' (L. Harvey 1990: 31).

10 In this regard we acknowledge the applicability to geography of Scharfstein's observation (1989: 48) that 'there is a strong suspicion, verified by innumerable examples, that the members of every tradition regard themselves as superior to every other'.

2
Research design

Although many have the impression that data collection is the major enterprise in research, this is not strictly correct. Preparation, Phase 1, takes the most time, and drawing conclusions and writing the report takes more time than data collection in most cases. Data collection itself takes the least time.

(Bouma 1993: 9)

There is no doubt that research design issues are central to good-quality research. Yet research design is a process rather than an event. The weight placed on 'setting up' a project before data collection is affected by the research approach followed, but whatever the approach the research process should be treated as a learning process. At one level this seems obvious, for a central element in research is discovery. If there is no prospect of finding something new, at the very least through evaluating whether a well-regarded theory 'works' in a new setting, then why bother? Once this is recognized, it should be apparent that there is a need to keep an open mind, so as to explore new insights and raise new questions. These can be unexpected when a project starts. Some methodologies, with induction as an overt case, have exploration built into their core fabric. Other methodologies, at least as specified in the idealized form in textbooks, seem little able to provide the adaptability a learning process demands. Most evidently in this regard, the deductive phases of logical positivism appear unbending. That stated, there is a wide gulf between idealism and practice. This is one reason why so many researchers have become disenchanted with the precepts of logical positivism: 'What I realise now, with the benefit of hindsight, is that the positivist paradigm of problem formulation, hypothesis, operationalisation and testing is not so much misleading as personally inoperable' (Newby 1977: 108). This is an important message, for there is a clear tendency amongst (positivist) researchers to justify the product of their explorations *a posteriori*.[1] Recognition of how common this is lies behind Vaus's observation (1991: 9) that: 'A basic difficulty when trying to describe how to do research is the gap between textbook accounts of how research should be done and how it actually is done.' In this chapter we seek to illustrate why there

1 It takes no genius to recognize this from the literature. Examine the output of multiple regression analyses that seek to 'explain' the incidence of a phenomenon, and ask why the 'explanatory' variables are so different, even for studies from a similar theoretical perspective. As an example, for a review of studies like this on local government policy, see Boyne and Powell (1991).

are good reasons for this. In doing so, our aim is to draw attention to the strategic choices that must be made in undertaking research. The intention is not to indicate what should be, but to highlight possibilities.

As a quick examination of the literature reveals, there are many books that focus explicitly on questions of research design (e.g. Campbell and Stanley 1963; Hakim 1987; Hedrick *et al.* 1993). We are not seeking to cram what they provide into one chapter. On the contrary, these texts show that when it comes to the practicalities of designing a research project you need to be well aware of the particular strengths and weaknesses of specific research methods. Hence this chapter is not about putting design principles into practice. Instead it is concerned with raising key issues that investigators should consider, irrespective of the precise methods they employ in a project. To contextualize what will follow we should make clear that we do not believe there is a template for good research, even if certain issues need to be given serious and careful attention in order to produce good work. How these issues are reconciled cannot be defined *a priori*:

> *The course that a piece of research actually takes will be peculiar to that piece of research: it is affected by the research topic, the technique of data collection, the experience and personality of the researcher, the politics of the research, the types of people or situation being studied, funding and so on. (Vaus 1991: 9)*

Researchers should be vigilant in seeking to understand links between philosophy and technique, with good researchers learning from past experience in utilizing research methods. Single research projects should be approached in a spirit of learning, just as research experience should accumulate to improve understanding.

This point is not only relevant for those who wish to pursue a research career. The major element of training in research is not oriented towards this. Few undergraduate or graduate students end up with research posts. Yet it is rare for graduate jobs not to involve research, as seen in the preparation of position papers for company decisions. More broadly, in our everyday lives we use research to evaluate ideas and conclusions. Reading newspaper reports falls under this rubric, just as the television news does (Glasgow University Media Group 1976, 1980, 1982). To put this point generally:

> *Considerable knowledge and experience are essential if one plans to conduct good research. Even the nonresearcher who only reads the results of studies would do well to understand basic design principles so that he or she can evaluate conclusions drawn from a study. Many times we see in published articles conclusions that seem not to follow from the study described. The reader who is unaware of design principles is at the mercy of an author who might well have overlooked a problem with the design of a study. (P. E. Spector 1993: 4)*

What this chapter examines are different components in designing and implementing a research project. The chapters that follow are about implementation as well, so the reader should think of this chapter in terms of research strategy. The chapters that follow look at the tactics of implementation and 'completion'. In dealing with strategic issues, the structure of this chapter is not meant to specify the order in which decisions have to be considered. It is true that we start with problem identification. But the sections that follow on purpose, styles, evidence, analysis, verification and presentation cannot be disentangled from the original spark that ignited interest in a research question. This does not mean that the research issue decides choices in research processes, but it would be mischievous to suggest that these choices are made independently of the problem posed. In similar vein, the purpose in undertaking a research project, and the kind of insight that is sought, have direct implications for what is accepted as evidence, alongside the type of analysis the data are subjected to (and so on).

Problem identification

For most academics who supervise undergraduate dissertations, student selection of dissertation topics is one of the most frustrating times in degree programmes. Perhaps there are institutions that provide students with a list of topics from which to choose a project. Where this is not the case, it is apparent that many students find it difficult to identify a research problem. This quandary is understandable. There is little in books on how to select a topic for investigation – even if some assist with study skills and dissertation projects (e.g. J. Bell 1993; Phillips and Pugh 1993; Parsons and Knight 1995; Cryer 1996; Kneale 1999). Advisers are prone to offer little advice on topic selection (at times hemmed in by examination rules, if a dissertation is meant to be the student's own work) or cause irritation by telling students the topic has to interest them so they must decide. Perhaps more commonly than either side wishes to recall, this last comment comes after a student has made a multitude of suggestions the adviser has either groaned over or simply rejected as 'unworkable'. Let us try to unpick this situation.

Forget about projects that are predefined, where employer or funding agent specifies the problem. Here the issues are different, for the weight that has to be placed on various 'interests' in problem specification are not the same as for projects the researcher selects. To be clear on this, in any project we can recognize a number of different interests. There is the researcher him or herself, who wants to find material of interest (otherwise investigating it is likely to be a chore). Then there is the significance of the issue investigated for stakeholders in the project (Andranovich and Riposa 1993). For some projects the stakeholders will be funding agencies. When data are collected directly from informants, informants are themselves stakeholders. Consideration of their interests involves more than questioning whether there is sufficient interest to encourage participation, for there are ethical issues concerning methods of data collection, data analysis and result presentation. To a lesser extent, people who read the work are stakeholders. Inevitably, research problem selection has to have an eye on eventual 'impact' (whether examination marks or building a research reputation). If a study follows a well-trodden path, without adding to well-established knowledge, the work is likely to generate little interest or excitement. Perhaps the investigator might be interested, for there are many who want to know more about their home place or a particular activity. But if the intention is to score in examination or professional terms, the temptation to ignore what interests others is self-indulgent and likely to be counterproductive. The critical point is to recognize linkages between research questions and target audiences. That said, it is critical that a topic captures the researcher's imagination, for this will sustain personal interest as the project progresses. The key is to find a topic that intrigues you as well as attracting others.

The sources from which this curiosity is derived are almost limitless. As Strauss and Corbin (1990: 35) put it: 'The minute one asks the question "but what if" and finds there is no answer, then one has a problem area.' Yet the need to enthuse others commonly leads advisers to direct students to explore 'the literature'. The intention behind this suggestion is to identify debates that interest you, questions that are unanswered, disputes over causes or impacts, issues on which little is known, and so on (Hedrick *et al.* 1993). The existing academic literature should be a source of inspiration by providing a theoretical context for a project, as well as indicating existing publications on a topic, potential data sources, possible analytical techniques and even a guide on the structure of result presentation. Of course, not all academic papers are equally helpful in this regard. Literature reviews or conceptual-theoretical papers can direct attention to questions, but provide less guidance on methods of analysis or possible data sources. Empirical research papers are better suited for this, with most providing a review of relevant concepts, theories and prior empirical results in the field in question. At one level this seems like a short cut to finding a topic to investigate, but some students feel uncomfortable with this approach. One reason appears to be a view that ideas cannot be culled from previous work because dissertations (and graduate theses) have to be original. This feeling seems to rest on a major misunderstanding of originality. If you think the corridors of your department are trampled by Einstein clones, with a full set of original ideas, then please have a cold shower and think again. Look at the 'stages' of the positivist scientific method. If deductive analysis is a key element in scientific progress, then a great deal of that progress is based on other people's hypotheses or theories. Whether we are looking at undergraduate, graduate or postdoctoral research, the meaning of originality is not that no one has thought of an idea or undertaken the work ever before. Rather it is that a project adds to previous work, perhaps by asking the same question in another setting, perhaps by introducing a consideration previous studies neglected.

Approaching project identification from this perspective places primacy on reviewing the literature (and/or perhaps policy debates). How this is achieved depends on familiarity with research in the field to be investigated. It also depends on knowledge of research abstracting services. But even if you have no access to sources like *Geo Abstracts* or online search engines like the *Web of Science* or *OCLC*, provided you can examine hard copies of journal runs you are well placed. We emphasize journals here, for most books provide review type material, rather than expositions of research projects. As noted above, the advantage of research papers is that they focus on a research problem, provide a literature review pertinent to that problem and indicate potential data sources or data collection methods. Reviews of literature focus more on the conclusions of studies. But how do you start the search process? You should start with an idea of the field you would like to work in, although initially your definition of this field might be broad. The aim is to narrow this field. First, identify research journals in the field that interests you (e.g. *Urban Geography* for urban topics, *Gender, Place and Culture* for feminist topics, *Environment and Planning D: Society and Space* for cultural topics, and so on). From this point, a starting place is to examine the contents of these journals. Use a hierarchical approach, start with the titles, looking for a topic that appeals. If one seems interesting, check the abstract of the article for more information. This process usually has to be repeated on many occasions, perhaps because the topic seems less interesting on reading the abstract or possibly because the approach used for data collection does not interest you or does not seem feasible in your context. It is important to recognize that a research project is a compromise between the desirable and the achievable. This means achieving a happy convergence between research problem and data availability. It is common to find students identifying a data set or empirical issue that interests them but not having a theoretical question to investigate. The end result is usually a descriptive account, with no purpose other than to précis a data set. Alternatively, many students have great theoretical ideas, but little appreciation of how to explore them. Commonly, this scenario is associated with projects that are not feasible, at least given the time and cost restraints a student faces. This means that, as abstracts are identified that interest you, you need to examine the paper in more detail, to establish if this kind of project is feasible for you. There are no hard-and-fast rules here. Creativity is a key element in judgements on what is feasible. Recall that you need to approach this process asking if you can add something to the existing literature.

In this regard, Ragin (1994) provides a useful guide to different types of investigation, by specifying seven (at times interrelated) research aims: (1) identifying general patterns or relations – including cross-national divergence (A. J. Fielding 1982; Sellers 2000); (2) testing or refining existing theories (Lees 1996; Clark and Hoffmann-Martinot 1998); (3) making predictions (Inglehart 1990; Marsden 1999); (4) interpreting culturally or historically significant phenomena (Ley 1974; J. C. Scott 1985; Addison 1994); (5) explaining social diversity (Massey 1995; Women and Geography Study Group 1997); (6) giving voice to those who are outside the mainstream (Philo 1992; Sibley 1995); or (7) advancing new theories (Tilly 1984; Beck 1992). In addition to seeking a contribution in one or other of these ways, the originality of a project can come from the methods it employs. For instance, it might be noted that studies on a particular topic have tended to use standardized questionnaires, which might prompt you to ask if more in-depth analysis might offer different insights (e.g. Forrest and Murie 1992; Gutting 1996). As one illustration, the masters degree of one of the authors focused on linkages between accessibility and consumer shopping decisions. At the time much of the literature on this issue used questionnaires, examining links between shopping places (and attitudes towards them), distance from home and the distribution of consumer spending. Rejecting this approach, this project used a diary to record shopping activity over a month followed by in-depth interviews to explore reasons for shopping decisions. Far from confirming that attitudes towards shopping centres and distance from home were important, it showed that attitudes towards shopping as an activity and time constraints on consumers were more influential (Hoggart 1978). Of course, the opposite point can be made. At times in-depth studies can become so engrossed in particularities that they fail to identify local specificities that make some forms of local action more likely than others (on one approach to explore this, see Hodson 1999). This reasoning provided one attraction for quantitative studies in the 1960s, which sought to move beyond the 1950s emphasis on the uniqueness of place (R. J. Johnston 1991), to develop theoretical accounts of geographical diversity. Such quantitative approaches have since been rightly criticized (Chapter 1), but we need to be careful here. Too commonly academics throw baby and bath water out of the same window, as new research fashions gain currency, when principles and notions are worth keeping. In this regard, there are some excellent critiques of the view that quantitative analysis has nothing to tell unless positivism is accepted (see Marchand 1974; R. A. Walker 1989). Sound understanding of methodological principles, alongside creatively blending method with research problem, are key factors in producing quality work (see also Philip 1998; Philo *et al.* 1998). A critical eye should be vigorously applied when asking whether a different research method could add insight on theoretical relationships.

All the above should not be taken to indicate that the existing literature must be trawled for research problems. In many cases there is no literature on problems that make interesting research enquiries. This is seen when new legislation is passed, with students commonly wanting to explore the impact of such legislation. The 1990 introduction of the so-called Poll Tax (the Community Charge) in England (1989 for Scotland) provides an obvious example. This tax set the same per capita tax on all residents in a local government area, irrespective of wealth or public services used. This had massive distributional consequences, as those in million pound homes, who had previously paid taxes based on property value, found themselves charged the same as those in run-down bedsits. Many student dissertations in geography sought to explore the distributional consequences of this new tax; did these vary across cities, what were the implications for different neighbourhoods within cities, and, given attempts by many to avoid the new tax, what was the impact on rates of electoral registration (a key index for officials seeking to identify non-payers, who were often prepared to give up voting rights to avoid payment)? Yet the fact these issues had not been investigated before did not mean the research literature was not relevant. Although offering no empirical analyses on this tax, the literature provided theoretical expectations and investigations on the impact of previous taxes. For example, there were expectations that political parties favour their own supporters, or alternatively seek to win marginal constituencies by awarding them special favours (Hoare 1983). Such expectations about the impact of new legislation provide guides for an investigation.

New topics?

It is important to caution about investigating 'new' topics. Most evident in this regard is the problem of data availability. Academic work is not journalism. In academic research emphasis is placed on securing as accurate a data set as possible. This can take a long time to collect or become available. By contrast, as Matthews's early commentary (1957) showed, time is a key consideration for journalists, who are pressured to 'get the story out', even if this can result in inaccuracy in reporting. Contrast this with the government statistics academics often use, which can take years to produce (often owing to the massive quantity of data involved). For the 1991 UK Census researchers had to wait a number of years before they could gain access to some data (Openshaw 1995). Although publication time has speeded up, this will be the same for the 2000 US Census and the 2001 UK Census (see e.g. Martin 2000). Information is even less likely to be available if it comes from private or non-profit organizations. This problem is lessened if data are collected by researchers themselves (for example, through interviews). But often this is not feasible, given time and cost restrictions.

Our general caution is that students who study very recent events, often in the mistaken belief that this makes their work more 'relevant', exciting or innovative, can generate considerable problems for themselves. Many geography lecturers have tired of students suggesting they want to study an issue that has yet to run its course. Take the example of the Poll Tax again. Before the Act of Parliament was passed many students sought to investigate its provisions. This could have been disastrous. What if you completed a study on the impact of the Poll Tax and, just as you are about to hand it in, the government changes its mind on a key provision you investigated. Disaster, what are you left with? A study of what the distributional consequences could have been had the government not changed its mind. Who cares? All right, there is merit in exploring the issue, but in terms of personal satisfaction this scenario would be very deflating. Let us assume no last-minute changes. Even so, studying issues as they unfold can raise serious limitations on an investigation (albeit this is not inevitable – see the section on Action Research in Chapter 7). Commonly, for example, participants in an unfolding issue are cautious about giving interviews or providing data. This is especially so if the issue is sensitive, but the sensitive nature of issues often draws students to them. The literature on sensitive decisions reveals that decision-makers are cautious and capable of manipulating interview situations (Healey and Rawlinson 1993; G. Walford 1994b). At the height of such sensitivity, in the run-up to a formal decision, reluctance to divulge information should be greater than normal. In such a setting researchers need to be aware that their efforts are more likely to be thwarted. In the context of dissertation work, with limited time to complete a project, this consideration has to be taken seriously.

Efforts in this direction are greatly aided if the questions investigated are specific. Given epistemological divergences, this does not mean you need a hypothesis. However, students often find that posing a question so it can be answered yes or no is helpful, even if they never expect a yes/no answer, given the bounty of qualifications that are likely. So rather than asking how women find working in the financial services sector, a more directed question would pose that women are less/more likely to find the environment in financial services militates against job promotion. Posing the question to enable a positive or negative response should provide a more direct link with theoretical expectations. Of course, when stated so simply, there is an implication of hypothesis testing. But, as the question is considered, a series of other positive/negative issues will become apparent that impact on the relationship specified. This opens up the topic, as recognition of the potential interplay and mutual disturbance of relationships can take the study into more uncertain and for many more interesting territory. Put simply, we need to step back from any appearance of hypothesis testing in a positive/negative formulation, by recognizing that all investigations are underpinned by a conceptual (theoretical) understanding, even if at times implicit. As Hedrick and associates (1993: 19) put it: 'All studies, whether it is acknowledged or not, are based on a conceptual framework that specifies the variables of interest and the expected relationships among them.' Our emphasis is to enhance student understanding of this. Reviews of the literature are fundamentally intended to ground ideas in a conceptual and theoretical framework that informs the project. Undertaking a study because a topic 'seems interesting' without this grounding is a recipe for a poor end product. Put simply, major pitfalls can usually be identified before a project starts to be implemented. Students should seriously evaluate potential pitfalls prior to embarking on data collection. Setting up the project in an informed manner, in terms of concepts, theory and methods, can save heartache, trauma and disappointment later.

In addition, researchers should be reflexive during project selection and implementation. As outlined in Chapter 1, this includes reflecting on the effect of their presence on the data collected. There are many issues involved in reflexive contemplation. Most discussions of these issues centre on the relationship between researcher and researched. It raises issues about the impact of the age, class, gender, political disposition and, amongst many others, race on research conclusions. These are serious questions we explore later. For present purposes, we draw attention to a different element of reflexivity, which is to ask if you are an appropriate person to undertake the research proposed. This is little explored in the literature. From the few commentaries that exist, we can glean that some investigators are ill-equipped for specific projects. Wilson (1992: 180) provides an illustrative commentary:

I would argue that as much damage has been done by poor academic quality as by poor ethics in relation to informants. Many researchers have contributed to negative stereotypes, incorrect but fashionable ideas, exciting but unfounded theories and over-simplified notions that have supported (directly or indirectly) ideologies, policies or programmes with negative results.

Perhaps this stresses the detrimental consequences of 'closed minds' more than we are seeking to do, but, in drawing attention to the damaging effects of poor research, Wilson provides an important message. What we ask is that researchers are honest with themselves about their capabilities and limitations. Today, in-depth qualitative research might be the flavour of the moment, but does this mean everyone can undertake this work in a proficient manner.[2] As individuals we all have strengths and weaknesses. Some are more theoretically inclined, some more empirically oriented. Some have a greater capacity to 'see behind the scenes', to deconstruct or abstract. It is futile to pretend we are equally capable in each dimension of academic endeavour. Similarly, if more than one researcher is involved in a project, it should not be assumed that the distribution of tasks will be easily agreed, or that each partner will contribute equally. Although academics appear reluctant to reveal personnel and personality problems associated with research projects, a few instructive contributions exist (see e.g. C. Bell 1977; D. W. Harvey 1992). If the aim is good research, researchers need to examine themselves (and their co-workers) critically, and be sure they can use a research method effectively. Many students do not do this. Most commonly they use the method that is the current 'flavour of the month'; what they often end up with is work that is not convincing, and that does not attain depth or quality to persuade readers. The paper by Baxter and Eyles (1997) on the inability of published qualitative studies to convince readers on rigour indicates that the same point applies for established researchers. In good part, this is because too many researchers confuse methodology and method. To refresh memories, the distinction we draw is that, taken in isolation, data collection and analysis are issues of method, whereas methodology embraces method in a manner that is cognizant of an underlying vision about the nature of 'reality' (namely, ontology) and the basis on which knowledge claims are made (namely, epistemology). Researchers too often naively assume particular methodologies necessitate specific research methods. This is a fatal underpinning of too many research studies.

---

2 Greele (1991) notes that the quality of oral history interviewing is very variable, yet few submit their work to public scrutiny. In similar vein, note the stricture of Daniels and Cosgrove (1988: 2): 'The iconographic approach consciously sought to conceptualize pictures as encoded texts to be deciphered by those cognizant of the culture as a whole in which they were produced.' How many of us can claim we understand 'culture as a whole'?

Summary: choosing a research topic

- When choosing a topic, use a funnel approach; start broadly, identifying areas of interest, then narrow this down.
- Use the research journals and research monographs to help identify a topic, not literature reviews and textbooks.
- Make sure the topic selected interests you and others.
- Remember a topic is a compromise between the desirable and the achievable – you need a research problem and the data to examine it.
- Be extra careful over studying issues that are unfolding, as data access can be a problem.
- Question yourself pointedly – is the research approach one that you are suited to.

Getting started

We are running ahead of ourselves here, for the reader might feel that the discussion has begun to drift into questions of research method. This is inevitable, as the identification of a research problem draws attention to other decisions that have to be made. Take, for instance, Bernard's articulation (1994: 103) of what constitutes a realistic approach to problem specification. This involves considering five questions. Does the topic interest you? Is the problem amenable to 'scientific' investigation? Are available resources adequate to explore the problem? Will the investigation raise ethical problems? Is the topic of theoretical interest? You might quibble with some of these provisions (some postmodernists would not see the need to be theoretical, for example), but the basic ideas merit attention. For one, it can be asked how those who are new to research can make such evaluations. How do you identify what is a researchable topic, explore whether ethical dilemmas might emerge, judge whether the issue is of broader interest or assess the resources you need? First and foremost this has to come from reviewing the literature, to gain insight from those who study similar problems or work in comparable locations. Secondly, there are advantages in discussing the proposal with those who have worked on similar issues. Andranovich and Riposa (1993) suggest there is merit in holding brainstorming sessions with colleagues, who might not have worked on similar topics, but who can raise issues for you to contemplate and assess, before deciding on the location and methods that best suit your purpose. They add that those with previous field experience are invaluable for bouncing ideas off (the point can be extended to those who have utilized a relevant data set or data collection method), especially in 'distant' sociocultural contexts (see e.g. Derman 1990). Such stipulations come with a health warning, for those who have undertaken previous research provide information from a particular perspective, which colours their views.[3]

3 As an illustration, Redfield (1930/1973) concluded that Tepoztlan in Mexico had a relatively homogeneous, isolated, well-integrated social structure that was slow to change. In a restudy of the same village, Lewis (1951) concluded that Redfield had misinterpreted social relations in Tepoztlan. Conditioned by a theoretical model that emphasized a socially stable 'folk culture', Redfield had overemphasized stability and downplayed the dynamism and diversity.

Information from other researchers must necessarily be filtered through the 'sieve' of research aims. It is one thing to take advice on good hotels or to listen to health tips, but researchers need to focus on the coherence of their project when evaluating advice. As Hedrick and associates (1993) note, the objective that should underscore the design of research projects is ensuring data collection and analysis are appropriately linked to the problem under investigation and the research agenda to be followed. In this regard, investigators need to ask if a proposed design is *credible* (capable of providing convincing conclusions), *directed* (targeted at the question at hand) and *feasible* (given cost and time constraints).

By implication this points to good research design depending on the problem under investigation. There is no such thing as good research design, viewed as an off-the-peg formula that can be trundled out for any project. There are more and less appropriate designs, depending on how the problem is posed and what insight is sought. We accept there is bad research design, although when it comes to evaluating the degree of 'badness' conclusions will be influenced by the problem at hand. But how can we assess goodness of purpose? The specification of fit-for-purpose criteria is not straightforward, for different problems lie at the heart of research projects. Even given this qualification, it is useful to think in terms of four issues when thinking about strategies for research project implementation. These are criteria for evaluating project 'success', assumptions and operationalizations built into a project, the utility of dissimilar data types, and the strategy for assessing meanings or relationships.

Summary: getting started on research design

- Use the expertise of others when developing a design but treat their advice critically.
- Ask if your research problem or question is credible, directed and feasible to investigate.
- In the abstract there is no such thing as good research design. Designs are more or less suited to research questions, approaches and contexts – they have fitness-for-purpose.

Evaluating project success

The range of research styles that analysts find acceptable is diverse. Contrasting with the (unattainable) quest for neutrality on the part of positivists, those following an Action Research strategy are concerned to change society or at least establish causation by gauging reactions when a 'stimulus' is deliberately introduced into a social situation. But deliberately injecting new information and actions into a social setting raises ethical questions. Unequal power relationships exist between researched and researcher, with the latter often seen to take from the researched and give little (if anything) in return. Various commentators have suggested more 'equal' treatment, perhaps with researchers revealing as much about themselves to the researched as vice versa (e.g. Oakley 1981).

Such sentiments are a long way from contriving to change a social situation to see how people react. This latter has more in common with enforced experimentation. This raises real ethical dilemmas. Note reports that suggest that the Amazon's Yanomami tribe was devastated by researchers deliberately bringing measles to this isolated tribe, with this controversy said to be tearing anthropology asunder (e.g. Cornwell 2000). But the division lines between what is and is not seen to be ethical are hard to draw. There might be doses of political correctness in some commentaries on power relations in interviewing, but there is rarely recognition that exposure to a researcher's values might be discomforting to interviewees. The seeming neutrality of positivism surprisingly has similarities with claims by some postmodernists that they primarily give 'voice' to informants (rather than imposing their values), with both raising questions about what is ethical research. Western (1986: 33) brings this to the forefront when discussing research on apartheid in South Africa:

> *If we are worth our salt as academics, as geographers of the city, then description and human sympathy and moral outrage are not in themselves enough. If they were, we would become little more than wordsmiths, quality journalists, or protagonists of the [politically] correct cause. What we must do is delve down to find that which has created pauperism . . . We must see if we can reveal causation . . .*

The strand in postmodern thought that seeks to draw out previously silent voices, letting voices 'speak for themselves', fearing an attempt to provide explanation will impose an 'outsider' perspective, is from this perspective a cop-out. Yet the case for explanation as an essential ingredient of good research is itself based on political assumptions about the merits of researchers helping the less advantaged or at least exposing mechanisms of societal organization, to provide a basis for changing society. Yet the academic literature offers many pronouncements suggesting that research is intended to change society, with little evidence of actual impact. Will an explanatory account really stimulate change more than a descriptive account that lays before an audience the plight or interpretative world of a social group?

Our sense is that seeing explanation as a criterion for good research implies that the explanation offered will be acted upon. This is a presumptuous claim. It has links to the assertion made by some academics that they undertake *applied research* – with the implication that applied research is more meritorious than 'pure' research. For us, this belief is nonsense. Applied research is research that is applied. The history of social science, like that for the physical sciences, the life sciences or indeed medicine, is that research output will have practical application to an unknown degree. Even if the intention is to establish evidence (or explanation) that can be applied, whether or not action follows depends significantly on the conclusions reached (see Cloke *et al.* 1997: ch 8). This is especially so for the social sciences, where the results of research are commonly regarded as 'political' by governments. The most overt and intense exemplification of this was the era of the Thatcher governments in the UK. Built on an antagonism towards social science, the Thatcher Government cut funding for social science research dramatically and even changed the name of its research council, to exclude the word 'science' from its motif. The emphasis in research fund allocation, as well as themes that were funded, changed towards subjects that might win government approval (Dalyell 1983; C. Bell 1984; more broadly Demeritt 2000). Conclusions from research projects were rejected or accepted depending on whether they were in line with the ideological persuasion of the government; not according to whether they explained or described issues accurately or innovatively. Insights on this process are readily available from newspaper reports. In 1994, for example, the minister responsible for immigration refused to publish Home Office research, as its messages contradicted the imagery the Conservative Government was promoting. The research 'undermine[d] popular conceptions of asylum seekers as poorly educated economic migrants by concluding that they have a wealth of skills that are not being employed to the detriment of the country' (Travis 1994: 2). Even if work is not blocked, it can be issued in such a way that few can get access to it (for example, few copies are printed). Brody's research (1971) on North Americans living on skid row provides a Canadian example. In this context, presenting results in a relatively descriptive manner, with an approach that is decidedly 'pure' in orientation, can have as much impact as explanatory accounts with a proclaimed intention of being 'applied'. In terms of changing society, if that is the aim, the likelihood that a research project leads to social change is extremely difficult to predict. Lees's work on the marginalization of youth in urban public space in Portland, Maine, provides a good example. Lees (1999) did not envision the broader attention that her research would receive from the city's youth (in the form of self-politicization), the city's main newspaper (taking her results to the Chief of Police to enlist his reaction for a front-page editorial) and radio (interviews alongside the chief of police for WMPG). Lees's research had the unexpected effect of enacting social change. First, highlighting the marginalization of youth from downtown public space in Portland fuelled the politicization of young people who dubbed themselves 'The Undesirables'. This led to sit-ins and rallies in downtown public spaces. Second, with heavy media coverage, city officials began to back down on their policies of marginalization (for example, curfews).

Our message here is twofold. First, the division between applied and pure research is confusing rather than illuminating. In some cases it is politically inspired, in an effort to promote certain types of work by vested interests. Overall it confuses the topic investigated with the application of results. Second, while it is critical that researchers have a purpose in their investigations, to proclaim the superiority of explanatory or descriptive accounts does little more than tell us about personal bias. That said, we add the caution that so-called descriptive studies inevitably rest on a specific ontology. They are not value free. Indeed, their capacity to conceal political messages is generally greater than for explanatory studies, as the theoretical models used in explanatory analyses point to the underpinnings of the study.

Summary: evaluating project success

- Ethical considerations should always be a factor in this equation, even if differences in values mean there are no easy answers to the question 'is this work unethical?'.

- The assumption that 'applied' research has greater worth than other research is questionable. The very definition of what is 'applied', is difficult to answer.

Explanation in social research

In most investigations, explanatory intent is embodied within the project. Surprisingly in this context, few studies or commentaries give much attention to what is meant by *explanation*. Likewise, few offer clear expositions of what provides an explanation. What some researchers find disturbing is what they see as a laxity in the manner some investigators use 'evidence' to provide explanation. In human geography, this problem appeared to increase as critical realism and the cultural turn challenged what some saw as the previous sureties of positivism. For studies utilizing a realist philosophy, 'certainties' about what supportive evidence might look like were challenged, as priority was given to theoretical explanation. Lee Harvey (1990: 7) summarizes the position aptly:

> *Despite its long history and concern with material reality, critical analysis of society has tended to be dominated by theoretical treatises. Empirical material is often taken for granted or even regarded as an encumbrance to the abstract theoretical analysis. There are, one suspects, a considerable number of critical commentators who regard empirical material with suspicion.*

At times evidence that contradicts a theoretical picture seems to be ignored.[4] With theory taking precedence, studies can be based on restricted empirical work (e.g. Duncan and Goodwin 1988) or overgeneralized empirical trends (see e.g. Worsley's critique 1980, of Wallerstein's world systems theory). The sense of discomfort empirical researchers experience over this has been heightened by postmodern research, where generalization and abstraction have been eschewed in favour of argument and explanation by illustrative example. Vaus (1991: 20) provides a pointed commentary:

> *The emphasis on basing theories on observation and evaluating them against further observations may seem to be common sense. However, it is not universally practised in sociology [or geography]. The practice of some sociologists [and geographers] involves the formulation of 'explanations' which are never systematically tested empirically. At best they use examples as proof. Examples, however, are a weak form of evidence, for regardless of the explanation we can find some examples to illustrate the argument. The key to empirical testing is looking for evidence which will disprove the theory, not simply to find supporting illustrations.*

4 A common claim is that these are 'contingent' conditions, with some commentators discarding 'contingent' relations as theoretically 'unimportant'.

There is no suggestion here that the insights such studies provide are not interesting. For example, the suggestive paper by Kinsman (1995) on the way the English countryside is intimidating for ethnic minorities, Stephen Daniels's interpretations (1992) of what watercolour paintings of Leeds tell us about industrialization, and Jackson's commentaries (1991, 1994) on masculinity in advertisements, all offer interesting interpretations. But the use of a limited number of selective examples raises the question of whether we should see their messages as anything other than 'unique'. A counter-argument is that such studies are not seeking a general picture but explore specific biases. Dressed in a particular garb, this line of argument would hold that, since everyone is unique, and writers cannot control how their work will be interpreted, it is futile to seek generalization. As such it is inappropriate to question whether the points raised through analysing specific examples are general trends. This we accept at one level, but reject at another as sleight of hand. Scharfstein's wry observation (1989: 48) on those who seek to 'tell it as it is' is pointed: 'The need to generalize is apparently irresistible – one says that generalization about some particular thing is wrong; and then, one's conscience pacified, one generalizes about it.' Even if the researcher does not fall into this trap, it would stretch the imagination not to recognize that readers often generalize from a particular study. This is particularly likely when there is little competing literature. Moreover, despite protestations to the contrary, most researchers are hoping others will 'generalize' about their work. After all, academic reputations are made by publishing work that is innovative and has broader messages than a single investigated (or empirical) issue.

In fairness to those who rely on examples to provide 'evidence', deciding on what constitutes an explanation is controversial. To explore this we start with Dey's words (1993: 40): 'To explain is to account for action, not just or necessarily through reference to actors' intentions. It requires the development of conceptual tools through which to apprehend the significance of social action and how actions interrelate.' Andy Pratt (1994) takes this point further by noting that, because causality is not empirically observable, this creates methodological difficulties for empiricist work. Pratt is incorrect here, for methodological difficulties are created for all approaches by this situation, not just empiricist ones. This essentially holds because causality is 'an explanatory frame imposed upon the data by the observer' (J. U. Marshall 1985: 121). Pratt might be content with the realist framework he proposes, as this provides the glue that links the observed with its explanation. But, as James (1984: 5) observed, the assumption that theory provides a convincing basis for explanation is questionable:

> *First, there is no independent standpoint from which to test claims; we must be content to rely on standards of evidence internal to theories themselves. Second, there is no neutral ground from which to compare theories, and we can only reject one from the standpoint of another.*

This is not meant to imply that explanation can be achieved without theory, for, as Chapter 1 made clear, data are theory dependent. Rather it is a reminder that there are internal inconsistencies in relying on theoretical explanation. Reliance on theory to brush aside the 'unexplained' is just as problematic as selectively choosing examples to illustrate a point. In both cases, it is beholden on the researcher to undertake a rigorous analysis in order to persuade others. At a philosophical level, there is no satisfactory solution to this conundrum. In practical terms, the researcher needs to have in the forefront of her or his mind that the strength of an investigation does not come from how satisfied the investigator is with it, but from the capacity to convince others.[5]

In so far as an investigation seeks to explain human behaviour, a key issue is the understanding of explanation used. Most commonly, analysts have seen causation as producing a change in behaviour. This seems simple enough but it raises complications. For a start, as Hellevik (1984) points out, over what time span can an effect occur? To hold that a cause must have an immediate effect contradicts accepted ideas on causation. One illustration is the distinction between what caused and what occasioned the start of the First World War. What occasioned it was the assassination of Archduke Ferdinand in Sarajevo. This led (fairly quickly) to war. But, if the underlying conditions of tension between European powers had not been there, this assassination would have been

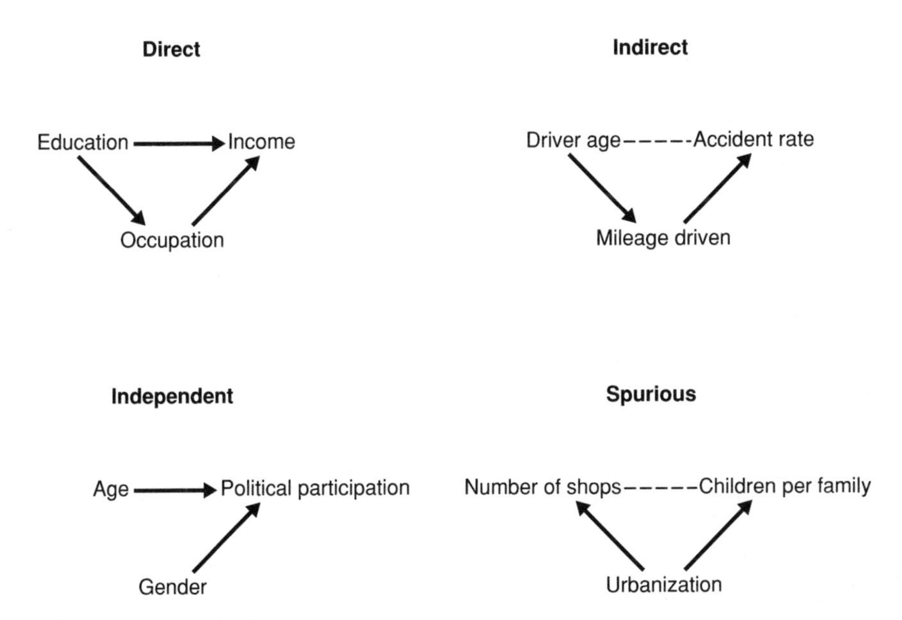

Figure 2.1 *Types of relationship between variables.*
Source: *Hellevik (1984).*

5 The implication of this is profound, as Longino (1990: 4) identifies, for it means that what is seen as good explanation is decided by personal values. This is one reason why some researchers accept selective examples as evidence in support of an argument.

unlikely to have provoked war (see e.g. A. J. P. Taylor 1963). After all, there have been many assassinations and most have not led to war. To turn this point around, underlying conditions of state rivalry had been present for many years but the absence of a critical spark kept conflagration in check. But, if we accept that causes can take a long time to bear fruit, then the provision of a direct link between cause and effect is made more difficult. It has to depend on argument (ultimately theorization) rather than observation of immediate effects (as positivism implies).

This point is evident when we recognize that causation need not have a direct effect. This is illustrated in Fig. 2.1. To take one example, while rates of road traffic accidents might have a relationship with driver age, it would be short-sighted not to take account of average distances driven each month when exploring the relationship. Of course, the range of potential intervening variables is immense. Analysts should appreciate the variety of potential causal forces that are at play. Here theoretical insight holds a key role in defining relevant variables. All well and good, but how is causation assessed? Dey (1993) provides an illuminating commentary, when referring to the identification of causal relationships as being akin to working on a jigsaw puzzle. Some pieces in the puzzle are so unique their place is quickly identifiable. For less obvious pieces in the puzzle, characteristics have to be assessed and need to be assigned to a category of potential contributors (classification). In many regards the process is the same for a qualitative description of divergent cultural meanings as for explaining human action. The objective is to look for natural breaks and transitions in the data that distinguish one meaning (or behaviour form) from another. This commonly involves exploring different ways of aggregating and splitting analytical categories. For analyses utilizing a deductive approach (hypothesis testing), the theory that underscores the analysis provides the key to what is examined. The procedure, if quantitative data are used, could be similar to that outlined in Box 1.4. For inductive analyses a similar approach could be used, but here the analyst would not be guided (as overtly at least) by an established set of expectations from (a single) theory. Here, for a guide to evaluating categories, the student is well advised to explore the rigorous methods of grounded theory* (Strauss and Corbin 1990). This embraces imagining the opposite, as seen, for example, in evaluating how events unfold from a position of domination as well as subordination. It involves maintaining an attitude of scepticism, of progressively narrowing the focus as the data are tested to exclude categories that do not distinguish meanings or behaviour forms.

Summary: the nature of explanation

- The temptation to generalize is common, even if this is in terms of hoping others will see broader applicability of a project. Analysts need to be careful in providing messages on the limits of their investigation.
- Explanation is 'imposed' by analysts through the lens of theory.
- As theory defines data, with independent evaluation of theoretical claims difficult and

rare, analysts need to persuade others of conclusions through analytical rigour and the support for their argument.
- There are important differences between what occasions and what causes.
- Classification of events, processes or actions is central to explanation, whatever the epistemological base of the study.

Operationalization

Fundamental to theory development are the definitions analysts impose or derive for key concepts. For quantitative studies such definitions are linked to how variables are measured. For qualitative analyses they constitute criteria for categories to distinguish investigated social groups and/or human actions. Operationalization is far from easy. Its difficulties are central to much of the dissatisfaction analysts express over evaluations of theory:

> *Measurement is one of the most difficult and most important tasks facing the quantitative researcher because so much depends on accurate measurement ... Because researchers usually hold fast to their theories they often blame their measures and complain about the difficulty of measuring social phenomena with precision. (Ragin 1994: 143)*

That many researchers have difficulties finding data to represent variables is clear. A quick review of the literature on attitude measurement provides a ready illustration (e.g. Oppenheim 1986). But the questions that have to be posed in thinking about operationalization are more profound than this. For one, questions have to be asked about the appropriate geographical area in which to investigate relationships. As one example, an issue that has attracted geographical interest is the so-called inverse care law (which is linked to the underclass hypothesis). This essentially posits that (public and private) service provision discriminate against those who are in the weakest socio-political position (two early analyses include Bradley *et al.* 1978 and Knox 1978). Evidence on this score has been contradictory, but Bolotin and Cingranelli (1983) have argued that a good reason for this is that analysts used inappropriate areas for their investigations. They argue that, when the hypothesis is concerned with service provision to residential areas, it is inappropriate to include central business districts in investigations. Effectively many services here are not oriented towards the residential population but are for offices or for the whole city. Analyses of Sassen's hypothesis (1991) that prevailing economic processes enhance social inequalities in global cities raise similar questions about territorial remits. Here many studies use arbitrary administrative boundaries for cities or metropolitan areas in their analyses (for a review, see Hamnett 1994). But can we make judgements about how economic change in London affects social inequalities if we ignore commuter field impacts beyond the capital's green belt?[6]

6 A similar question of appropriate study sites arises in research by those from wealthy nations working in poorer nations. As Gow (1990) notes, site selection here reveals spatial bias (urban or tarmacked, roadside locations), project bias (showpieces favoured), informant bias (more professionals and elites), seasonal bias (dry seasons have less disease) and temporal bias (the 'tourist' nature of short visits), amongst others.

Even if there is no question mark against the appropriateness of geographical areas, a common operationalization problem is concept definition. Social scientists use certain words regularly and seemingly with certainty, yet even core concepts lack agreed understanding. An early signifier of this was Hillery's review (1955), which found 94 different definitions of the concept 'community'. To name a few, noteworthy social science concepts that are subject to definitional disagreement include social class (G. Marshall *et al.* 1988; Crompton 1991), power (Lukes 1974; Foucault 1982), ethnicity (Ahmad 1999; or see ethnicity-ethnocentric entries in McDowell and Sharp 1999), economic development (Perroux 1983) or homelessness (Widdowfield 1999). Yet conceptual definitions can have major implications for how a variable is measured or human behaviour interpreted. This does not mean we should fix on one definition and have done with it. Concepts are abstractions, not artefacts. They are not 'out there' waiting to be identified but have to be constructed (or for post-structuralists, deconstructed), using a particular theoretical framework. As such, how a concept is defined will owe much to theoretical emphasis (albeit more than one theoretical perspective can be involved here). Yet how a variable is measured or behaviour is interpreted also depends on data availability. Here compromise is an inevitable accompaniment to an investigation.

Validity notions as research criteria

Fundamental to such compromise is the degree to which deviation from the 'ideal' brings research results into doubt. This is associated with the extent to which construct validity is compromised. Expressed in the context of quantitative variables, construct validity is the extent to which the operationalization of a variable does 'measure what it is designed to measure' (Spector 1993: 8). Stated more generally, representation of a concept is valid to the extent it accurately reflects the concept it is purported to characterize. Yet researchers often assume their concepts are measured or represented validly, with little evaluation of this assumption (Wasserman and Faust 1994: 58), even though many potential causes of invalidity exist. Campbell and Stanley (1963) provide primary examples:

- *Instrument reactivity*. The classic example of this is referred to as the 'Hawthorne Effect', which was first identified in studies of industrial organization (Roethlisberger and Dickson 1939), and has such profound potential impacts on medical research. The Hawthorne Effect refers to those participating in a study behaving differently because they are being investigated. Most obviously this is seen when medical researchers compare a group that receives a new drug with another that receives a dummy, neutral substance. Whether for psychological reasons or not, those who receive the dummy substance can have a notable rate of medical improvement. As it is a reaction to being studied, the Hawthorne Effect has broader implications than reacting to a specific measurement instrument.

- *History*. This is when change occurs that has nothing to do with the study but impacts on research conclusions. Consider assessing a campaign to provide a more positive image for politicians. Having run a questionnaire survey before the campaign and being set up to undertake the second as it nears its end, the effort to assess the impact of the campaign could be blown off course by a stream of revelations of impropriety by politicians. In the UK one is reminded of how the Major Government's 'back-to-basics' campaign (characterized by some as back to Victorian values) was badly dented by revelations about 'improper' behaviour by Conservative MPs.
- *Unreliability of instruments*. This refers to 'measuring' (or representing) concepts in a manner that does not reflect their 'real' essence. One example that has become legendary is assuming that formal positions of authority can be equated with the exercise of political power. For the USA, Hunter (1953) exposed the fallibility of this idea by showing that elected city leaders in Atlanta, Georgia, did not exert critical power over major decisions, with unelected business leaders having decisive influence. The revelation that political position cannot be assumed to equate with political power was revolutionary at the time – one of those studies that can rightly be described as having 'launched a thousand theses'.
- *Differential subject loss*. This is a common problem for studies that revisit a panel of informants over a period of time (panel studies). Maintaining contact with panel members is time consuming and expensive. In the UK the most well-known panel studies are the British Cohort Studies (see Davies and Dale 1994). A key here is that those who either refuse to continue or cannot be found might be 'exceptional' in some way, so changes in identified behaviour or values might owe something to subject loss, rather than real social change (see e.g. Boruch and Pearson 1988).
- *Bias in the assignment of those investigated to analytical groups*. This largely refers to comparisons made between two (or more) groups. Examples of this problem make regular appearances in undergraduate dissertations. Take, for example, investigations that have sought to explore Ross Davies's hypothesis (1968) that the number and quality of retail services in neighbourhoods vary with the average income of neighbourhood residents. The central issue of this hypothesis is that income differences determine service provision, so the areas selected for comparison should be similar apart from income divergence. In smaller cities, areas that are comparable in this sense are not necessarily easy to find. For example, many neighbourhoods lie on a main road, so retail services pick up passing trade, which means that they are not 'equivalents' to relatively isolated residential belts with a shopping centre at their heart. Comparison is made more difficult by the variety of dimensions along which places differ. As a result compromise is inevitable. But this should not be stretched to the key dimension of the study (in this case the nature of shopping centre catchment areas).
- *Instruments change over time*. There are many aspects to this. One is that informants respond to instruments in different ways over time. A good reason is that, once

exposed to an instrument, they react to it. So, once a survey question has been asked, informants might think about that question after the interview. Having reflected, if asked the same question even a few days later, the response might be different. In effect, for this respondent, the nature of the instrument has changed. Foddy (1993) provides an array of examples of this.

Given the potential for error resulting from such problems, it is little surprise to find commentators recommending that 'by using well-known and generally accepted indicators, the validity of your description is enhanced' (Andranovich and Riposa 1993: 49; see also Vaus 1991). This advice is useful for quantitative work, in that more established indicators of concepts or ways of interpreting concepts have been subject to prior investigation, so problems of usage should be identified in the literature. Yet established indicators cannot be assumed *a priori* to be helpful. It has to be recognized that something of an industry can attach itself to some indicators, which could encourage neglect of its weaknesses. The SAT and GRE tests that are so beloved as standardized entrance examinations for US universities are transparent examples (Bernard 1994), with anyone who has seen the questions asked in geography test papers soon questioning whether this can identify good geography graduate students. IQ tests are similarly endowed by some people with grand pretensions. In each of these cases it might not be that the tests are invalid as such, more that their use in particular contexts is questionable. In this regard, analysts have usefully reminded us that ideas on concept validity are not neutral with respect to theoretical models (P. E. Spector 1993), political perspective (Plummer 1983) or epistemological stance. This should remind us that concept validity is not independent of the underlying framework or the purpose of an investigation.

This provides one argument in favour of in-depth, qualitative analyses, in which the meaning of actions can be explored through primary data collection. But validity is not straightforward even for this kind of work. Nevertheless, such approaches have the joint advantage of not relying on secondary data, which only provide often proxy measures of concepts (Hedrick *et al.* 1993), while enabling direct investigation of meanings and understandings in people's behaviour. By contrast, quantitative analysis, even if primary data are used, necessitates imposing a degree of uniformity on variables that removes nuances of interpretation and understanding. The response to this, which holds whether qualitative or quantitative approaches are used, is to seek cross-validation by exploring alternative ways of interpreting or measuring concepts. As will be described below, this commonly involves the triangulation* of data sources or interpretative mechanisms. Whatever the approach used, concept validity is more likely to be enhanced by checking interpretations and results using a variety of data sources.

Reliability or validity?

Do not assume this means that interpretations or results will neatly coalesce to signify the sanctity of one perspective. Human society is too messy for this. Throughout analysts must be looking to build an argument to convince others about their conclusions. As Chapter 1 should have made clear, although there might be broad agreement amongst people about some aspects of human life, there is no single 'reality' waiting 'out there' for researchers to find. Moreover, as eloquently articulated in grounded theory (Glaser and Strauss 1967), society is subject to change that makes for uncertainty over the repeatability of social events (Toffler 1970). As numerous analysts have commented, this should place higher value on validity than the positivist call for reliability (e.g. Plummer 1983). This arises in part because informants 'learn' through the process of being involved in a research project. Dorst (1989: 204) makes an amusing observation in this regard, which is troubling for those who see the repeatability of research results as an essential component of theory verification:

> *It is increasingly common to hear from ethnographers anecdotes about textual ambush at the field site, for example, the anthropologist whose key informant checks his cultural facts by producing from the hut a previously published ethnography, the very one the field worker has relied on to formulate his own questions.*

What we see here is circularity between research and human behaviour. The first research study informs those who were studied, who change or reinterpret their behaviour in the light of the research results. But one has to ask if this is really an analytical problem (it might well be seen as an ethical one, but that is another issue). If human behaviour has changed or if informants describe themselves using categories that do not fit their actual behaviour and beliefs, this should be revealed in analysis. Dorst seems to imply that we should take what we are told by informants as a true interpretation (so that those who describe themselves using the categories employed by others are 'cheating'). This is rather simplistic, for good research involves cross-checking information and evaluating inconsistencies. Take, for example, Maquis's comparison (1970) of interview responses with medical records. Here only 12–17 per cent of hospital episodes, 23–26 per cent of visits to doctors and just 50 per cent of chronic or acute conditions in medical records were reported by interviewees. Ask which you think is more likely to be accurate? Then ask if exploring unevenness in response is not important in its own right. Inconsistency is not a problem as such. It requires exploration. The key is not to rely on single data sources but to seek cross-validation.

Objectivity and validity

It should be clear from the above that validity is not the same as objectivity, for analysts can employ concepts validly within the context of theoretical positions that take a singular view on the world. This is something geographers have made much comment on. As Duncan and Ley (1993: 8) articulate it, 'by definition all representations are inextricably intertwined with the theory-laden categories of the research', so that: 'What is seen as avoidable bias by the positivists is acknowledged by the hermeneutician as an inescapable part of the formation of knowledge'. Such realizations led Myrdal (1969) to argue that research analysts should make their values explicit in their writing. Even in 2001 relatively few do this in journal articles, whether because of editorial policy over article length or reviewer antagonism towards 'personal views'. Although values are easy to identify when reading some publications, in others they are concealed, often under a guise of 'objectivity'. For Harding (1991: 143), there are inherent power relationships in such claims to objectivity: '[the] objectivist justifications of science are useful to dominant groups that, consciously or not, do not really intend to "play fair" anyway.' These points are not ones we disagree with. Yet we caution over their interpretation. There is a tendency in geographical circles, implicit most of the time but sometimes overt, to turn criticism of ideas about value neutrality into a claim that one-sided investigations are acceptable. This could lead to the malpractice of selectively choosing examples to support pre-established positions, without examining other interpretations or conflicting evidence.

Here we see a link between validity and a specific kind of objectivity. In a sense we accept the spirit of Ragin's ideas (1994) that social researchers should be 'objective' in that they need to be cautious: (*a*) not to whitewash; (*b*) to seek out 'good' and 'bad'; (*c*) to be wary of how people rationalize what they do; (*d*) to maintain scepticism over evidence; and (*e*) to examine events from different viewpoints. Although expressed in naive fashion in this list, what this asks is for researchers self-consciously to review ideas and values, seeking to establish if they hold up when other viewpoints are taken into account. Following Harding (1991: 151), this vision has become associated with the concept of strong objectivity (as opposed to weak objectivity, which, like positivism, defines objectivity solely from one point of view):

> *To enact or operationalise the directive of strong objectivity is to value the Other's perspective and to pass over in thought into the social condition that creates it – not in order to stay there, to 'go native' or merge the self with the Other, but in order to look back at the self in all its cultural particularity from a distant, critical and objectifying location.*

This vision of objectivity has been brought into prominence by writings on feminism, but it has a broader base, as seen in the 'scientific' interpretation of Cunningham (1973: 4):

> *it is possible for an inquiry to be objective if, and only if, (a) it is possible for its descriptions and explanations of a subject-matter to reveal the actual nature of that subject-matter, where 'actual nature' means 'the qualities and relations of the subject-matter as they exist independently of an inquirer's thoughts and desires regarding them', and (b) it is not possible*

> *for two inquirers holding rival theories about the same subject-matter and having complete knowledge of each other's theories (including the grounds for holding them) both to be justified in adhering to their theories.*

This definition is harder to comply with than Harding's, for stipulations like 'complete knowledge' are unreasonable and the idea that a subject-matter is independent of thought and desire is contentious (even if some accept it is feasible if most agree). However, comparing these two views on objectivity is not our intention. Rather we wish to emphasize three points. First, while Cunningham and Harding approach objectivity from dissimilar philosophical stances, they both conclude that objectivity is strengthened by examining more than one perspective. Secondly, rather than rejecting objectivity, as many relativists do, these analysts seek strategies to convince others of the validity of their work (e.g. Rocheleau 1995). Finally, these commentators highlight the inappropriateness of associating objectivity (or validity for that matter) with positivism. Objectivity, as self-critical evaluation of evidence, is to be encouraged. Thus, while you might be horrified by sexist tendencies in research agendas and practices, to conclude from this that subjectivity must be embraced runs the danger of failing to convince those who need to be won over to non-sexist practices (Eichler 1988). Objectivity in a positivist sense might be unattainable, but in the strong objectivity terms of Harding (1991) it is desirable. Unabashed subjectivity might produce internally coherent and (for some) politically correct accounts, but it can fail in a central test for researchers, which is to persuade others, most especially those who hold other views.

Qualitative or quantitative?

When it comes to persuasion, there is little doubt that research methods have a role to play. Investigators have long recognized that government officials are more inclined to accept quantitative data that give an overview and are inclined to dismiss as anecdotal qualitative evidence, no matter how often they themselves rely on anecdotes for policy justifications (see e.g. Green and MacColl 1987). Yet the association in many geographers' minds of quantitative research with positivism has cast a shadow over this analytical approach. Turning away from this approach is not simply a function of epistemological dislike, but arises for ethical reasons. In feminist research, for example, many adherents turned towards qualitative research in reaction to sexist biases in much quantitative data (e.g. Oakley and Oakley 1979), with a conviction that through closer relations with informants research could become less exploitative and more empowering for informants (Oakley 1981; McDowell 1992b). Although revealing a less overt ethical concern, realist researchers tend to collude with this view; for abstractions regarding structural forces, alongside the 'chaos' of contingent effects, is seen to militate against quantitative investigation. The rationale for a qualitative approach can also be theoretical. Thus, as Ragin (1994: 40) noted: 'The significance of most historical phenomena derives from their atypicality, the fact that they are dramatically nonroutine, and from their impact on who we are today.' How would one quantitatively study the causes of the French Revolution or the US Civil War?[7] By comparison, on the positivists' side, discomfort is found amongst those who think it unlikely (if not impossible) for qualitative work to meet desired objectivity and reliability standards. These entrenched differences are reminiscent of the battlefields of the First World War, with analysts trying to ignore, or at best look over the parapet at, 'the enemy'. Occasional movements forward by one side are quickly followed by retreat, at times bloody in the face of stiff resistance from the 'other side'. Stalemate is the norm.

A major reason for this is the interpretation each side places on the work of others. On one side there is an unfortunate tendency to see quantitative research as conformist. This charge has been made in the context of easier compliance with UK research council student stipulations (Cantley 1992), of qualitative data being messier and more expensive to collect, so that quantitative analysis should be the default (McCracken 1988: 59), and of quantitative work having more 'respectability' (Gittins 1979). Those with an inner desire to see themselves as 'radical' might, not surprisingly, be pulled towards qualitative work. Associated with this, the literature offers reminders of distinctive goals and capacities for quantitative and qualitative work. Thus, McCracken (1988: 16) holds that:

> *The quantitative goal is to isolate and define categories as precisely as possible before the study is undertaken, and then to determine, again with great precision, the relationship between them. The qualitative goal, on the other hand is often to isolate and define categories during the process of research. The qualitative investigator expects the nature and definition of analytic categories to change in the course of a project.*

This view is unduly restrictive and, given the strong emphasis on probability and fuzzy sets in much quantitative analysis, misrepresents the precision that is expected – albeit we accept this description for some work (for global climate change models, see Demeritt 2001a). Added to which, the implication that quantitative research is well thought out before analysis starts, with little change once it has begun, ignores the haphazard nature of many quantitative studies (for commentaries on a long recognized pattern, see J. S. Armstrong 1967 and Gould 1970). Such work is commonly critiqued for engaging in trawling exercises, with justification for research steps *ex post facto* rather than *a priori* (Berry 1971). Moreover, as Ragin (1994) notes, you cannot distinguish qualitative from quantitative work by the use of words or numbers, for both use them (see e.g. Siegel 1956). What more accurately distinguishes qualitative from quantitative work is a qualitative focus on a large number of attributes (and their linkages) over a

7 This is not to claim that associated conditions cannot be analysed in a quantitative manner, for quantitative approaches can offer much to history (Tuma 1994; see also Isaac *et al.* 1998).

relatively small number of cases (whether people, factories or nations). Quantitative work, by contrast, usually examines a larger number of cases but focuses on fewer attributes. Quantitative work is better equipped to highlight general trends (across a limited range of attributes). Qualitative work is capable of more nuanced understanding, drawing out social distinctions and diverse cultural meanings (Ragin 1994).

It follows that, rather than being in conflict, qualitative and quantitative approaches should be complementary (see e.g. Hodson 1999). Advocates from the political left made precisely this point in expressing exasperation over blindness towards quantitative analysis in political economy research (Marchand 1974; R. A. Walker 1989). More recently, feminist and cultural geographers have made similar points (Philip 1998; Philo *et al.* 1998; Barnes and Hannah 2001). These commentaries have been accompanied by a number of studies that have incorporated both qualitative and quantitative data (e.g. Hanson and Pratt 1995; Rocheleau 1995). One example of this is the work by Ley (2000) on immigration into Vancouver, which utilizes in-depth interviews with 24 Chinese immigrants (or couples) to identify rationalities and experiences, while drawing on the Canadian Government's longitudinal immigration database, which links immigrant landing forms with subsequent annual tax information, to provide a broader evaluation of trends. In this case, the material used links effectively, but we caution that the coexistence of qualitative and quantitative analysis does not mean the two inevitably integrate in an effective way.

Summary: operationalizing research projects

- Analysts must be clear about what (geographical and social) units of analysis are most appropriate for the research question investigated.
- Ideas on the validity of key research concepts are influenced by theory, study purpose and epistemology.
- Reliability, in the sense of the repeatability of results, is a questionable goal for social research.
- Objectivity, in the sense of neutrality, is not possible, and would in any event be a limited goal. A stronger sense of objectivity can be achieved by securing the acceptance of those adopting other perspectives.
- High-quality results are secured by cross-validation, as in utilizing different concept formulations, opening analysis to other perspectives (as Dear and Flusty 1998 have it, 'rehearsing the break') and drawing on different types of data.
- Linked to this, quantitative and qualitative data are not oppositional, but offer dissimilar insights.

Multi-method and multi-level analysis

As we began to think about this book, one of the authors attended a one-day conference organized by the Population Geography Research Group of the Royal Geographical Society/Institute of British Geographers on multi-method analysis

(McKendrick 1995). The day was disappointing in the manner multi-method analysis was approached. In particular, there was confusion in delegates' and speakers' comments over the difference between multi-method and multi-level analysis. This is an important distinction, with the question of 'levels' being particularly important owing to the ecological fallacy* of transposing conclusions from one analytical level to another. What is meant by an analytical level is the level at which observation units are aggregated. Using urban research as a framework, Andranovich and Riposa (1993) specified these levels as individual people (households could form the next level), neighbourhoods, cities, regions, national systems of cities and the world system of cities. The ecological fallacy involves undertaking research through the lens of one of these levels, then assuming that the results apply to other levels. The first author of a major work to popularize this problem was W. S. Robinson (1950: 352), who spelt out that analysts commonly use certain information because it is more available, when they are interested in a different analytical level:

> *In each study which uses ecological correlations [meaning data aggregated above the level of individuals, so electoral wards, neighbourhoods or cities], the obvious purpose is to discover something about the behavior of individuals. Ecological correlations are used simply because correlations between individuals are not available. In each instance, however, the substitution is made tacitly rather than explicitly.*

Today few researchers slip into the trap of assuming that associations at one level can be transposed to another. A major reason is the work of Robinson himself. For example, when analysing the incidence of illiteracy amongst individuals in the USA, Robinson reported that a correlation of 0.118 showed a weak tendency for those who were illiterate to be born outside the USA. Examining the percentage of each state's population that was illiterate and the percentage born outside the USA, the correlation was –0.526. At the state level, places with a high percentage of foreign-born residents had lower rates of illiteracy. To take the state-level analysis and assume this fitted individuals would be incorrect. Illiteracy was higher where more of the population was born in the USA, but this does not mean that those born in the USA were less likely to be illiterate.

More recent expositions of the danger of confusing relationships across geographical levels are seen in cautions about confusing the incidence of socio-economic deprivation at the personal and neighbourhood level (Fieldhouse and Tye 1996). This confusion creates regular criticism over government policies; where the temptation to concentrate government spending, whether for electoral, public-image or cost-saving reasons, commonly results in funding allocations to those who are not 'deserving', while many supposed targets of the policy are ineligible because they live in the 'wrong' area.

Indicative of mismatches between individual and area-based representations of specific social conditions or processes, Dunleavy (1977) reported that in the urban renewal campaigns of the 1960s and early 1970s, 52 per cent of dwellings demolished as 'slums' in the London Borough of Newham were fit for human habitation. Similarly, policies to help low-income residents upgrade their housing were 'captured' by professional and managerial workers, who used public subsidies to gentrify inner-city areas (Hamnett 1973; Balchin 1979). Funds targeted at poorer areas disproportionately benefited richer people.

The fact we cannot transpose results from one level to another indicates that studies that explore relationships at two levels are not multi-method. Multi-method implies utilizing more than one method to explore the same problem. Studies that examine, say, relationships between voting and social class using data for geographical aggregates (like wards or constituencies) are not examining the same problem as those who look at electors themselves (e.g. Peake 1986). Material from these two levels can inform one another (see e.g. Crenson 1971), and analyses using areal data can contextualize in-depth investigations (see e.g. Hanson and Pratt 1995). But analyses that use different analytical levels are not examining the same issues.

Using multiple methods

It is not controversial to state that research quality should be enhanced by multi-method investigations. The literature is replete with commentaries that make this point (e.g. Yin 1984; L. Harvey 1990; van Meter 1994). Most evidently, the advantages of multi-method approaches are asserted on account of the capacity to undertake 'triangulation'. What is meant by triangulation is the use of a series of complementary methods in order to gain deeper insight on a research problem. The advantage of using complementary methods is that they enhance capacities for interpreting meaning and behaviour. This is because the insight gained can strengthen confidence in conclusions by providing multiple routes to the same result. Providing a rather different justification, investigators in rural Third World settings have been drawn to Participatory Rural Appraisal (R. Chambers 1994a, b, c). This emphasizes the need for multiple methods to be used, as well as local residents' involvement in an investigation (Box 2.1). Here consciousness of inadequacies in formal databases (see e.g. Gow 1990), relationships between investigators and researched that can lead to mis-reporting (see e.g. Derman 1990), and the difficulties differences in language, culture or social standing pose for interpreting local events (Howard 1994; F. M. Smith 1996), all caution against reliance on single sources (see e.g. Gow 1990). The coupling of a strong ethical dimension with a shrewd appreciation of the limitations in data collection is a distinguishing feature of Participatory Rural Appraisal. Central to this approach is recognition that research is a learning process, whereby uncertainties arising from a lack of correspondence between data sources provide a focus for further investigation – much like grounded theory (Glaser and Strauss 1967). Triangulation in this framework involves checking and progressive learning, with

Box 2.1 Menu of methods used in Participatory Rural Appraisal

Significant principles in Participatory Rural Appraisal are related to the behaviour and attitudes of outside facilitators. These include not rushing into the work, 'handing over the stick' to local people and being self-critically aware. As Robert Chambers (1994a, b, c) has made clear, amongst the reasons that account for the popularity and research strengths of the approach are the analytical abilities of local people, given the catalyst of a relaxed rapport for the facilitator, which are expressed through a series of methods of data collection. The combination of a wide variety of data collection mechanisms is regarded as significantly enhancing the reliability and validity of the information obtained, especially given the potential of cross-checking conclusions suggested by different sources, alongside improved understanding gained through local people sharing in data interpretation. The many data sources that are commonly employed in Participatory Rural Appraisal include:

- secondary data sources (census data, police records on crime of traffic accidents, etc.);
- semi-structured interviews with a selection of local residents;
- interviews with key informants;
- focused discussions with groups of local residents;
- engaging in village activities;
- watching villagers undertake their usual activities (i.e. direct observation);
- mapping, with residents, local conditions (tree species, garden plots, etc.);
- interpreting air photos;
- transect walks (e.g. to establish land-use patterns);
- trend and change analysis;
- oral histories (e.g. of cropping patterns, fuels used, etc.);
- maintaining seasonal calendars (e.g. of cultivation activities);
- daily use of time records;
- livelihood analyses (exploring family budgets, responses to crises, etc.);
- institutional (or chapati or venn) diagramming (identifying role of institutions in lives);
- analysing differences (asking why and what makes for differences);
- matrix scoring (e.g. using seeds to score);
- ranking and sorting (e.g. identifying wealth or status groups);
- stories, portraits and case studies of village life;
- drama, games and role plays (includes playing out scenarios in workshops);
- brainstorming sessions about ideas (e.g. on preliminary research findings.

Source: R. Chambers (1994a, b, c); Mikkelsen (1995).

approximations from single data sources refined through a plural investigative strategy (Chambers 1994b).

Readers might detect differences in the vision of what triangulation constitutes across the last paragraph. This would be in keeping with the messages in the literature. In some cases the meaning given to triangulation is too shallow. For one, Plummer's statement (1983: 95), that it is 'a mixture of participant observation and almost casual chatting with notes taken', is flawed in that it prescribes specific methods to triangulation, in its implication that it involves casual data collection, and in assuming that use of different elements in the same method constitutes triangulation.[8] What constitutes the core of 'triangulation' is the idea that complementary methods add insight on a research question. We accept that this notion can be fine-tuned, and have a lot of sympathy with the Participatory Rural Appraisal idea that research is a learning process. However, the latter is a broader issue. It should be embraced irrespective of triangulation issues. Moving forward from this position, there are two issues to confront before we can feel comfortable with introducing triangulation ideas into research projects.

Styles of triangulation

Asking for the deployment of complementary data collection methods might have a simple ring to it, but the implementation of this requirement is often complex. Both Burgess (1984) and Mikkelsen (1995) spot this, when specifying different types of triangulation. For Burgess these are data, investigator, theory and method. Mikkelsen accepts these four but adds academic discipline. Put simply, the argument is that we can have more faith if researchers call on different types of data, use more than one investigator (with enhanced effect if they are not from the same discipline), interpret evidence through different theoretical lenses and use more than one data collection technique. This might appear reasonable, but we need to qualify the positivist tone in this message. This tone is seen in the implication that conclusions are strengthened if they are confirmed from different theoretical perspectives. This message embodies an underlying assumption that there is a 'reality' to be identified. As Chapter 1 made clear, some accept this perspective, but others reject it. This could leave the impression that triangulation is not relevant for their research. But there is no doubt that investigations can be strengthened by researchers offering different perspectives on the same phenomenon (see Lees 1994, on using both Marxist economic and postmodern cultural explanations to understand the gentrification process). If researchers disagree in their interpretation, this should prompt questioning of positions, in turn prompting analysis of points of disagreement. This does not mean that agreement will be reached on one view, nor does it mean that a lack of agreement points to poor research.

A great many research projects explore a number of different theoretical perspectives, but their intention is not to seek one 'true' view. Rather it is to evaluate which theory provides a more convincing account. This is not triangulation, in the sense of using complementary methods to gain deeper insight. However, the underlying idea that conviction about a perspective is enhanced by evaluation of alternative interpretations is sound. This practice is central to many research projects. Where feasible, its introduction into research design is highly desirable. The same point can be

---

8 Participant observation involves being a 'full' member of a group, so casual chatting is an aspect of learning about a community, social group or organization (Chapter 7). Because 'chatting' is part of participant observation, it offers no 'protection' against inherent biases in this observational approach. It hardly qualifies as a 'complementary' method.

made about using different researchers to explore the same issue, although this is more difficult to achieve, owing to cost and time considerations. It should also be noted that, despite advantages in principle, multiple research participants generate particular problems (see e.g. C. Bell, 1977; D. W. Harvey, 1992). At the same time, especially in an environment in which high prestige is bestowed on research grant winning, there can be uneven power relationships between researchers, with one as grant-holder and the other(s) as employee(s). How this works out in practice will depend in part on personalities. We do not wish to dwell on this point or make great pronouncements. What we need to note is that practices that seem advantageous in principle can prove troublesome on implementation.

In general terms, the most common understanding of triangulation is the employment of different data sources and collection procedures to examine the same research issue. In our view we need to be careful in our understanding of what this means. Reading Burgess (1984) and Mikkelsen (1995), as two examples, we are left with a sense of disappointment by the implication that using data at different scales of geographical aggregation and for different time periods is sound triangulation practice. We have already made this point clear for geographical scale, so let us turn to time differences. First of all we should recognize that investigations utilizing data from one point in time (namely, cross-sectional data) can be used to explore associations between human conditions, but only hint at causal relationships. As causality relates to what leads to change (or thwarts change), we need to explore relationships over time. We make this point even if a project seeks to provide an in-depth, qualitative understanding of a social group at one point in time. Accounting for social 'form' requires an understanding of process, simply because people, groups, institutions and places are always in the process of becoming. They have never 'become', for, even if they appear to be 'static', form and habitual action must be reproduced. Comparisons across time periods might identify similar or dissimilar behaviour, but they do not constitute the same empirical situations, because processes in the intervening period have either deepened or weakened prior social processes.

Readers might object that, provided there is a satisfactory theory to underscore an analysis, data for one time point provides a satisfactory base for theory evaluation. Here we beg to disagree, in terms of theory evaluation. By exploring what is a temporal relationship in a static time framework, we might get results that fit theoretical predictions, but these will be based on false reasoning. Let us illustrate with a hypothetical example, in that we do not know the true direction of causation. The starting point is the expectation that parties of the political left are more committed to providing housing for governmental (or subsidized) low-income housing. This issue has been explored by a number of researchers in the UK (e.g. Boyne and Powell 1991). The general expectation is that control of local government by the Labour Party leads to higher rates of public housing construction – at least until the Thatcher governments imposed huge cutbacks in council house building (Boyne and Powell 1993). With Labour largely controlling urban local authorities, this relationship is said to hold for towns and cities in particular. In the 1970s statistical evidence showed that Labour control was allied to more council-owned accommodation in cities. But what is the direction of causation? During the Second World War there was considerable damage to the urban fabric as a result of bombing. After the war the national government decided that public housing would be dominant. In the late 1940s private house building was insignificant, with private-sector completions surpassing the public sector only in 1959 (UK Office of National Statistics, annual (a)). With war damage as a central concern, it was not surprising that a great deal of building took place in inner-city areas. Moreover, even if they escaped bomb damage, inner cities were the focus of massive slum clearance programmes (Merrett 1979). There is evidence that suburban Conservative governments were reluctant to accept council houses, so building was 'overconcentrated' in inner-city areas (e.g. Young and Kramer 1978). But this is icing on the cake, rather than a fundamental determinant of the geography of public housing. To illustrate, during the Thatcher Government years from 1980 up to the 1988 Housing Act (which effectively removed the public-sector role in housing provision), while 310 114 public housing completions were recorded, this was mitigated by 127 443 demolitions. The legal requirement that councils rehouse those in housing need would have focused much rebuilding into areas of heavy demolition.[9] The fact the Labour Party tends to govern councils with a large public housing stock might owe more to the distribution of war damage and poor housing than to political differences. Cross-sectional analysis might suggest Labour favours public housing. Temporal investigation might point more to factors outside local control – like the priority given to slum clearance in national government policy, the distribution of older dwellings, low incomes and so on. To have the potential for identifying causal relations, change over time needs to be explored, with a variety of ways available in which this can be achieved (Box 2.2).

What all this points to is the need for care in thinking through what triangulation means. It is not having different data sources as such but using *complementary* sources. The word 'complementary' is critical, for the data used must be capable of addressing the question investigated. It should not just draw attention to the same human condition (at different geographical or temporal scales). Ideally, as later chapters show, complementary data sources should utilize different means of information collection, since every mode of data collection has built-in biases and limitations. As a result, data collected in dissimilar ways have the potential to highlight different aspects of a research problem, so providing greater cross-validation in research interpretations.

---

9 In the 1960s, when public-sector construction was higher, 1 275 241 new-sector housing units were built, with 592 314 demolished (1961–70). This means that 46.4% of new build replaced demolished units, compared with 41.1% for 1980–8.

Box 2.2 Experimental research designs

One-group designs

Pre-test post-test	$0_1 \; X \; 0_2$
Interrupted time-series	$0_1 \; 0_2 \; 0_3 \; X \; 0_4 \; 0_5 \; 0_6$
Time-series	$0_1 \; 0_2 \; 0_3 \; 0_4 \; 0_5 \; 0_6$
Correlational design	$0 \; X$

Two-group designs

Two group	X_1	0_1	
	X_2	0_2	
Multiple group post-test	X_1	0_1	
	|	|	
	X_n	0_n	
Multiple group pre-test post-test	0_1	X_1	0_2
	|	|	|
	0_{n1}	X_n	0_{n2}
Ex post facto	X_1	0_1	
	X_2	0_1	
Multiple group time-series	$0_1 \; 0_2 \; 0_3 \; X_1 \; 0_4 \; 0_5 \; 0_6$		
	$0_1 \; 0_2 \; 0_3 \; X_2 \; 0_4 \; 0_5 \; 0_6$		

Key: 0 refers to an observation, X to an 'event' whose impact is being assessed.

Source: P. E. Spector (1993).

Inconsistency between methods

This all sounds well and good, but the use of different data sets has the potential to yield contradictory results. Midanik (1982) points to this, in noting inconsistency in interviewee reports and other data sources on alcohol consumption. Rabow and Neuman (1984) produce similar conclusions, in an examination of the effect of the opening of a liquor store on a residential area. Contrasting the empty bottles found in people's garbage cans with questionnaire evidence, they found the questionnaire survey seriously underestimated alcohol consumption (albeit the project had to be abandoned because so many refused to have their garbage cans checked!). However, as in the case of patients not having detailed knowledge of their own surgery (Maquis 1970), such inconsistencies are informative. Of course, the suggestions such inconsistencies pose might not be those that most interest the researcher. Reaching the conclusion that people are rather conservative in reporting alcohol consumption might be amusing, but can be troublesome for studies of consumer purchasing.

In simple terms there is no real answer to the problem of inconsistencies in conclusions drawn from different data sources. This is because there are so many different implications from such findings. At one level there is the situation in which one data source is taken to be more accurate than another. An example would be checking interview responses against documentary sources in which you have a lot of confidence. But the key to making best use of multi-method approaches is not to assume automatically that a problem of method exists. This should be examined but should be approached as one of a number of possible sources of inconsistency. A powerful prospect is that inconsistencies lead to a more penetrating consideration of the concepts and theorizations that guide a study. Good research relies on investigators having an open mind towards changing fundamental precepts in their research approach, as inconsistencies unsettle the comfort blanket that familiar concepts and theories provide. Certainly, coordinated multi-method approaches strengthen key results by confirming them using data sources that offer dissimilar insights. At the same time, they can discomfort common assumptions, which have regularly relied on definitions and mainstream theorizations that are embedded in evaluations using one type of data. These are the ultimate strengths of multi-method approaches.

Summary: using multi-method appraoches

- In so far as they offer complementary insights on a research problem, multi-method approaches enhance prospects of developing arguments with supporting evidence that convince others.
- Multi-method does not mean multi-scale. Seeming consistencies and inconsistencies between the two can be informative and lead to deeper understanding, but they are addressing different issues.
- Multi-method analysis can highlight important inconsistencies in results. These should led to deeper questioning, of method, of results and of theorization.

Designing a project

We started this chapter with a quotation from Bouma and we wish to conclude it by drawing again on this source to emphasize two fundamental points about research design. The first has been implicit in what we have said already but is worth stating overtly. This is simply that: 'In order to succeed, research must be guided by a clear statement of the problem or issue to be addressed' (Bouma 1993: 35). Woolly objectives lead to woolly outcomes. But add to this a second critical point. This is especially important for student projects, for it is another issue students can find difficult to grasp, and they consequently find setting up their project more unsettling and nerve wracking than it should be. This is to bear in mind that: 'It is better to answer a small question than to leave a large one unanswered' (Bouma 1993: 10–11). In other words, be sure the objective is attainable and do not over-complicate the question such that the demands of research methods are beyond your capacity or resources.

What this chapter has not addressed is what the demands of research methods are. This is what we turn to now, with the rest of this book focusing on the particularities and peculiarities of research employing different types of data. What we have sought to provide in this chapter is a framework of issues to consider when putting together a project. As we will show in the chapters that follow, epistemological frameworks impose themselves in articulating the contribution research methods can offer on theoretical questions. An essential need in designing research projects is to understand the epistemological foundation of the approach you adopt. This not only has implications for what is interpreted as data but also specifies how the methods you employ offer insight on the problem at hand. Social science research is about argument, not proof. The researcher's aim is to convince others. For this the research project needs to have coherence in epistemology, theory and method. It also needs quality evidence to support the arguments made if readers are to be convinced. As we will show in the chapters that follow, since all research methods have peculiar weaknesses and strengths (just as epistemological positions do), single method approaches have question marks placed against them. Research strategies that creatively employ diverse methods in an attempt to circumvent particular weaknesses enhance the prospect of deepening our knowledge. For some projects coherent multi-method strategies are not easy to achieve but they are a goal worth striving for.

3
By the book? Using published data

The regular use of secondary information often develops a healthy skepticism about information provided by others.

(Stewart and Kamins 1993: 17)

In the jargon of research methods, published data are secondary data. The designation 'secondary' indicates that researchers are removed from the data collection process. Interpreted as data collected by the investigator him or herself, primary data might give off a warm glow that convinces a reader of their sanctity, but primary data are shot through with imponderables and doubts, just like secondary data. A key issue is whether data are adequate for investigating a particular research question. In this regard primary data collection is often portrayed as more advantageous (even pliable), as the researcher has more 'control' over what information is gathered, and how 'events' are interpreted. Lacking this 'control', secondary data analysis imposes different strictures on the researcher. One of the first books to examine this form of data made this point clearly: 'One of the advantages of secondary analysis is that it forces the researcher to think more closely about the theoretical aims and substantive issues of the study rather than the practical and methodological problems of collating new data' (Hakim 1982: 16).

Yet the most thought many researchers give to secondary data seems to be deciding whether or not to use it. Forget for the moment using paid interviewers to undertake questionnaire surveys on your behalf. This might qualify as a form of secondary data if the analyst has not engaged in the interviews, but you as researcher had a hand in deciding what information to gather, how to represent key variables, how to collect and how to collate data. By contrast, for secondary data, information is out of your control, even if 'authoritatively' collected. In this circumstance, it might appear that the most appropriate insights this book could offer is how to analyse secondary data. This interpretation would be wrong. If you treat information as 'facts', you make a grave mistake. Had we wanted to present a book in this genre, we could have added to a long list of barely distinguishable geography texts on statistical analysis (or other data manipulation methods). Had we wished to warn that statistical information can manipulate readings, humorous offerings make this point well (e.g. D. Huff 1954; Runyon 1981). Because we need to appreciate what numbers and texts reveal and conceal, we need to understand the nuances of secondary data. This is especially so for published information, since the process of publication might seem to provide legitimacy.

But the social nature of official statistics needs to be appreciated (Miles and Irvine 1979). As Mercer (1984: 168) expressed it: 'statistics do not, in some mysterious way, emanate directly from the social condition they appear to describe ... between the two lie the assumptions, conceptions and priorities of the state and the social order, a large, complex and imperfectly functioning bureaucracy' (see also Desrosières 1996). This is seen in male-centred biases in categories used and topics presented in official statistics (Oakley and Oakley 1979; Madge *et al.* 1997). The empiricist view that official statistics are objective facts is challenged by recognition that these outputs emerge from subjective judgements that reflect organizational priorities. There is a politics to what information is presented to the public, how it is presented and what use can be made of it (e.g. Dorling and Simpson 1999). This does not mean data do not reflect social conditions, nor that they are not powerful research tools. It does mean that they conceal elements of the population and, taken by themselves, direct attention towards certain issues and away from others.

This might seem surprising, given that the most systematically gathered and published information comes from governments. There are various reasons for this. For one, why should private companies go to the expense of publishing data about their own activities? Unless there are commercial reasons,[1] such outputs would be a drain on profits. Of course governmental regulation might necessitate the release of information, but is publication necessary? With the dawning of the Internet, putting information into the public realm is cheaper, but are companies happy to divulge material on themselves? Generally, private sources offer limited information in 'published' form. Commercial considerations will probably direct web-trawlers towards certain activities and away from others. This does not mean there are no valuable data on the web for researchers. When these are available, they are likely to be restricted, as with the Nationwide Building Society regional house price data, which draw on mortgage information from this society alone (http://www.nationwide.co.uk/).[2] Governmental agencies are likely to present a more comprehensive picture. Theories of the state predict this; either in the form of

---

1 One obvious example is the provision of information so others can assess performance or provision for customers. Examples include the quarterly *Hambro Company guide* on the performance of stock-market companies and telephone directories.

2 This chapter does not explore WWW data availability specifically. Commentary on using the Internet is offered in Chapter 5 and Chapter 7, with one listing of data 'online' in Talbot (1992). But note that the Internet provides a mixed bag of data. Many official agencies have useful sites. Notable examples include the US Bureau of Census (http://www.census.gov/) and the UK's Office of National Statistics (http://www.statistics.gov.uk/), with many countries having similar sites (as with Saudi Arabia, which offers a bounty of data: http://www.saudinf.com/). Eurostat, the EU data service, also has a useful site (http://europa.eu.int/comm/eurostat). A variety of private-sector and interest-group sites also exist. These might tell you more about the vested interests of the organization than a topic under investigation.

pluralist theory articulating a state impetus to understand and then to improve society, or in Marxist theory, which emphasizes how the state must intervene to thwart social unrest, underpin profit accumulation and provide communal resources (Dunleavy and O'Leary 1987). Whatever perspective you take, Higgs (1996) provides an indication of the forces that propel state agencies. Writing on the nineteenth century, he notes almost indiscriminate state data gathering in an effort to reveal 'the state of the nation'. This was propelled by contemporary associations between social conditions and health. Here concerns about the differential reproduction of social classes were influential, alongside a belief that radical opposition to the economic system was based on ignorance, which could be dispelled if the true economic and social structure of the nation was visible (see also Cullen 1975; Tooze 1998; MacKenzie 1999). Yet demand for more statistical returns also arose from fears about economic competition from abroad, much as expansion of French official statistics was linked to policies of national and imperial expansion (Woolf 1984). Quite apart from state resources generally exceeding those of other national organizations, making this the 'logical' institution to collect data, the dictates of electoral accountability press governments to identify and respond to socio-economic (or recently environmental) potentialities and problems.

This does not mean that state agencies have a free hand in deciding what information is reported or how it is reported. Official data are embedded in, and partly determine, national cultural norms. The representations that official reports provide consequently 'emerge' from political compromises, between levels of government, agencies within government and negotiations with a range of social actors, such as employer associations, civil servants, politicians, trade unions and researchers (Desrosières 1996). The presentation of official information is political not simply on account of these compromises, but also from instrumental use of the information. This is seen in 'the narrative of nationhood' being enlivened and embellished through visions of national progress, as recorded through official statistics, and as consumed daily by the newspaper-reading public from the nineteenth century to the present (Tooze 1998). This should not be read to imply that conspiracy lies behind the collection and presentation of official information. Nor does it suggest that compromises over official information are fixed. Demand for information changes over time, with resulting outcomes, with what information is gathered and what is made widely available, reflecting historically specific social conjunctures. Thus, one cause of rapid growth in state information collection has been war. This was apparent in the First World War, but the efforts of Albert Speer in keeping German manufacturing in operation in the face of heavy Allied bombing offers a more classic example of the centrality of data in governmental decision making (Tooze 1998). Perhaps not unexpectedly, given the power of economic interests, the economic sphere received early attention in public information gathering (even if at times indirectly, as with efforts to collect information related to health, given the adverse effect illness had on labour productivity). But as time progressed broader domains came under the governmental umbrella. Thus, in the Second World War, the UK Government found it needed more social information, owing to a new deployment of men and women, a mobilization of voluntary activity, rationing and so on (L. Moss 1991). More recently we have seen a different set of pressures, with the OECD and the UN taking an active role in promoting greater information availability and the standardization of information, as a consequence of globalization pressures.[3]

There is an important implication in the above messages. This arises from the compromises states make on what information to collect and present. Put simply, it is to be expected that such compromises reflect power relations in society. Offering a simple taster of this, Glover (1996) notes that if feminist views held sway in the way official statistics are gathered, collated and presented, then nurses might not be grouped under one occupational category but have as many subcategories as engineers. As another example, information on farm income (in the UK and EU) is woefully inadequate, with the vested interest of the farm community and agriculture ministry bureaucrats leading resistance to more valid income measures (see e.g. B. E. Hill 1999). At this point we do not want to start debate about the character of such biases. Evidence on this score should become abundantly clear as the chapter progresses. But the question of bias in what is and is not reported, how material is presented, and who has access to that information has troubled researchers for some time. This is not surprising, given criticisms levelled at official statistics. To take one example, Dorling and Simpson (1999: 4) hold that in the UK we do not have the kind of statistics that enable us to understand and improve life, citing the following inadequacies.

- There is a paucity of statistics on gender, with children commonly treated as a household property.
- Work done in the home is not measured, only paid work.
- Unemployment figures do not cover those who want jobs.
- Health statistics do not cover health-care needs.
- Housing statistics do not assess homelessness or housing need.
- Poverty statistics do not identify the number of poor people.
- Economic and transport data neglect social and environmental costs.
- Only since the 1990s have there been statistics on racially motivated assault.

Despite such weaknesses, the analytical potential of official data is clear, even for radical critiques of 'the Establishment'. If you recognize that Karl Marx used governmental statistics and official 'Blue Book' reports, or that Emile Durkheim based his path-breaking *Suicide* (1952) on official statistics and unpublished reports in the Ministry of Justice (Macdonald and Tipton 1993), the centrality of official data in social science theorizing becomes clear. One reason for inconsistency between contributions to social theorizing and critiques of official data is the criticism that such data are socially constructed. But all

3 Offering one indication of this increase, Tooze (1998) notes that national income estimates only existed for 39 nations in 1945, but for 93 in 1955 and more than 130 in 1969.

data are socially constructed in one sense or another. What is important for quality research is to understand how the 'creation' of a data set infiltrates its core assumptions, silences and emphases. In a nutshell, we need to be conscious of serious limitations in using official statistics, alongside their powerful advantages. A balance must be struck. This should be an informed balance.

Summary: The nature of official statistics

- These data are socially constructed, so they embody the emphases and biases of their generation and the interests of powerful social groups.
- The strength of these data sources arises from the comprehensiveness of their coverage, by population and topic.
- The magnitude of statistical information now available from private and public sources is immense, growing and deepening.
- Researchers need to be conscious of rationalities for the expansion of official statistical accounts in order to understand their emphases and silences.

What else are we going to use?

So the reader can contextualize comments in this chapter, we should first point out that material explored in the next chapter can also be said to be 'secondary data'. To distinguish what we report here from what is found there, we have made the primary focus of this chapter statistical counts. These might be lists (such as directories of retail outlets, like those provided by Kelly's in the UK and Dun and Bradstreet in the USA), where the researcher composes the numerical representations used in analysis. Alternatively, the information might already be aggregated when published, as with census reports. As the breadth of secondary information is so broad, there is no sharp line between what we cover in this chapter and the next. There our attention is devoted more to discursive accounts, although the distinction is not rigid. Many statistical reports have extensive commentaries, exploring regular, mostly annual, outputs from the UK Government (for example, the *Digest of environmental statistics*, *Family spending*, the *Family resource survey*, the *General household survey*, the *Labour force survey*, *Living in Britain*, *Local housing statistics*, the *National food survey*, the *New earnings survey*, *Regional trends*, *Road traffic statistics* and *Social trends*; for a short review of some of these, see Owen 1999). Likewise, in the midst of discursive material we regularly find statistical tables (as seen in minutes of local government meetings). To clarify further, we are not seeking to distinguish between published and unpublished material in a hard-and-fast manner in these two chapters. For instance, census reports are available as published documents, but such information also comes from computer-based files, and, if the data were collected sufficiently long ago, individual census returns can now be examined. We do not wish to draw a line between these types of reports, as they are the same data, presented differently. As such the primary criterion for inclusion in this chapter is whether data are expected to go into the public realm in a 'published' form. If so, these are included, with the emphasis on reports that provide data – such that these reports tend to be statistical in nature – rather than describing or justifying organizational policies (that is, they are not policy documents or policy reviews).

When viewed in this light, and given this numerical bias, the central advantage of published data is that we often cannot get the information another way. In some cases this is due to access issues. The annual reports published by companies, local governments, housing associations and so on might come across as dry, superficial accounts, but the chances of the same organizations allowing researchers to rummage through their confidential records is often slim. It is not just that organizations want control of the information they release, but that at times they need to preserve the confidence of those they work with. Unless we are living somewhere above Cloud Nine without realizing it, we doubt the citizenry would relish researchers being able to publish details on them from medical files. Reporting on large numbers of individual events or people, published accounts are prone to be flat in this regard. Occasionally official agencies allow us glimpses of what lies behind aggregate accounts, by exploring individual-level data. One example is applications to construct or extend buildings in the UK. Here there is a blandness to local planning authority reports that inform us about the proportion of building applications accepted or rejected, the percentage approved with conditions, or the number of dwellings built. Yet analysis of single applications can reveal biases that favour construction in certain sites or by certain people or organizations (see e.g. Buller and Hoggart 1986; Gilg and Kelly 1997). Such insights should remind us that official reports commonly come across as 'factual' rather than analytical.[4] Such reports leave a lot to be desired. Yet it is often through the presentation of simple statistical (or other) accounts that researchers are attracted to issues. Take, for example, evidence on divergent occupational mobility amongst ethnic minority groups (see e.g. Heath and McMahon 1997; Modood 1997), or the uneven incidence of monetary gain and pain following housing market booms and busts (Hamnett 1993; Dorling 1994), or gender differences in the impact of migration on occupational mobility (Fielding and Halford 1999), or the impact of parental housing tenure on children's (adult) housing (Lyons and Simister 2000) or how change in government spending and taxation bring regionally uneven income gains and losses (R. J. Johnston 1979; Hamnett 1997).

What lies behind the provocations that published accounts provide is the relative ease with which analysts can grasp trends, deviations and seeming inconsistencies (if only at a

4 This has received media attention. As one example, Melanie Phillips and associates (1989: 21) comment on 'political' intervention in official UK reports in the 1980s: 'the figures are published but the accompanying commentary is truncated. The recent disablement surveys by OPCS [Office of Population Censuses and Surveys] were written under the instruction that the commentary should contain no value judgements. This rules out completely any use of words such as poverty or disadvantage, for example. The text could make no attempt to tease out the policy implications of the findings. And once again the Treasury not only combed the text but insisted that certain words were changed'. (See also M. Dean 1995.)

surface level). This capacity is enhanced by the amount of data presented, whether articulated as the number of informants, the number of places or the time period covered. Even in a seemingly 'descriptive' format, there is enormous power in information on comparative 'performance'. Fundamentally, this arises from the magnitude and breadth of state involvement in information gathering and presentation. It might be that the British are prone to do things on a shoestring, but the 1991 Population Census still cost the UK Government £135 million (Raper *et al.* 1992). Add an increment for a population that is four times greater, then throw in a good dose of American largesse, and the figure for the US Census was $2.6 billion in 1990, with an anticipated cost for 2000 of $6.8 billion (*The Economist*, 1 April 2000, 47). Described by *The Economist* as the federal government's largest peacetime logistical task, the magnitude of the US task, in collecting, checking, collating and publishing, is well beyond the capacity or inclination of private organizations.

The issue is not just data volume but also range (see e.g. Goyer and Domschke 1983; Domschke and Goyer 1986). The vastness of the US response is highlighted by the size of the monthly supplement to the *American statistical index*, which lists new US government publications (Congressional Information Service, monthly).[5] Yet national governments are not alone in providing outpourings of statistical information (let alone reports). International organizations are of major importance also, especially at the national level. Publications that provide a broad range of indicators include the United Nation's (annual) *Statistical yearbook* and the World Bank's (annual) *World development report*. Among the more familiar (annual) reports with a narrower theme are those on fisheries, forestry and farm production (and agricultural trade) from the UN's Food and Agriculture Organization (FAO), or the UN's (annual) *Industrial commodity statistics yearbook*, with data on the earth's ecosystems available in the World Resources Institute's (bi-annual) *World resources*. General economic indicators are found in the Organization for Economic Cooperation and Development's *OECD in figures* (OECD, annual) or the UN's (monthly) *Bulletin of statistics*. Amongst the publications giving data on trade there is the International Monetary Fund's *Directional trade statistics yearbook* (IMF, annual) and the UN's (annual) *International trade statistics yearbook*. Educational information can be found in UNESCO's (annual) *Statistical yearbook*, which carries a more social tone, albeit with an economic imperative; data on health problems and facilities are available in the World Health Organization's *World health statistics annual* (WHO, annual); visitor information is presented in the World Tourism Organization's *Yearbook of tourism statistics* (WTO, annual). All this plus enormous output from the European Commission (see Eurostat (http://europa.eu.int/comm/eurostat/)). Not all such information comes in regular reports, for many compendiums are extracted from

5 In terms of five- or 10-yearly census reports, the US Bureau of Census provides nationwide statistical documentation on agriculture, county business patterns, government, housing, manufacturing, mineral industries, population, service industries, transportation, communication and utilities, and wholesale trade.

varied sources, so offering ease of access to material on themes (e.g. McGillivray and Scammon 1994; Simmons 1999). In some of these cases the compilations offer valuable historical runs that identify changing national circumstances stretching back over the centuries (e.g. Mitchell 1988, 1998a, b, c; D. Dodd 1993; Smith and Horton 1995).

Although the volume of printed material from the UK government is less than from US sources, or other European nations, even 15 years ago the national UK administration was reported to produce about 160 kilometres of shelf space of documentation each year (Hamshere 1987). The volume of information is vast, with the add-on of national organizations publishing information – sometimes primarily for their members but with availability to the public, as with the Chartered Institute of Public Finance and Accountancy reports on local government in England and Wales (Box 3.1)). Moreover, huge amounts of information are available from private and quasi-private sector organizations. One example is the 'country report' series of the Economist Intelligence Unit, which offers in-depth political and economic examinations of 200 countries on CD-ROM, the World Wide Web and in print (albeit for a hefty price – if libraries do not

Box 3.1 First date for publication of local government statistical series by the Chartered Institute of Public Finance and Accountanct (CIPFA)

These annual statistical profiles offer data on each local government unit that administers the functions identified. There have been minor changes in titles of some series, with the most substantial noted below. CIPFA changed its name from the Institute of Municipal Treasurers and Accountants (IMTA) in 1973. This list provides a good indication of the wealth of data made available by some non-government units.

Series	First year issued
Administration of justice statistics	1977
Archive service statistics	1988
Capital expenditure and treasury management statistics[a]	1947
Cemeteries statistics	1979
Charges for leisure services statistics	1978
Council tax demands and precepts	1992
Crematoria statistics	1956
Direct service organization statistics	1982
Education statistics	1948
Environmental health statistics	1982
Finance and general statistics[b]	1945
Fire service statistics	1948
Highways and transport statistics	1976
Homelessness statistics	1977

Housing rent arrears and benefit statistics	1994
Housing rent statistics	1968
Housing revenue account statistics	1948
Leisure and recreation statistics	1976
Local authority assets statistics	1996
Local authority superannuation and fund investment statistics	1966
Local government comparative statistics	1981
Local government trend statistics	1999
Personal social service statistics	1951
Planning and development statistics	1976
Police statistics	1949
Probation service statistics	1974
Public library statistics	1960
Revenue collection statistics[c]	1933
Trading standards statistics	1975
Waste collection and disposal statistics	1976

[a]Original title *Return of outstanding debt.*
[b]Original title *Return of rates.*
[c]Original title *Rate collection.*

Source: CIPFA Statistical Information Service.

have copies, see http://www.store.eiu.com/cr_offer.asp/). Often data on private and quasi-private organizations are not provided as sources for researchers but as aids for a business sector or for customers.[6] Each of these aids has different problems in terms of its representativeness, accuracy and regularity of publication.

The range of such materials is so extensive we cannot attempt a comprehensive list, for many are ephemeral outputs (for example, London's *Time Out* can offer insight on, say, the location and incidence of gay and lesbian bars). Other outputs appear regularly. For example, the performance and location of corporate headquarters for the largest manufacturing firms and banks in the world are available in an annual *Fortune* magazine survey (*The Times 1000* offers a UK list). Figures on crime in the USA are subject to detailed reports from the FBI (annual). UK local government election results are presented in collations by Rallings and Thrasher (annual), with a historical collation (and opinion poll results) in Rallings and Thrasher (2000). When there is sufficient interest, private organizations, non-governmental organizations or individuals are prone to compile data of considerable value to geographers.[7] As one example, various compendia exist that

6 Official statistics can be used in innovative ways to gain insight on research questions. Thus Gallent and Tewdwr-Jones (2000: ch 1) note second-home ownership levels can be estimated from spending figures in the UK's *Family spending* (UK Office of National Statistics, annual (b)).

7 For example, the British Geological Survey produces reports that can be applied in economic analyses, environmental research and sustainable development planning, as in the Survey's *World mineral statistics, Individual commodity tables and United Kingdom minerals yearbook*. On these, see http://www.british-geological-survey.co.uk/.

allow comparative judgements on city living standards (e.g. Boyer and Savageau 1989). Regular newspaper reports on the cost of living in cities or their relative attractiveness as living spaces indicate the potency of such compilations for businesses and (some) householders. There might be fierce disagreement over the measures included in such indices, but as these publications have material on cultural, economic, environmental, political and social conditions in cities, they can be valuable sources.

Yet the variety of topics covered and the range of themes in data compilations mean that analysts need assistance in searching for sources (especially given uncertainty over the reliability of sources). Directories on sources, as well as commentaries on information reliability, should be referred to early. Analysts should recognize that even national censuses have uneven coverage and reliability. Commentaries on unfamiliar censuses should be explored early – whether or not they involve cross-national comparison (e.g. Goyer and Domschke 1983; Domschke and Goyer 1986). Excellent analyses exist on the politics of official statistics (e.g. M. J. Anderson 1982; Alonso and Starr 1987), as well as on biases in state mind-frames favouring numerical accounts (e.g.

Box 3.2 Regular statistical reports on housing in England and Wales

Chartered Institute of Public Finance and Accountancy (CIPFA), *Homelessness actuals, Housing rent arrears and benefits actuals, Housing rents actuals, Housing revenue account combined actuals and estimates, Planning and development combined actuals and estimates.*

This array of annual reports provide information on public-sector housing for every local housing authority. The range of information available is vast, with the number of variables increasing significantly over time. CIPFA sources to contextualize housing policy and performance measures include the annual Local government comparative statistics and Local government trends. Information on the range of CIPFA publications is at http://www.cipfa.org.uk/.

Council of Mortgage Lenders: *Housing finance.*

The quarterly publication contains commentaries/articles and a statistical appendix. The latter has information on housing starts and completions, house prices, interest rates, building society loans and financial statistics. A few tables have regional information, but most is at the national scale.

London Federation of Housing Associations 1996, *LFHA directory 1996.*

This is an irregular publication, listing non-profit housing association dwellings. The report lists associations with a remit to provide specific services (e.g. social service provision), caters for particular populations (e.g. ethnic minorities) that are active builders, and shows how many dwellings each association has in every borough. The previous report to this was produced by the London Housing Associations Council in 1989.

London Research Centre, *London housing statistics*.

Offers a wealth of detail, by borough and across the metropolitan area, on private, public and non-profit sectors. Amongst the information on each London borough are data on new build, sales, local authority waiting lists, local authority lettings, ethnic monitoring, racial harassment, empty properties, sale of council dwellings, housing association rents, rent arrears, house prices, housing benefit receipts, national housing subsidies for councils and new housing investment.

National Assembly for Wales, *Welsh housing statistics*.

Provides a range of information on each local authority with housing responsibilities in Wales. This annual publication has data on renovations, new build, the age of stock, demolitions, homelessness, social housing, council house sales, lettings and vacancies, housing finance and rents.

National Housing Federation and Housing Corporation, *CORE statistics*.

Offers information on housing associations. The COntinuous REcording system has been developed by the National Housing Federation (NHF) and the Housing Corporation. There are two primary publications – the *Annual digest* and the *CORE bulletin* (plus a *General needs bulletin*, a *Sales bulletin* and a *Supported bulletin*, some quarterly). The *General needs bulletin* has rents and income by region, with analyses of affordability. Supplementary tables are provided on household type, ethnic origin, housing benefit eligibility, homelessness and sources of referral. The *Sales bulletin* reports on housing costs for those with low-cost home ownership, including house prices, mortgages and shared ownership rents by size and type of property. The *Supported bulletin* records lettings and rents, tenant profiles and tenant incomes. More information is available at http://www.core.ac.uk/.

Nationwide Building Society, *House price index*.

Formerly presented as a published document, in the late 1990s this became available on the Internet. The address is http://www.nationwide.co.uk/hpi/. The data are at a regional level on house prices, including historical runs, for different types of houses. Data are derived from homes with mortgages from the Nationwide Building Society. The web site explains the methodology.

UK Department of the Environment, Transport and the Regions, *Housing and construction statistics*.

This quarterly publication provides regional information on housing starts and completions, dwelling prices, building society advances, mortgages, housing renewal grants, slum clearance, orders received by building contractors, and the number and value of dwellings sold by local authorities. It is issued in two parts each quarter.

UK Department of the Environment, Transport and the Regions, *Local housing statistics*.

Reporting on every local authority in England and Wales, this quarterly report includes information on housing starts, homes under construction and dwelling completions. This is provided for the private sector, the public sector and non-profit housing. Some years have data on demolitions, homelessness applications and local authority responses to them.

UK Department of the Environment, Transport and the Regions, *Households found accommodation under the homeless provisions of the 1985 Housing Act, England.*
A quarterly account with information for all local housing authorities in England and Wales. Figures are presented on applications made on the grounds of being homeless, applications accepted, rejected and under consideration, plus the number provided with permanent and temporary accommodation. Commentary on the problems of homelessness statistics can be found in Widdowfield (1999).

UK Department of the Environment, Transport and the Regions irregular, *English housing condition survey.*
Data have been collected in 1967, 1971, 1976, 1981, 1986, 1991 and 1996. The survey is based on a sample of dwellings, with no fine-grained geographical distinctions in data presentation. The information indicates state of repair and quality of living conditions (including abandonment), with broad distinctions between regions, rural and urban zones and housing tenures. Trends in physical condition can be inferred.

UK Office of National Statistics Social Survey Division, *Housing in England.*
Compiled annually, this includes information on trends in tenure, recent movers, social renters, owner occupiers, private renters, household formation by the young, and even views about the area in which houses are located.

Valuation Office, *Property market report.*
A bi-annual report with information on selected places to 'represent' regions in the UK, with data on house prices, housing types, age of dwellings and dwelling size. As an illustration, for a region with few places to represent it, the North-East of England presents values for Durham, Middlesborough, Morpeth, Newcastle-upon-Tyne and Sunderland.

Poovey 1998). In terms of coverage, Box 3.2 offers one insight. This is an illustrative example, to show how governmental, private-sector and voluntary-sector publications provide varying (but often complementary) entry points for numerical information on a specific research field – in this case, housing in England and Wales. Similar lists of key data sources can be run up for other activity spheres by noting sources used in research papers on a theme.

What of subnational units?

There are implicit messages about data availability in Box 3.2 to recognize. The first concerns the geographical scale at which data are reported. There is a wealth of information at the national level. One indication is the number of international organizations that provide annual (or more regular) statistics for nations of the world, even if there is a long way to go in securing harmony in cross-national measurement (e.g. Cheshire *et al.* 1996; Singleton 1999). Below the national level a multitude of figures and tabulations again exist (e.g. European Commission 1995a, b, 1999), at least for basic administrative areas within nations. Significantly here, reporting units tend to be meaningful for administrators, but commonly not for single research projects. Illustrative of this, for non-profit housing associations in the UK, organizations often provide data by association, rather than by place. Since associations can manage dwellings in a large number of localities (see e.g. Kirby 1985), the absence of a geographical index means researchers can face a lot of work establishing which associations operate in a study area (e.g. Richmond 1985). Fortunately, this task is occasionally eased by local directories (e.g. London Federation of Housing Associations 1996). The key message is that researchers cannot assume data will be provided for the geographical unit they are interested in. Moreover, government agencies can use different units to present figures. Thus, the monthly UK publication *Employment Gazette* lists the number of unemployed by employment office area. Other than by special tabulation, comparable information from other governmental agencies is not available for these units (they do not coincide with local government areas, for example). A further consideration is unevenness of coverage across geographical areas. This is readily seen in the information provided by the US Bureau of Census on substate areas, with many data returns restricted to large cities and metropolitan areas (as exemplified in the wealth of data in the US Bureau of Census's *County and city databook* or in the block statistics in the same organization's *Housing census*).[8] Let us not confuse this issue with that of confidentiality. It is to be expected that government agencies will be assiduous in ensuring data are not available in a form that enables individuals to be identified, so statistics for small areas are more difficult to obtain than for larger zones. But often information is published for relatively small population units (like municipalities) within metropolitan areas, when the same information is not published for places with a larger population if they are independent cities or low-density rural areas (e.g. US Bureau of Census *State and metropolitan databook* or the US Department of Commerce publication on *Metropolitan area exports*).

The importance of administrative units should not be forgotten here, for the ease of obtaining data is greatly reduced if areal units are not formal administrative zones. This does not mean data are not available in unpublished form, but it increases the chances this will be the case. In making this statement we advise the reader to exercise caution in its interpretation, for there are cultural divergences in data availability. This is exemplified in studies stimulated by the inverse-care law. In this case we find a marked disparity between the UK and the USA in the availability of sub-administrative area data. While US data are readily available on (say) municipal service provision to neighbourhoods (Lineberry 1975; Bolotin and Cingranelli 1983), or even citizen contacts with their council by neighbourhood (e.g. B. D. Jones *et al.* 1977), in the UK Government officials have seen less utility in exploring distributional issues below the administrative

8 We use the US Bureau of Census for illustration. This pattern applies in many countries. For England and Wales, for example, early publications on local government expenditure and taxation were available annually for cities designated as county boroughs, but not for other areas (UK Ministry of Health, annual).

level, so neighbourhood level data are sparse (Webster and Stewart 1974). As a result, UK studies of 'territorial justice' largely compare local government districts (e.g. Boyne and Powell 1991). Some interesting UK studies were instigated by political antagonism in the 1980s (e.g. Pattie 1986), and some thought-provoking studies explore intra-authority issues (e.g. Hebbert 1991), but there is a lack of empirical depth in the UK literature, given the absence of information on service provision to neighbourhoods (e.g. Whitehead 1983). An obvious advantage of governmental statistics is their broad coverage. This does not help if data are presented for inappropriate areal units. This is not just an issue for researchers, but also how people are primed to view their own society. A pointed example occurred in the early 1990s when one of the authors was asked to take part in a BBC radio broadcast about rising rural crime. Sensing a new data source of interest, the BBC contact was asked about the geographical units for the data. On being told this was police force areas (which cover a county or even more than one county), the comment was made that this could not indicate that rural crime was rising. Even if Kent had seen a sharp rise in crime, this could be accounted for by Chatham, Dover, Gillingham, Maidstone, Tonbridge and Tunbridge Wells, with no rise in countryside areas. This misrepresentation could have been avoided had distinction been made between metropolitan centres like London and Manchester and non-metropolitan counties with no major city, for this was the real contrast the programme had identified. Certainly, referring to a 'rural' crime wave might be discarded as poor understanding of data, or even journalistic hype resulting in a too common misreporting of events (Glasgow University Media Group 1976, 1980, 1982). For our immediate purposes it does not matter which is the case. What should be noted is that an issue the media still projects as 'fact' cannot be easily substantiated owing to data non-availability. This does not mean that important studies of rural crime in the UK cannot be undertaken. Detailed studies of a small number of places provide an obvious example (e.g. Yarwood and Gardner 2000). However, politicians and officials reveal a preference for 'systematic evidence' rather than case studies, as well as quantitative measures rather than qualitative interpretation (see e.g. Cantley 1992; Poovey 1998). One result is that local studies and qualitative investigations can be cast aside as 'anecdotal' (e.g. see the discussion on presenting research to government officials in Cloke *et al.* 1997). This creates particular problems when studying issues like crime, whose incidence has been conjured into a mythological standing by attention-grabbing (or more accurately ratings-inspired) media reports (see e.g. Box 1983). With the media so intent on 'selling its story', data presented at inappropriate geographical scales can have important implications for images of societal problems.

Let us extend this point in an important direction. Recall that this section has an underlying question; namely, what else can we use? The message is that, for many issues, governments and other large organizations are at times the only institutions with the capacity to provide accounts with a broad coverage (by social agent, place or over time). In that context, (published) data from such organizations commonly provide the only source on general trends. The previous paragraph hinted that just because the data are there does not mean they should be used. Put simply, data must be appropriate for the research task at hand; not used simply because they cast a light vaguely in the direction of the question posed. Analysts must be rigorous in limiting the temptation to 'let the data speak'. In more conceptual terms, the reader should recall discussion of the ecological fallacy and of validity in Chapter 2, in order to place our comments appropriately.

Are subnational units comparable?

We want to go further in our consideration of geographical units for published data. In particular it is necessary to recognize that incorrect inferences can also be drawn because comparison is made between units that are not equivalent. Cheshire and associates (1996) draw this out when discussing difficulties in pan-European research on cities and regions. Focusing on Eurostat data, which try to offer some uniformity in areal and variable measurement across nations (Langevin *et al.* 1992; Eurostat 1999), this paper notes that the administrative units used for reporting data are inadequate for comparative investigation:

> *The existing set of official regions for European policy purposes conform to no common conceptual definition of 'region' but represent a hotchpotch of historical accident, local identity and national concern – or lack of concern – for consistency or logic in defining subnational administrative units. Because of the absence of systematic logic in defining Europe's official regions, official analyses of 'spatial disparities' confound truly spatial differences with differences due to social and economic stratification which, because of residential segregation, happen to have a spatial manifestation, and with the product of (changing) commuting patterns. (Cheshire et al. 1996: 43)*

In comparison with the USA, an inability to examine socio-economic conditions and trends in cities or city regions in Europe leads to an inability to evaluate satisfactorily competing explanations for local economic performance, to identify whether regional/national elements are critical in such performances or to see if trends in European cities reveal common components. This problematic is not simply due to divergence in statistical accounting, for there are important distinctions in the structure of 'city regions'.[9] Caution is needed in comparing jurisdictions with a dissimilar logic in

9 While accretions to 'the city' tend to be contiguous to built-up areas in France, green belt policies in the UK mean outward urban expansion is dispersed. As a result, the geographical principles of city regions differ across Europe (Cheshire *et al.* 1996). Reflecting such differences, state agencies have resisted efforts to encourage uniformity in data presentation (Desrosières 1996), which complicates sound policy formulation at the EU level.

procedures of data collection and presentation. The same applies for the definition of variables, but for the present it is the geography of data representation that concerns us.

What of temporal change in subnational units?

This concern has to be extended to the temporal dimension. Very obviously this is seen when major transformations occur in administrative units. Thus, in 1974 the local government system in England and Wales was reorganized, with a vastly reduced number of local authority districts (this restructuring occurred in 1965 for Greater London). In most cases the new districts were comprised of more than one old district, with many new units crossing the boundaries of old ones. As a consequence you need a special volume to compare post-1974 districts with those that existed before (UK Office of Population Censuses and Surveys 1975). Although 1971 Census returns are available for both local government systems, examination of trends stretching back beyond 1971 has been made difficult (even were variable definitions consistent). But our real concern is not with substantial transformations. These are 'in-your-face' events that researchers are unlikely to miss. More insidious are multiple small changes, which shift boundaries marginally. Again publications outlining changes can help, although maps are likely to be required if exact impacts are needed (Lipman 1949). A critical problem is when researchers do not expect such adjustments. If this occurs, temporal trends might be assumed for a location that is a different place.[10] How to handle this is not straightforward. In some cases analyses have been restricted to places that experience only small boundary changes. A key issue is that boundary changes are unlikely to make random impacts on population structures. This is apparent for the USA, where annexations that appear 'logical' in terms of the daily lives of residents can fail to be annexed for racial or socio-economic reasons (see e.g. Aiken 1987), just as annexations can occur that appear illogical on these grounds (e.g. Miller 1988). Irrespective of country, changes in the geographical reporting units can be less easy to recognize for areas with no administrative responsibility. Most evident in this regard, especially as they provide the smallest geographical units for census reports, are the enumeration districts (EDs) that demarcate the areas for officers who collect census information. Data for EDs are commonly available for researchers, in photocopy, microfiche or computer file form. Undertaking single year analyses with these units is not a problem, even if confidentiality concerns result in adjustments in data presentation.

10 Lipman (1949: 185–6) reports that from 1929 to 1937 boundaries were changed in 50 of the 83 county boroughs (cities) in England and Wales. The total land area affected was 45 170 hectares gained and 312 hectares lost. The net gain of 44 859 acres was twice the size of Birmingham. Analysts can reach incorrect conclusions if raw figures are taken for granted. For example, in 1934 Eaton Bray Rural District in Bedfordshire (1931 population 3935) was absorbed into Luton Rural District (RDC). At the same time, Toddington parish (1931 population 2500) was transferred from Ampthill RDC to Luton RDC. This was a substantial population increase for Luton RDC, which had 9666 residents in 1931. For 1951 Luton RDC reported 19 634 residents, with a good share coming from these boundary changes.

For these small areas, units for data presentation are also not necessarily ideal for research purposes. As Tim Butler (1997: 70) points out:

> *despite the rhetoric of the OPCS [the then Office of Population Censuses and Surveys] about the Census areas reflecting natural boundaries, they rarely do – at least in the inner city. Enumeration districts do largely reflect such natural boundaries but ward boundaries are fixed by local authorities, often on political grounds.*

Even more serious is variation over time (for example, compare the socio-economic classification of wards in Wallace *et al.* 1995, with that in Webber 1978). As Higgs (1996: 30) draws out for nineteenth-century UK censuses, it is not just that changes to enumeration districts make temporal comparison difficult, but that it is difficult to give general guidance on changes. Rather than resulting from a set of principles, changes appear to occur owing to local circumstances, change in population or the conscientiousness of local officers. In more recent decades this has been less true, as UK censuses have tried to use consistent enumeration districts since 1971. Moreover, from 2001, tabulations of UK census material will be possible using postcode information, so areal aggregates can be compiled for specific purposes (Martin 2000). However, investigations of small areas that seek trends over time have to commit long hours to ensuring comparability of reporting areas (for example, by merging units). Failure to do so means comparisons of unlike units are made, so bringing the validity of a study into doubt.

None of this should turn geographers away from governmental (or other large organizational) published statistics. We highlight the fallibility of incautious utilization of spatial units, alongside the problem of assuming consistency in units over time. There is no point in belabouring these issues. The real and potential problems caused by them should be explored prior to committing to a particular research approach. Accepting this, we come back to recognition that the magnitude, reliability and breadth of data provided by 'official' organizations make a bountiful resource for the careful researcher. If you wish to put this idea in numerical terms, then simply note the observation of Raper and colleagues (1992) that the UK censuses provided 1571 cells of tabular information for each area in 1971, 4400 cells in 1981, and between 8000 and 14 000 cells, depending on which area you are interested in, for 1991.

Summary: what else can we use?

- The volume of official material is immense, with much that can be used as research material not designed specifically for this purpose.
- Regularity of publication and extensiveness of coverage mean there can be little incentive to produce alternatives to single official sources.
- Data are often available on sources that do not focus specifically on the central theme of a research project.

- There are marked differences across nations in how data are reported, in topics covered, and in reporting areal units, with confidentiality concerns complicating data presentation.
- Over time, a further problem is shifts in the boundaries of areal units, which complicates temporal comparisons.

Then why be cautious?

Critics of governmental and other large organizational information are primarily concerned about their silences, their worrying tendency to be taken as indisputable 'facts' and the mistaken belief that they reflect the population, rather than part of it. For a start, we have to recognize that seemingly 'objective' statistics are socially constructed. Put simply, they embody the value dispositions and mindsets of those who produce them. As Miles and Irvine (1979) make clear, this does not mean we should be interpreting 'official' reports and statistics through the lens of conspiracy theory. This offers a rather convoluted explanation for official statistics, raising awkward questions about why official figures regularly embarrass governments. Rather than accepting simplistic accounts, a more nuanced understanding of official statistics is called for.

What about content?

As Hakim (1982) notes, a distinguishing feature of government statistical information is that it covers non-sensitive issues, for which data can be collected from the population at large in a reliable manner. This statement is easy to find support for. By way of illustration, fears about civil liberties and privacy led to the 1981 Netherlands Census being abandoned, while official statistics in Denmark rely on files that link data from separate agencies, as it has had no census since 1970 (Hall and Hall 1995). In Germany strong protests against the census led to six years of delay before the 1987 enumeration.[11] Go back to the instigation of census taking in the UK and you find fears about confidentiality and intrusiveness were potent worries. Thus 1750s proposals for a census were defeated as an attack on liberty (Lawton 1978). Only with growing fears about population growth, which owed much to Malthus (1798/1998), did views change, with the first census in 1801. Even then the first four UK censuses were little more than numerical counts, with an accretion of demands for information leading to an expansion of content and greater professionalization in census taking from 1841 (Higgs 1996). Only from this year were data gathered on individual members of households. Under the UK's 100-year rule, from 2002 the public can scrutinize information on individual household members from enumerators' books for 1841–1901 (for the USA, this kind of information is available from 1850; for census information availability by state, see Lainhart 1992).

11 One consequence is that some European nations use sample censuses to supplement population registration systems (Langevin *et al.* 1992).

Controversy over the information governments obtain and release on their citizens – even if in aggregated form – extends beyond whether data should be collected, to include what kind of data is acceptable. In the UK immigration from former African, Asian and Caribbean colonies is an issue that has stirred the political blood more vigorously than many policy debates. Perhaps it comes as no surprise therefore that questions of immigration and race have proved controversial. In the 1961 UK Census, for example, a significant proportion failed to complete the question on nationality (Benjamin 1970). For 1971 officials allocated an ethnicity to informants, based on their name, birthplace and parental birthplaces. Controversy over immigration in the late 1970s, especially linked to the hostile stance the incoming Conservative Government was taking, resulted in fierce antagonism towards 'race' and immigration questions in pre-tests of the 1981 Census, with fears that such statistics could contribute to (anti-)immigration laws (Barn 1994). As a consequence, researchers interested in ethnicity in 1981 have little to go on except birthplace; so UK-born black and Asian populations are 'hidden'. With self-definition of ethnicity in the 1991 Census, another change occurred.[12] Similar controversies arose in the 1971 pre-test question on income. The poor response rate achieved here was matched by a 1972 sample survey on income that secured information (even then some only partial) from half the respondents. This might come across as a strange hang-up, for the US Census regularly reports income information, whereas the UK still has no question on this topic. Indeed, even though it has such information from tax returns, UK governments have not been willing to release income data at anything other than the grossest regional level. Yet, despite irritating researchers with these huge, rather meaningless spatial units, interesting research can be conducted (e.g. R. J. Johnston 1979; Hamnett 1997). We do not wish to propose hypotheses on the inclusions and exclusions of items in government statistics in different countries. If the collection of statistics is controversial, expect compromises in deciding what is gathered. Coming up with longer questionnaires is a sure way to antagonize those who provide information, as research on questionnaire design shows (e.g. Nichols 1991). This means that judgements have to be made about what is and is not a central data requirement. Such judgements will be made with an eye to governmental information needs, in order to respond to economic and social trends (if only for political reasons). As such, a compromise is to be expected between the interests of powerful groups and the general population (Dunleavy and O'Leary 1987). Not unexpectedly, there are intriguing disparities between states in terms of the themes and coverage of governmental statistics.

12 Comparison of 1971 and 1991 classifications using Longitudinal Study data, which traces the same anonymous people across censuses, reveals that 87.4% of those who defined themselves as Black Caribbean in 1991 were categorized in this manner by officials in 1971, with just 60.5% who defined themselves as being from the Indian subcontinent in 1991 classed in this category in 1971 (these figures were computed by LS staff for Keith Hoggart).

In order to quell any inclination to interpret government data collection as a process of compromise between outside forces, the interests of governments should be acknowledged. This point is picked out by Moss (1991: 105), who headed the UK's first government social survey unit:

> *The dominance of the political factor in the fortunes of the [Government Social] Survey is well illustrated by the contrast between the early and later years of its history; between growing recognition and integration in the government apparatus and the later, very pointed attack.*

This attack owed something to a sense of disquiet in governmental circles towards 'social' science and social data (Sharpe 1978), but this sentiment was intensified by antagonism to the social sciences within the Thatcher governments (Dalyell 1983). Accompanying these sentiments was a desire to cut burdens on the private sector (which form filling is regularly reported to be by the political right), alongside a desire to roll back the state – with cuts in expenditure on data collection forming one element in this trimming. But if this sounds like something the private sector will warm to, think again. Thus the Statistical Users' Council, which represents the Confederation of British Industry (CBI) and the Trades Union Congress (TUC), is reported to have been perturbed by reductions in official statistical information, with its chairperson expressing the view that the cuts appeared to be 'pure dogma' (Harper 1990). Whatever the claims made by business owners about the burden of reporting to government agencies, their representative organizations want data for lobbying purposes, as well as to evaluate government policy. Transparently revealing why this might be, in a 1995 dispute between the UK Government and the CBI over the accuracy of economic indicators, newspaper readers were reminded that:

> *While both the CBI and the CSO [the Central Statistical Office, now the Office of National Statistics] acknowledged that the two sets of data [of each organization] sometimes deviated, the Government is well aware of the dangers of misreading economic runes. In 1988, the Treasury believed its own data pointing to a post-Black Monday slackening in activity, rather than the buoyant CBI surveys, and as a result failed to act quickly enough to stifle the Lawson boom. (Elliott 1995: 21)*

Lest readers assume that governments respond only to possible flaws in their assessments of how the economy operates, it is worth noting media reports on adjustments to how poverty is recorded, after Institute of Fiscal Studies revelations that the system of recording exaggerated income gains for the lowest paid (Brindle 1991). In this case, change led to recognition that poverty was more rife than the then government claimed.

The essential messages here are twofold. On the one hand, uneven weight is placed by national governments on different economic, social and environmental indicators in information gathering and reporting. These differences are not simply short-term quirks, but are built into the fabric of society, such that their investigation constitutes a valid field of social science enquiry (Desrosières 1996). On the other hand, governments are faced with competing pressures, from a variety of groups, to collect and publish information, as well as to ensure it is accurate, efficiently acquired and easy to understand. How varied such pressures are is indicated by Martin Walker (1990a: 23):

> *Although the US Census Bureau is probably the most intrusive and inquisitive state organisation outside the KGB, they have spared us from the pet food industry, which has been lobbying hard for decades to get the census to ask every family whether they had a dog or cat or goldfish ... Psychiatrists begged the Census Bureau to include two questions: do you dream? And if so, do you dream in colour? They were refused. Sexologists thought this would be the perfect opportunity to find out how often people have sex, but even with the 72-year guarantee of confidentiality, this was thought too intrusive ...*

Governments do not always respond to pressure. But, as we will see below, political considerations do influence whether change is accepted or rejected, albeit decisions are set in a framework in which information has become increasingly important for society to operate (Castells 1989, 2000). This has seen enormous growth in the range, regularity, detail and precision with which governmental data are collected and presented. The same point can be extended to many private and non-governmental organizations, as seen in Box 3.1.

Is the population covered?

If one of the critical advantages of factual reports by large private- and public-sector organizations arises from their ability to gather information on a scale most of us cannot conceive, then a key issue in evaluating utility is the comprehensiveness of cover. Here there are no simple answers, partly on account of the variety of data types and partly as a result of change over time. As already noted, for the 1801–31 UK censuses little more was involved than numerical counts, although some enumerators did exceed their duties, with the patchy survivors of these censuses available in some local record offices (Chapman 1998). Here you must dig in the expectation that treasure will not be found, but in the hope it will. After 1841, as with any census, unevenness in coverage and accuracy occurred. Higgs (1996: 13) provides an example of one enumerator whose records can be expected to have a high degree of accuracy, although he signifies that rural enumerators often had considerable local knowledge, which enhanced the quality of their returns. In the case of Crosby Ravensworth in Westmoreland, Higgs notes that the local schoolmaster was the enumerator in each of 1841, 1851 and 1861.

Examination of the schoolmaster's diary reveals a high level of local involvement. He attended parish meetings, collected tithes and church rates, filled out income-tax forms, witnessed the wills of neighbours and collected tax returns. Not only might we expect less day-to-day involvement in local affairs amongst urban enumerators, but we can anticipate that, as statistical services became more professionalized, this level of involvement lessened. This idea prompts two thoughts. The first is to emphasize the dangers of using one example to assume a general pattern of urban/rural (or other) differences. This danger is suggested by examining the census records of ancestors of one of the authors (George Hoggart and Jane Hall), who lived in small villages throughout their lives. As Box 3.3 shows, even within tiny communities, enumerators entered names incorrectly in two of the nineteenth-century censuses for which records are available. Not all rural enumerators had a sound knowledge of the local population. The second thought relates to the professionalization of statistical services, which has continued as the decades have progressed, partly as a result of research on information gathering and questionnaire design, which has improved technical capacities, and partly as a result of more, better-trained staff. The importance of professionalization is well illustrated by the US Census. Here the full census is regarded as less accurate than sample censuses by Bureau of Census employees. A key reason is the gigantic scale of the full census, which far outstrips the capacity of census staff to undertake, so temporary workers are used as enumerators. This is a problem that is common to virtually all censuses. It can be expected to affect the results obtained, but not in ways that are known with precision.

Box 3.3 Misreporting basic data in nineteenth-century UK censuses

This box reports on George Hoggart and his wife, the former Jane Hall, the great-great-great-grandparents of one of the authors. On the night of the 1841 Census, George Hoggart was away from home, hence he is not recorded. This list is from enumerators' returns. The information covers name, age, birthplace [not for 1841] and paid work (if relevant). All locations are in North Yorkshire. 'Source' refers to the relevant enumeration book.

1841 Census [location: Great Broughton – East Side; source: HO107 1257]

Name	Age	Birthplace	Paid work
Hoggart, Jane	20 years		farm labourer's wife
Hoggart, Margaret	6 months		

1851 Census [location: Ingleby Greenhow; source: HO107 2376]

Name	Age	Birthplace	Paid work
Hoggard, George	36 years	Battersby	farm labourer
Hoggard, Jane	28 years	Bilsdale	
Hoggard, Margaret	10 years	Bilsdale	
Hoggard, Thomas	7 years	Broughton	
Hoggard, Hannah	4 years	Broughton	
Hoggard, Elizabeth	2 years	Broughton	

1861 Census [location: Broughton; source: RG9 3657]

Name	Age	Birthplace	Paid work
Hoggart, George	48 years	Battersby	railway platelayer
Hoggart, Jane	40 years	Bilsdale	

Hoggart, Margaret	21 years	Bilsdale
Hoggart, Thomas	18 years	Broughton
Hoggart, Elizabeth	12 years	Broughton
Hoggart, Mary A.	10 years	Battersby
Hoggart, John	7 years	Ingleby Greenhow
Hoggart, Jane	4 years	Ingleby Greenhow
Hoggart, Sarah	2 years	Ingleby Greenhow
Hoggart, Ada	4 months	Ingleby Greenhow [granddaughter]

1871 Census [location: Ingleby Greenhow; source: RG10 4859]

Hoggart, George	57 years	Battersby agricultural labourer
Hoggart, Jane	48 years	Bilsdale
Hoggart, Sarah	12 years	Ingleby Greenhow
Hoggart, William	8 years	Ingleby Greenhow
Hall, Ada	10 years	Ingleby Greenhow [granddaughter]
Hall, William [blind]	7 years	Ingleby Greenhow [grandson]

1881 Census [location: 9 Battersby Junction; source: RG11 4864]

Hoggarth, George	65 years	Battersby
Hoggarth, Jane	58 years	Bilsdale
Hoggarth, James	9 years	Ingleby Greenhow [grandson]
Hoggarth, Sarah	7 years	Stokesley [granddaughter]
Hoggarth, Henry	6 years	Ingleby Greenhow [granddaughter]

1891 Census [location: Rose Cottage, Easby; source: RG12 4022]

Hoggart, George	78 years	Battersby
Hoggart, Jane	68 years	Bilsdale – East Side

The inconsistent spelling of the surname is significant, as this is the same family across these years. Similar misspellings are recorded for George and Jane's son Thomas. His 1843 birth certificate lists him as Hoggatt (hence complicating finding him in Family Record Centre lists of births), with Mary Jane, the first daughter of Thomas and Ellen Hoggart (née Mudd), listed as Hogart on her 29 March 1869 birth certificate. At the time of Thomas Hoggart's marriage to Ellen Mudd on 17 January 1869, his surname and that of his father [George] was entered as Hoggett, with his father listed as 'John', a farm labourer. Cross-checking all possibilities (including potential misspellings of Hoggart), alongside family associations for Ellen Mudd, reveals that the birth and marriage certificate entries are misspellings at the time of registration. For example, in the 1871 Census, Thomas Hoggart, aged 28, is shown to be married to Ellen Hoggart, with daughters Mary Jane (2 years) and Ann (10 months). The birthplaces of the parents (Broughton and Yarm-on-Tees) and of Mary Jane (Yarm-on-Tees) are the same in the 1871 Census as birth certificate information for these three people. As the names Hoggart and Mudd are relatively uncommon, even on a national scale, a universal tracing is possible to ensure inconsistencies are spelling errors not different people. The ease with which registration errors occurred is suggested on both birth and marriage certificates, where the entry 'the mark of …' regularly appears, indicating that George Hoggart, Thomas Hoggart, Jane Hall and Ellen Mudd were illiterate. Verbal communication with members of the family only stretches back to grandparents, but these (and current) family members pronounce their surname hogg-ett [or even hogg-utt] rather than ho-gart. If the same applied in earlier generations, this might explain registrar and enumerator errors in transcribing the name from verbal sound to written representation.

Readers might note the inconsistency in the ages across census decades. As Higgs (1996: 78) makes clear: 'The reading of ages is without doubt one of the most problematic features of the manuscript census returns.' In part this arises because enumerators were told to record ages within five-years periods in 1841. Errors also occur owing to people's shaky knowledge of their age or, in the absence of official records and with illiteracy rampant, because good reasons existed for disguising ages with little fear of contradiction. As an illustration of the inconsistency, in a comparison of returns for Preston, Lancashire, Higgs (1996: 79) found that only 53 per cent of his sample had an age 10 years older in 1861 than in 1851 (albeit only 4 per cent were adrift by more than two years)

For whatever reasons, it has been estimated that early censuses missed large numbers of people whom they should have recorded. Thus, under-enumeration in nineteenth-century US censuses was estimated at around 15 per cent for those with a high chance of being enumerated – namely, those recorded in the previous census, eligible voters, legislators or property-owners (Steckel 1991: 593). Detailed comparisons of different localities suggest under-enumeration of 11.7–23.1 per cent for 1850, 9.2–16.3 per cent for 1860 and 12.5–17.8 per cent for 1870, with one estimate for three Georgia counties in 1880 placing the under-recording at 34.9 per cent (Steckel 1991, 588). Zoom to the present and omissions are less severe, but still worrying. Thus, *The Economist* (1 April 2000, 47) reports that 1.6 per cent of the population was missed from the 1990 US Census, with the 2000 count expected to be 1.75 per cent down.[13] There is also unevenness in under-reporting, with 4.4 per cent said to have been missed from the black population. Given the geography of ethnicity in the USA, this translates into a marked under-recording for inner cities. Newspaper headlines capture the outrage this creates, as seen in: 'NY up in arms at "wrong" census' (Tran 1990) or 'US cities refuse to go down for count' (M. Walker 1990c). Significant in this regard is the difficulty presented by pressure from vested interests not to reach a reconciliation on under-reporting totals (e.g. Darga 1999). For one thing, in terms of political parties, the Republicans see adjustment as harming their supporters and their re-election prospects, given that federal funds are distributed using census totals and the same figures are employed for Congressional district realignment.[14]

Across the Atlantic the situation is no better. Thus, Hall and Hall (1995) report that 2.2 per cent were missing from the 1991 UK Census; up from 0.45 per cent in 1981. Under-enumeration was heaviest amongst young men in their early 20s and those over 85 years, as well as being higher in metropolitan areas (Simpson 1996). This pattern of geographical unevenness is not new. Offering a slightly higher figure than Hall and Hall's, Higgs (1996: 117) reports that the 1981 Census missed 0.5 per cent of the population, but this value rose to 2.75 per cent for inner London. Ethnic distinctions were noteworthy, as in the USA. Thus, computational adjustments at the ward level to bring undercounts closer to 'reality' suggest a multiplier of 1.021 for whites, 1.038 for the black-Caribbean population, 1.042 for the Chinese and 1.052 for black-Africans (Simpson 1996). As with the USA, there are monetary and electoral consequences of

13 These figures are from US Bureau of Census sample surveys after the full census. Figures suggest 10 million were missed from the 1990 Census, with six million double counted (S. A. Holmes 1998). As happens often with government statistics, these figures have been adjusted over time (test this point by looking at runs of economic statistics – such as annual import figures or GDP figures – and trace how values for the same year change in annual editions of the UK Office of National Statistics, *Annual abstract of statistics*, or the US Bureau of Census, *Statistical abstract of the United States*). For the 1990 US Census, *The Economist* (20 July 1991, 39) had earlier reported that 2.1 per cent of the US population were missed, with 4.8 per cent for blacks, 5.0 per cent for native Americans and 5.2 per cent for Hispanics.

under-recording, which discriminate against areas with concentrations of socio-economic deprivation. If surveys are correct, then a key reason for this under-recording is distrust of government (Maier 1995).[15] This was evident in the marked downward shift in UK response rates from 1981–1991. While not wishing to discount other explanations, a key reason people wished to avoid being captured in the census was avoidance of the Poll Tax.[16] With this tax abolished, a lesser rate of undercounting is expected to occur in the 2001 UK Census.

A critical message is that we need to acknowledge the context in which the census is undertaken. Attitudes change over time. They are not consistent for population groups. Even short-term events could impact on perceptions of worth or threats from data gathering. The case of UK farmers illustrates the point. Here, prior to joining the EU, an annual, usually consensual, 'negotiation' took place between the National Farmers' Union (NFU) and the Ministry of Agriculture, Fisheries and Food (MAFF) that determined price supports for the next year (see e.g. Grant 1995). In this harmonious environment, it comes as no real surprise to find Zarkovich (1965) reporting that the collaboration of UK farmers with the agricultural census meant that annual postal questionnaires yielded very reliable results. Contrast this with Oldenburg's observation (1989: 191) on the Indian city of Lucknow. Here, as data for the 1869 Census were collected by the city police, a rumour swept the city that the aim was to count virgin girls so that they could be abducted for the pleasure of European soldiers and bureaucrats. The impact on undercounting, in a colonial city with little history of census taking, and coming not many years after the Indian Rebellion of 1857, is not difficult to imagine.

But what can be done about under-recording? Being realistic, there is not much that can be done, save to be conscious of omissions. This consciousness is not something to be kept in the back of the mind, for it has profound epistemological implications. For one thing, simple counts, let alone how variables are measured or which are included and excluded, bring into question the validity of certain analytical approaches. Explore the foundations of parametric statistical tests and you find underlying assumptions of unbiased data collection and interval scale measurement (see e.g. Ebdon 1985; O'Brien 1992). Can we really hold to these assumptions when censuses have built-in biases that

14 The 1980 undercount lost New York City $50 million in federal aid a year (Maier 1995: 9).

15 An interesting insight on the 1991 UK Census appears in the 25 July 1991 edition of the *Guardian* ('Why the census doesn't make sense', p. 21). This reports on the experiences of one of the army of temporary enumerators, revealing the difficulties of finding some people and getting completed forms from others. As Maier (1995) reports, enumerators generally feel there is growing suspicion of authority, growing distrust of statistics, growing insecurity over casual visitors and a pervasive sense of disenfranchisement that affects people's willingness to respond to efforts to collect information.

16 The Poll Tax, or Community Charge as the Conservative Government preferred to call it, levied the same charge on all householders irrespective of wealth or service usage – with some reduced payments, as for students (Travers 1989). This replaced a tax based on property value. This regressive tax proved extremely unpopular, leading to riots in the streets of London, alongside efforts to avoid payment, which involved ensuring you were 'lost' from the census and the electoral register, so disenfranchising yourself (see e.g. Hall and Hall 1995; Dorling *et al.* 1996).

result in uneven population coverage? This brings into question the positivist's visions of neutral, objective analysis – such a thing cannot exist if the data are biased in their own right. Yet it is often the case that, warts and all, official statistics are the most readily accessible, and often felt to be the most reliable, data. Our message is simple. Do not assume this means they are accurate; explore whether other sources can be used that might add insight on issues under investigation. Such sources will undoubtedly have their own biases, but inconsistencies between data sets should bring these biases to the forefront, so helping to sensitize the researcher to inadequacies.

Are there other biases in the data?

At one level the obvious answer to this question is yes. Most evidently this is seen in older records. For example, while information in enumerators' books is open to the public after 100 years in the UK (72 years in the USA), this does not mean you will find the information you want. For one thing, there have been a series of losses. Thus, for 1841, Kensington, Paddington, Golden Lane and Whitecross subdistricts are missing, plus parts of Essex and Kent. For 1851 Salford and Manchester returns have been damaged by water. Belgravia and Woolwich Arsenal are missing from the 1861 records. And, with Irish independence in 1922, the UK Government destroyed many enumerators' books (Lawton 1978). The same picture can be painted elsewhere, as with fire destroying many 1890 US Census records (J. Scott 1990: 101). But these absences are small compared with the grand scale of census data collection.

These holes are 'easy' to identify but this is not the limit of potential bias. While enumerators' books give researchers access to raw data, the latitude this provides is limited, as these books have gone through the filter of the enumerator. Anticipating that enumerators impose their own value judgements on the data is something readers should be familiar with. In the jargon, this is referred to as the action of a street-level bureaucrat (see e.g. Lipsky 1980; Lowe and Ward 1997). It is easier to draw out how street-level officials affect census taking using examples from the past. In one way government revelations about the past are unintended. Once information on the past is released in raw format, it can be deconstructed so that inherent biases and tendencies are brought into focus. In another way, governments are less concerned about insights from the past, as the messages are deniable. Take note when a review of government policy criticizes existing practice. How many seconds into an interview before the relevant government minister is telling us that the review refers to the situation of *x* months ago and that things have changed now – the government has responded? In terms of the particularities of the case, this message might be correct, but fundamental tenets of operations are not easy to change, even over long time periods. The simple message is that biases identified in the (recent) past often have much to tell us about contemporary affairs.

Here Hindess (1973) brings out an important point, for he notes that, as social statistics are collated from numerous observers, we cannot interpret them adequately without reference to background expectations. Illustrative of this, Higgs (1996: 94) reports that: 'On occasion the feelings of the enumerators shine through their returns. In Limehouse in 1871, for example, one enumerator described every prostitute as "fallen" in the occupational column.' Do we imagine that the sparse information on the male visitors to these prostitutes on census night can be put down to lack of familiarity with clients? Uneven enumerator inputs also appear in assumptions about occupation. Nineteenth-century returns reveal that male full-time work seemed to be reported reasonably accurately, but the quality of information on female paid work is questionable. For one thing, women in the nineteenth century often worked on a casual basis or part-time, with this information rarely appearing in census returns. Work undertaken at home – such as making garments for sale – was poorly recorded. Moreover, while multiple job holding was common (see e.g. Hussey 1997), as seen in the annual transformation of thousands of London dockworkers into Kent farm labourers at harvest time, the recording of such occupational mixes is hardly exhaustive (Higgs 1996: 97). As for unemployment, here information was seldom provided (this became a census item in the USA in 1930). None of this should be taken to imply that enumerators were deliberately misreporting. The more relevant point is that they would have used their judgement about what was important and what was not. As a consequence, even if access to original logbooks is obtained, expect omissions to be uneven.

Also note that systematic inconsistencies exist, as when instructions to enumerators were changed. For instance, in 1891 enumerators were informed that they were not to record the size of the farm and the number employed on it, although these had been recorded at previous UK censuses. More troubling were instructions on how those who worked on a farm were to be recorded. Thus, general servants, many of them women, who undertook some farm labour were not to be recorded as farmworkers (Higgs 1996: 160). There is little doubt that the recording of occupations has created difficulties for public officials. As economies change, new categories of labour emerge, sometimes without due regard from official statisticians. As an illustration, it was not until 1881 that definite instructions were given on how to code warehouse operatives in the UK Census. Add to this, as economies change occupational titles mean different things. Thus, in the UK Registrar General's classification of occupations, 'clerk' shifted from Class I in 1911 to Class II in 1921 and then to Class III in 1931, where it has stayed ever since. This shift reflected a different vision of what a clerk was, with interpretations in the nineteenth century biased toward clerks in holy orders whereas by the 1930s the understanding was closer to that held at the start of the twenty-first century (J. Scott 1990: 213).[17] Significantly, clarifications better to reflect 'reality' can make geographical distinctions. Thus, in 1871, census clerks were instructed that carters were to be referred to as carmen in large commercial and manufacturing towns, but in 'ordinary' provincial towns

17 C. Davies (1980) makes the same point about nurses, showing how interpretations shifted from a domestic servant to a medical profession.

as agricultural labourers (Higgs 1996: 163). Realistically, 'the general conclusion to be drawn from this analysis is that the published occupational returns from the censuses should be used with caution' (Higgs 1996: 166). The continuing potency of this commentary is seen in assessments of what some see as the deteriorating reliability of national economic statistics. For instance, an *Economist* commentary (27 May 2000, 18) adds to the charge that governments have starved national statistical offices of funds. It goes on to caution that even with more funding things are unlikely to improve significantly:

> *Deregulation (notably the scrapping of capital controls), the shift from manufacturing to services, and rapid technological change are all partly to blame. It is ironic that, in the information age, knowledge about economic activity has become ever more uncertain.*

Here we shift from individual decisions about recording data that have quirky results, to recognition that governments have not adapted to current economic conditions. But, when governments try to adapt to 'new realities', this creates analytical problems, as concepts with the same name mean different things over time. We see this in the way the UK Census categorizes households.[18] It is not just government decisions that make investigation of change difficult, but societal change. Even so, there is a sense that governments tend to be lethargic in responding to new social realities. In the UK Census, for example, it was only in 1991 that cohabiting (opposite sex) partners were able to specify this relationship. Prior to this there was no distinction between opposite sex householders who did or did not share such a relationship. That household composition measures are still unsatisfactory is evident in the way the labels cohabiting, divorced, single and widowed fail to capture the meaning of relationships for individuals. The relationship between same-sex cohabiting adults is as visible as coal on a moonless midnight in the UK Census (Barn 1994). Of course, decisions on the representation of a census item can reflect issues or controversies at the time a census was in preparation. Even then, the interpretation of social phenomena will draw on systematic biases in governmental values, for national data sets are constructed within a particular epistemological context. This is illustrated by exploring how key concepts are defined.

18 In the 1971 Census a household was a person living alone or a group living at the same address with common housekeeping (interpreted as sharing a meal a day). For 1981 common housekeeping was extended to include sharing a living room, even if no meal was shared (Penhale 1990: 29–30). The difference may not be large, but an accumulation of small adjustments across a range of variables raises the question, is change definitional or substantive?

Social class

As understood in theories of social stratification, the concept of social class in the UK Census is a measure of social status more than class. In the UK scheme, occupations are placed into classes based on judgements made by the Registrar General's staff. These assessments derive from a biological theory of nineteenth-century society that portrayed a hierarchy of inherited natural abilities. These were taken to be reflected in the skill level of occupations (Marshall *et al.* 1988: 19). While there is continuing debate about theoretically informed representations of social class (see e.g. Edgell 1993), the Registrar General's scheme is a long way from an accepted theoretical scheme.[19] Its unspoken assumptions ring loud when comparison is made with the class representations in other censuses. For example, in contrast with the lack of distinction between the public and private sectors in the UK, the French national statistical service (INSEE) uses a socio-professional classification of occupations with critical distinctions between the private and public sectors, as well as between employees and employed. With the definition of a head of household used to define the social class standing of its members, only in 1991 did the UK Census provide an opportunity for the head to be self-defined. Prior to that date, save for female-only households, the head was a male (on sexism in official statistics, see Oakley and Oakley 1979). Theoretically, this is problematical, as the imposition of a social hierarchy based on male-focused occupations can be misleading, given that gender divisions of labour imply distinctive social arrangements (Crompton 1991). This point is illustrated by Eichler (1988: 95), who reports that, by reclassifying female socio-economic standing so that it does not rely on male measures, different links are found between socio-economic standing and political action (see also Peake 1986). The verification of theoretical propositions can thereby be confounded by assumptions in published data.

For the 2001 Census, the UK public is to be rewarded with a new measure of socio-economic standing. The nineteenth-century biological model is to be no more, for the new scheme portrays class as dependent primarily on position in the labour market, rather than subjective notions of job skill. Hence, people will be divided according to their employment contract. Higher professional and managerial workers will have long-term contracts that result in rewards for work being both monetary and in the form of employment perks (which raises questions about those in new industries, noting the varying fortunes of dot.com companies). At the other end of the scale, routine occupations (which mainly comprise manual workers) are characterized by the short-term exchange of labour for money. This scheme might well infuriate analysts who are intent on exploring longer-term trends. Whether it helps address gender distinctions is also to be seen. However, according to *The Economist* (3 June 2000, 31), unlike the

19 Owing to poor theoretical grounding, the Registrar General's categories are imprecise, so coding errors abound (Marshall *et al.* 1988: 20). A check on 1966 returns showed 10.7 per cent of occupations were wrongly classified.

previous classification, the new scheme appears to correlate strongly with differences in income levels.

Ethnicity

Asking ethnicity questions is somewhat akin to walking with bare feet on broken glass – no matter how elegantly you cross the space, no matter how fast or slowly you move, somewhere along the line you know there is going to be pain. That the categorization of ethnic groups has proved troublesome for census officials is very apparent. In some cases, this might arise from the speed with which populations have grown, although changing official assumptions about what it means to belong to an ethnic group cannot be discounted. Take the US Census treatment of the Hispanic population as one illustration. In 1930 the Hispanic population was classed as 'other non-white'. In 1940 Hispanics were 'persons of Spanish mother tongue'. For 1950 and 1960 this was changed to a 'white person of Spanish surname'. In 1970 you needed a Spanish surname and Spanish mother tongue. By 1980 and 1990 there was a separate question enabling people to identify themselves as Hispanic (Maier 1995). As with many countries, in the UK Census rather narrow assumptions prevail about the nature of ethnicity. In 1991, for example, the nine prime categories were white, black-Caribbean, black-African, black-other, Indian, Pakistani, Bangladeshi, Chinese and 'other'. For white people, this means Northern Irish Catholics and Protestants living on 'the mainland' belong to the same ethnic group, as do the Scottish, the Welsh, Manx folk and Channel Islanders. Polish Jews, Czech émigrés, US service personnel and a horde of European corporate expats, if white, apparently share the same ethnicity. Do you buy this? The saga does not end here. Consider the following US stipulation about how children of 'mixed' parents should be classified: 'When the husband is white and the wife is not, the child is assigned the wife's race. When the husband is not white, the child is assigned to the husband's race' (Maier 1995: 15). This measure was dropped in 1989, when assignment was by mother, irrespective of father. Its impact on official statistics is nonetheless one to note, for this old measure appears to have underestimated infant mortality for the non-white population, while changes in definition pose difficulties comparing change over time. The struggle agencies have in deciding how to represent ethnic identity is reflected by both the 2000 US Census and the 2001 UK Census introducing new ethnicity categorizations. The more complex is the USA, where people will be able to choose more than one category, with 63 combinations proposed. Where is the broken glass in all this? Well some black groups are reported to be concerned that the new procedure will dilute their power, with some eyebrows raised at the federal government decision to list those who check 'white' and a minority group as belonging to that minority for civil-rights purposes (see e.g. *The Economist*, 1 April 2000, 48).

The range of concepts that are handled in state statistics in a way that suggests they need careful treatment is extensive. Most evident is the silence that exists on unpaid work. This leads to an inability to use most official publications for effective investigation of women's lives, whether the investigation relates to activities within the home or to unpaid community work (see e.g. A. K. Davies 1988). Also of note are crime figures, which are really measures of reported crime, so that, as the social stigma of certain crimes lessens or the potential gains for victims rise (for example, insurance), the likelihood of reporting swells. Throw into the pot the potential for political massaging. Thus, in Gardiner's study (1968) of traffic violations in Massachusetts he found substantial variety in reporting levels owing to differential municipal policies on the vigour with which traffic codes should be enforced. This local-level input into the construction of crime statistics is even more overt for suicide figures, which emerge from the social rules and practices that coroners and the police use to reach 'a socially acceptable judgement' (J. M. Atkinson 1978). But it is not just the providers of information who distort figures, but those being reported on or affected by their end product. Thus, in Spain many apartments that are rented out for income are recorded as second homes, so the owner does not have to pay tax on rental income owing to a quirk in the law (Barke and France 1988). Providing an even stronger caution against blind acceptance of official figures, Bennett (1985) provides the remarkable story of second-home registration in Austria, which puts results from the 1981 national census in serious doubt. Here local government funds were distributed such that places with many second homes received extra cash and those that provided the primary residence of second-home owners lost income. The response of local governments in which many second-home owners had their main residence was to offer cash incentives to register the second home as a main residence. In this way, Vienna is reported to have secured an extra $2 million in federal funds.

Temporal changes

This brings us back to the critical point that 'official' statistics, whether from the public, private or non-governmental sector, are socially constructed. They embody the cultural norms of the producer, as well as reflecting pressures from interested parties. As Macdonald and Tipton (1993: 188) put it:

> *they are produced on the basis of certain ideas, theories commonly accepted, taken-for-granted principles, which means that while they are perfectly correct [sometimes] – given certain socially accepted norms – they do not have the objectivity of, say, a measure of atmospheric pressure recorded on a barometer.*

Yet a key message the reader should capture is that coverage, definitions and meanings in official statistics are not unchanging. The 'norms' behind their production, that structure their presentation, have to be reproduced, whether in a consensual or a conflictual environment. In this process a multitude of small and intermediate scale decisions have to be made that bear on the ethos and trend in reporting. It would be startling to think that agencies do not take some of the opportunities afforded by such decisions to orient outcomes to favour themselves.

The central message is that we cannot assume consistency over time in the ethos of official information. This is not meant to be a comment about Nazi Germany or the former Soviet Union, where official information was distorted in an overt and outrageous manner (albeit many citizens of these nations might not have been aware of this). The issue is more about subtle adjustments that reorient the content of available information, and consequently official interpretations of events. To put the picture straight, in some cases official versions of events reflect 'public' understanding of social phenomena. A possible example is the announcement by the UK Government that it will change the manner in which it assesses national progress, so less weight is given to economic output and more attention is afforded to 'quality of life'; in which the 'barometer includes measures ranging from greenhouse gas emissions, road traffic levels and wildlife populations to public investment and length of healthy life' (Ward 1998: 2). Such adjustments might seem welcome, about time too, but changes are charged with political colouring. Thus, around the time the Blair Government was informing the media that it would provide a more 'humanist' view of national progress, it also announced that figures on the economy were to be adjusted. No longer would GDP be presented as simple numbers but would be conditioned by income distribution and quality of life. Again, the rationale is reasonable. For one thing, under simple GDP-based measures of wealth, oil spills are seen as a boost to the economy rather than detracting from it. Similarly, the old measure implies that a pound is worth the same for poor and rich people, when an extra pound is worth much more to a poor person than to a wealthy one. All this might seem reasonable. The sting in the tail is that it appears these readjustments are expected to lead to a noteworthy reduction in the official rate of economic progress under the Thatcher and Major governments (Elliott and Denny 1998). One hardly imagines the Conservative Party will be thrilled by such a rewriting of their record in office. No doubt we will continue to see more incidents of Conservative and Labour politicians chanting inconsistent statistics across the floor of the House of Commons; only this time they will be using not just different base years to show their party in the best light but dissimilar measures of what economic growth is.

To accept that a change in how statistics are presented rewrites history is one thing. Rewriting to distort is another. Again, draw back on any temptation to think of authoritarian regimes. The point we are making is about governments (or companies) in countries most people accept as 'democratic'. Our real message is to be cautious, for definitional and reporting changes can have major impacts on how economies and societies are presented. These need not occur in a single blow but can debilitate through a thousand cuts. In recent UK history, a noteworthy instance of this was the manipulation of unemployment statistics by the Thatcher governments. A portion of the changes that were made is demonstrated in Box 3.4. Providing a measure of the cumulative impact of disguising 'real' unemployment, Hutton (1992) argued that, while the official unemployment rate in September 1992 was 2 843 000 people, the more accurate figure was 3 988 000. Actual unemployment was estimated to be 40 per cent more than the official figure – with the trend of massaging down the unemployment toll continuing in the Major years (see e.g. Thomas 1995; Elliott and Thomas 1997). Not that changes in definition are required for governments to affect figures. As one example, the computation of inflation in the UK is based on monthly adjustments in the price of a fixed 'basket of purchases', so if governments wish to 'keep inflation down' they can influence inflation figures by keeping price rises low for items in 'the basket' (for example, differential tax rates, subsidies for services and so on). If items prove 'awkward', as with the price of housing in the UK, governments can exclude them (UK inflation figures are often quoted now with and without housing costs – the without figure usually being lower). Of course, the effort to 'distort' policy to influence official statistics might be less important than implementing sound policies; so we should not rush into the assumption that such 'distortions' are widespread. Nonetheless, the capacity to adjust official accounts is valued by public agencies – as seen in the Conservative Government's efforts to censor the 25th anniversary history of the Government Statistical Service's *Social Trends* (Dean 1995) and the Blair Government's succumbing to pressure from Whitehall departments not to hand over provision of official statistics to an independent organization, despite doing precisely this for a major aspect of economic control, in giving interest rate determination to the Bank of England (Denny 1998).

Hawthorne Effects

But the basis for official statistics being untrustworthy is not all about national policy. It also has much to do with the street-level bureaucrats. Widely recognized in this regard is a tendency for officials in communist Eastern Europe to report artificially high production volumes in order to appease regional or national head offices (see e.g. Nove 1975, 1977). Closer to home we find an abundance of 'fiddles' in the computation of education statistics. It perhaps comes as little surprise when a government announces that teachers will be paid 'by student results' to find reports that school teachers are 'routinely writing the coursework which counts towards GCSE results' (*Guardian*, 11 July 2000, 7). With schools embroiled in competition to attract the best students, the publication of statistics on examination performance has likewise become a target for creative accounting. Most widely reported is the use of expulsions from school to remove students who do not perform well, with a 450 per cent rise in expulsions for schools in England and Wales between 1990 and 1995 (Chaudhary 1998). Even without action by school officials, educational statistics can be distorted by the policy that seeks to improve educational performance. As one example, Nick Davies (2000a) reports on the case of the Phoenix School in Hammersmith. Official inspectors' reports indicate that, five years after this school went into the 'special-measures' category (on account of poor examination results), the leadership of its head was excellent and 60 per cent of the classes inspectors observed were graded good, very good or excellent. Yet this

Box 3.4 Adjusting official statistics for political ends?

In the first 10 years of Thatcher governments, the official definition of unemployment was adjusted 24 times. According to computations by the Unemployment Unit, a research and pressure group, only once did this increase the number of people recorded as unemployed.

Date	Change	Impact (000)
Oct. 1979	Benefit payment every two weeks	20 gain
Oct. 1979	Compensating seasonally adjusted figures	20 loss
Feb. 1981	First special employment and training measures	495 loss
July–Oct. 1981	Change to seasonally adjusted figures	20 loss
July 1981	Men over 60 unemployed for a year have more money if not registered for work	30 loss
July 1982	Unemployment benefit taxed, encouraging single parents to shift to supplementary benefit	? loss
Oct. 1982	Unemployment defined as claiming benefit, not those registering with the government as seeking work	c.200 loss
Apr. 1983	Men over 60 on benefit no longer need to sign on at benefit offices to get national insurance credits	107 loss
June 1983	Men over 60 allowed longer-term supplementary benefit as soon as qualify for supplementary benefit	54 loss
June 1983	School-leavers unable to claim benefit until September	(June–Aug.) 100–200 loss
Oct. 1984	Community programme restricted to unemployment benefit claimants	29 loss
July 1985	Reconciliation of Northern Ireland records	5 loss
July 1985	Payment of unemployment benefit in arrears	?
Mar. 1986	Delay in publication of statistics by two weeks	c. 50 loss
June 1986	Unemployment share no longer = unemployed/employed but unemployed/(unemployed + employed)	1.0–1.5% loss
June 1986	Re-interview jobless with tighter available-for-work test	200–300 loss
Oct. 1986	Removal of partial unemployment payments if national insurance payments are low	30 loss
Oct. 1986	Voluntary deduction to benefit extended from 6 to 13 weeks	2–3 loss
Apr. 1988	Voluntary deduction to benefit extended from 13 to 26 weeks	12 loss
Apr. 1988	Unemployed able to claim income support while working part-time	small loss
June 1988	New larger denominator to compute unemployment rate	0.1% loss
Sept. 1988	All 16 and 17 year olds denied unemployment benefit	90–120 loss
Oct. 1988	Contributions for short-term benefits amended	38 loss
Oct. 1988	Age limit for abating unemployment benefit to occupational pensions falls from 60 to 55	30 loss

Source: Guardian, 15 Mar. 1989, 21.

school saw its share of pupils scoring A–C grades in GCSE examinations fall from 17 per cent to 5 per cent in just two years. The trigger it seems was the measure the UK Government proclaimed would improve standards:

> *the school was damaged directly by being put into special measures in 1994. This triggered a rash of publicity which caused an immediate flight of teachers and of parents of motivated children. This left classes to be taught by supply teachers and drained many of the most able children from the new intake ... (N. Davies 2000a: 7)*

Universities do not come out of this smelling of roses. Reports come in of gamesmanship in university Teaching Quality Assessments (TQA), with the score (which could mistakenly be taken by parents, teachers and students as some form of 'objective' measure) seemingly being accorded more importance than learning (Baty 1999a). With so much money attached to its outcome, the UK's Research Assessment Exercise (RAE) is not unexpectedly tarnished by similar institutional attempts to raise scores above 'real' levels. Media reports have identified steps taken by Exeter and Loughborough in this direction, but they are unlikely to be the only universities to use ruses to exclude staff from RAE counts – the aim being to eliminate those who will lower research ratings and raise the proportion of staff who appear to be 'research active' (Baty 1999b; Goddard 2000). Readers might recognize this as Peter's Principle – if you set up a criterion that is deemed desirable, people will adjust their behaviour to score in terms of how performance is measured. Rather than reflecting 'reality', the outcome is determined by the skill with which agents boost scores by manipulation of inputs into their computation. Anyone who approaches figures like school exam results, TQA scores or RAE grades in a simplistic manner, not recognizing the potency of such 'fudges', is likely to perpetrate a significant disservice to social science understanding.

Without belabouring the point, we should again draw the reader's attention to the epistemological implications of this. Those who start from the standpoint that society and research outputs are socially constructed will hardly be troubled. Here a key issue is the deconstruction of representations, such that drawing out structures and agency responses that create the imaginary worlds of test scores and quality ratings is grist for the mill. By contrast, those of a positivist bend should feel discomfort. It is true that only the most confused would think there is no measurement 'error' in numerical data. Statistical analysis provides for this, with much theoretical work on how much error can be 'allowed' for valid conclusions to be drawn – this applying whether the distribution of values is assumed to be normal or not (see e.g. Siegel 1956). Yet the statistical basis for applying inferential statistical tests must at the very least be approached with caution, given the materials presented above. Most evidently this is seen in assuming that error terms are randomly distributed, or at least are not systematically biased. Can we really assume this for governmental manipulations of data sources, for state restrictions on researcher access

Box 3.5 Fogel and Engerman on the economics of US slavery

In 1974 the (1993) Nobel Prize winner Robert Fogel published a controversial book with Stanley Engerman, entitled *Time on the cross* (Fogel and Engerman 1974/1989). This significant book is now part of intellectual history. It challenged dominant ideas in the 1960s and early 1970s concerning slavery as an economic institution, and offered a quantitative analysis of the slave-based economy. To grasp the profound challenge this book presented at the time (and for many populist views today), it is worth quoting their main conclusions. These were:

- that slavery was highly profitable – holding slaves was rational economically;
- that slavery at the time of the US Civil War was not moribund economically;
- that on the eve of the Civil War slave-holders anticipated unprecedented prosperity;
- that slave-based agriculture was 35 per cent more efficient than northern family farming;
- that on average slaves were hard-working and more efficient than their white counterparts;
- that slaves working in manufacturing had a comparable productivity performance with free workers;
- that the family was the basic organizational form in slavery, so it was in the interest of the slave-owner to maintain the family stability; not to break it up or threaten stability through sexual exploitation;
- that the material conditions of adult slaves compared favourably with free industrial workers (this was not the case for children, for women were made to work, so depriving the foetus of nutrition, which resulted in higher infant and 1–5-year-old mortality rates);
- that rates of economic exploitation for slaves were less than commonly presumed, with a typical slave receiving 90 per cent of the income he or she produced;
- that the economy of the antebellum South was growing rapidly, with per capita increases for 1840–1860 faster than the rest of the country, and with per capita incomes in 1860 at a level Italy did not achieve until after the Second World War.

Antagonism towards the book had three sources. First, it used quantitative methods, when history as a discipline was dominated by qualitative approaches. This was seen as an affront – in effect a knee-jerk reaction from those holding to a vision that methodological purity is required to gain 'knowledge'.

Second, the conclusions of the study were criticized. Significantly, Fogel and Engerman's findings are now accepted by many researchers on slavery in the USA, albeit some have been fine-tuned through further analysis. One problem was that previous researchers had commonly failed to provide empirical backing for basic assumptions. Starting from the premise that slavery was evil, they found evidence of sexual abuse, economic inefficiency and so on from personal accounts. They devoted insufficient attention to general trends, which would have shown that these personal accounts exaggerated their incidence. Even with far from perfect numerical accounts, Fogel and Engerman showed that qualitative accounts emphasized 'the special' or those with 'appealing attributes', much as investigations using participant observation do when compared with questionnaire surveys (e.g. Becker and Geer 1957).

Finally, *Time on the cross* was criticized for not paying sufficient attention to the moral issues of slavery. This was recognized by the authors in the (1989) new edition of the book, who indicated that they had intended to investigate an empirical question rather than provide a commentary on the slave system. This draws out an important general point, which has two elements. The first is that researchers need to be aware that their moral (or methodological) biases can cloud efforts to investigate empirical situations. The second is that empirical studies are embedded in a political framework, whether this relates to the distribution of societal goods and bads, or ethical or moralistic issues. The two are inseparable.

to information or for the scores and ratings bestowed on educational institutions?[20] Set against the concerns on this point, we do not wish to give the impression that this is a whole-hearted endorsement of social constructivism. At least for some of its adherents social constructivism seems to do little more than expose biases in statistical (or other) representations. Issues of materiality, like the poor dying from malnutrition, are given less weight in such analyses than concern for its representation in media, academic or governmental circles (Gregson 1995). That published information is socially constructed is undoubtedly the case. What is needed is to go beyond this to explore the potentialities of using published resources to expose key social processes. As Box 3.5 explains, investigations using such data can provide path-breaking insights on key questions in social research, as well as offering the capacity to challenge unquestioned assumptions in qualitative approaches that might hold for moralistic or political reasons rather than because empirical evidence supports them. Add to which, if estimates show 40 per cent are dying of malnutrition, how productive is it to dwell on 'misrepresentations' that might lower or raise this figure by 10 per cent?

Summary: what biases are there?

- Biases in official statistics do not result just from state priorities but from the capacity of the state to secure information from the public. This cannot be guaranteed.
- The same forces mean that the accuracy of official statistics cannot be assured, even when state or private-sector agents seek to maximize their accuracy.
- Biases that are clear in official statistics are those that lean towards the more easily counted, with major areas of social life consequently often excluded.
- Because governments and major private-sector organizations have significant resources, it is easy to neglect the magnitude of under-reporting from the reports they produce. These can be large.
- As official statistics are important mechanisms for evaluating governments and their performance, they are political outputs, being subject to massaging by politicians and to manipulation by state officials who wish to show their agency in a better light. They are not neutral but socially reactive representations.
- Although not necessarily overt, the potential for conflict over such figures has a bearing on the way concepts are defined. These can have little similarity with social science theory.
- As society changes, the meaning of variables in official statistics also changes. Change in the way variables are measured also introduces inconsistencies over time.

Can we improve on limitations?

One of the ironies of social science research is that a great deal of theorization is either concerned directly with social change or else requires that social change be explored to

20 While our commentary on street-level manipulation of statistical returns looks only at education, the practice is much wider.

assess theoretical propositions. Yet, when researchers use secondary data, information limitations often confine investigations, especially quantitative ones, to cross-sectional studies. This means inferences about process rely heavily on unsubstantiated theory (Davies and Dale 1994). In the case of voting decisions, for example, there are those who hold that voters become more conservative as they progress through their life course, while others hold that successive cohorts of young people have their views formed at an early age with relatively minor realignments thereafter (see e.g. Greer and Greer 1976). There are two theoretical dimensions to explaining voting differences here. A cross-sectional analysis can only confound the two. Examination of behaviour over time is essential to assess these competing hypotheses. That such differences make a crucial analytical and substantive impact is evinced from explorations of the impact of rioting on US city governments. In studies that correlated the incidence of rioting with measures of fiscal policy at one point in time, the common conclusion was that there is little difference between riot and non-riot cities – much as analysts have found that socio-economic characteristics do not distinguish riot from non-riot cities; (see e.g. Keith and Peach 1983). Yet, comparison of changing city fiscal conditions in riot and non-riot cities reveals that riot cities received disproportionately more federal grants, as well as seeing (somewhat) greater spending increases for services that strengthened municipal capacities to react to future riots (see e.g. Hoggart 1990). As the discussion throughout this chapter has shown, either pointedly or by implication, securing reliable time-series data is a key flaw in official statistics. Certainly, data changes occur because society moves on. We do not really need to expend great efforts identifying the incidence of cholera in England today, unlike in the nineteenth century (Kearns 1985). At other times, changes in definition, reporting style or coverage owe less to societal change than to political considerations (see e.g. Box 3.4).

This does not mean there are no data series that offer valuable insights on change in society. Amongst some of the more well-known examples published by the UK Office of National Statistics are the General Household Survey (published as *Living in Britain*), the Family Expenditure Survey (published as *Family spending*), the Labour Force Survey and the British Household Panel Study (Davies and Dale 1994; D. Rose *et al.* 1994). In the USA the General Social Survey is particularly notable for providing a long and consistent record (Davis and Smith 1992). The precise form these surveys take varies. Some utilize a panel of respondents who are revisited. These tend to be rare, partly owing to the effort involved in organizing and administering them – although national cohort studies were started in the UK in 1958, 1967 and 1970 (Davies and Dale 1994).

---

21 The US National Opinion Research Center survey began in 1972, with more than 25 000 respondents by the year 2000 involved, covering 1500 plus questions. These are varied over time, although key questions are replicated (Davis and Smith 1992). The same is true for the Commission of the European Communities reports on citizen attitudes, which are published in *Eurobarometer* (CEC, annual). In addition to single-topic reports (see e.g. CEC, 1999), consistent themes in *Eurobarometer* are views on European integration and EU policies.

In addition, panel studies can have a conditioning effect on participants, who are more prepared to respond to issues the second time around (Lee 1993).[21] This can be partly addressed through a rotating survey, in which a proportion of panel members are changed each time (Nichols 1991). In other cases, whether because of logistics or methodological considerations, the survey is an annual one but its participants are new each time (Fink and Kosecoff 1985). This latter approach, which is used by the UK's General Household Survey, the Family Expenditure Survey, the Labour Force Survey and, less regularly, the Housing Condition Survey (Davies and Dale 1994), is effectively a cross-sectional survey repeated across years. Beyond the government-funded realm, such surveys are less regular, but various surveys help identify change in public values. In a single-nation context, the annual *British social attitudes* report provides one output, albeit the topics covered are not consistent across the years, even if topics are revisited. On an international level, those familiar with the work of Inglehart (1977, 1990, 1997) will be well aware of multi-nation attitudinal surveys conducted over some years. Linking measures of political value to local government policies, the Fiscal Austerity and Urban Innovation (FAUI) programme likewise draws on questionnaire returns from municipal leaders in more than 7000 localities in over 30 nations, with many nations surveyed more than once (see Clark and Hoffmann-Martinot 1998: 168–91). With the exception of the last-mentioned survey, one problem with such surveys is the weakness of their geographical indexing. In many cases these surveys involve such small numbers that geographical discriminations can only be sought at the most superficial level.

More researcher control?

A point to set against this is that the census (and other official material) became more readily available for researchers in the last decades of the twentieth century (see e.g. Cole 1994). This is seen in the availability of Small Area Statistics (SAS) from the UK Census. Improvements have taken place in the variables that can be called on, as well as the ease with which data can be extracted (for example, via the MIMAS service at the University of Manchester). Payments for individual photocopied sheets or hours sat transcribing from microfiche, which were requirements for 1971 SAS users, have been transformed through the ability to read directly from a computer file, draw off a CD-ROM or use networked access (Rees 1995).

Even more powerful is the capacity to manipulate original census data. There are two notable vehicles through which this can be achieved in the UK. The first, which is a major resource on account of its capacity to trace census information for the same individuals for every UK Census since 1971, is the Longitudinal Study (LS). This data source has been used with considerable effect by geographers in exploring social, housing and demographic change (see e.g. Hamnett and Randolph 1988; A. J. Fielding 1992; Hoggart 1997). As a research resource, the LS has great capacity, for not only is census information included in the database, but also information from medical and other records (Hattersley and Cresser 1995). In addition, the variables that are available to researchers on the LS are greater than those available in the census, since raw data here have been recoded, so more theoretical categories can be employed. A good example is social class, with the Registrar General's classification available alongside others, like that of Goldthorpe (1987). Although considerations of anonymity place restrictions on securing data for very small areas, fine-scaled analyses can still be undertaken. Moreover, the user can create variables specifically for an investigation, out of raw LS data. As the raw data are drawn from a 1 per cent sample of the total census population (selected by undisclosed birth dates), this provides the LS with data for a single census of approximately half a million individuals, with more than 800 000 in the 1971–91 data set. Moreover, as this data set is linked to significant medical projects, such as evaluating the life trajectories of cancer patients (see e.g. Murphy *et al.* 1990; Kogevinas *et al.* 1991), resources are expended in ensuring the data are 'clean' and user friendly. As the beauty of this data set is its unique longitudinal framework, efforts are also made to iron out inconsistencies in definition that exist across censuses (albeit, as users are well aware, some remain). There is little doubt that this is a data source of significance for the future (on the British Household Panel Study, see also D. Rose *et al.* 1994).

The same can be said for the Sample of Anonymized Records (SAR). Reflecting the enthusiasm with which some researchers have responded to records like the SAR, Middleton (1995: 353) proclaims that: 'Probably for the first time, the social statistician has access to a large, national, representative sample of individual data about people in Britain.' Again this is a data source that census officials are keen to see used, so materials can be downloaded and installed on local unix hosts. Stripped of names and addresses, this data set comes in two forms, although so far only for the 1991 Census. The first is a 1 per cent sample of households, the second a 2 per cent sample of individuals (Rees 1995). Confidentiality again restrains the size of geographical zones for which data can be reported, although tables can be secured for local authorities with at least 100 000 residents. The power of these new statistics for theoretical investigation should not be played down. As Openshaw and Turton (1996) point out, if a table is required that is not provided in Small Area Statistics and Local Base Statistics formats, this can probably be created from the SAR. There is great flexibility in this data source, as with the LS, although both suffer from general problems with official statistics – for example, the SAR has a notable under-recording of young males, which is probably linked to Poll Tax avoidance (Openshaw and Turton 1996).

The UK is not the only country with records available like this. In Italy, for example, raw coded data from the census can be released without names and addresses. The city of Bologna also has a version of the LS. In France also 0.25 per cent of the population has been linked across the 1982 and 1990 censuses (see e.g. Cribier and Kych 1993; R. Hall *et al.* 1997).

Summary: can we improve on limitations?

- Through demands being placed on large organizations for information-based decisions, more information is becoming available
- This involves including more innovative data sets, some as time-series, alongside materials covering new realms of human activity

Appreciating the balance

As these last examples suggest, it is difficult to generalize about published data sources. Not simply are there enormous differences between private, public and voluntary sectors, but also within these sectors. As one illustration, in 2001 only a small number of building societies in the UK provide easy access, even through web sites, to (regional) house price information; most do not. Moreover, the range of potential data sources that emerge from organizations and associations, as exemplified by the disparate selection and uneven longevity of privately published directories (see e.g. Shaw 1982), works against any easy treatment. Even defining what a published source is creates problems. Essentially this arises because of the innovative ways researchers can utilize information sources. Anyone looking for exemplary and thought-provoking examples to illustrate this need look no further than Webb and associates' classic *Unobtrusive measures* (1966/2000). With the range of possible sources so diverse, the aim of this chapter has been to explore potential gains and limitations associated with the use of published accounts. We have focused on public-sector documents in the main because they generally seek comprehensiveness. As we have stressed from the outset, a major advantage of published accounts is that they provide a short cut to obtaining 'the general picture' or at least as much of the general picture as the data allow interpreters to draw. Because the principles on which compilations are put together can vary, researchers are advised to start by exploring commentaries on potential sources. As with many other aspects of documentary evidence, John Scott's *A matter of record* (1990) provides an excellent guide to themes and dimensions of distinction. As one instance, he draws out nicely the way some directories are inclusive, others exclusive, some allow potential entrants to opt out, others do not. Particularly powerful is Scott's reminder that the list of directories (that is, data sources) is almost endless, with the potential to gain insight on research questions greatly enhanced by overlapping sources. As one illustration, there are divergent sources on the Establishment in the UK, such as *Burke's peerage and baronetage* (irregular), *Debrett's peerage, baronetage, knightage and companionage* (irregular), the *Directory of directors* (annual) and *Who's who* (annual), with the *Social register*, the *Million dollar directory* and *Who's who in America* (annual) offering companion volumes across the Atlantic. Beyond general commentaries like Scott's, investigate commentaries on directories for information on data collection, accuracy, content and so on. There are a vast number of publications of this sort. Some have a general remit (e.g.

Finlay 1981; Whyte and Whyte 1981; G. Shaw 1982; Richardson and James 1983; Porter 1990; P. Edwards 1993; Drake and Finnegan 1994), others examine specific topics (e.g. E. L. Jones *et al.* 1984; Southall *et al.* 1994) or single places (e.g. Szucs 1986; Creaton 1998). One weakness in these accounts is that the guidance offered is stronger on historical than contemporary sources. This bias is less noteworthy for official statistics. Although useful guides on historical resources do exist (e.g. Lawton 1978; Higgs 1996), there are valuable contemporary guides (e.g. Rhind 1983; Slattery 1986; Dale and Marsh 1993; Openshaw 1995; US Department of Justice, Office of Justice Programs, Bureau of Justice Statistics 1999). Our purpose has not been to replicate the content of these guides, but to use (primarily) official statistical sources to draw out issues of principle that researchers should consider when using published records.

Our primary message is that these sources are biased, with these biases, in our view, bringing into question the validity of research approaches that assume randomness in measurement error or the objective nature of official counts. This point is critical. Epistemological understanding in social science is more refined today than in the past, so we should not unquestioningly accept the foundation stones on which published reports are built. For most official statistics it was positivist ideas that provided the early underpinning (Hindess 1973; Bulmer 1984). The bias towards so-called objective facts in census taking draws on this assumption, without a concomitant appreciation that this vision of 'objectivity' is socially biased. In a nutshell, the foundations of published accounts need to be approached as political documents, irrespective of whether they are issued by private, public or voluntary organizations. Viewing this way, researchers need to appreciate that they reflect a particular model of society. If social science is to make a positive contribution to societal change, researchers need to be conscious of the basis on which evidence is accepted by those who are able to influence change substantively. Experience shows that two kinds of information are recognized in this context. The first, which political activists and lobbyists will be well aware of, but which largely seems to reinforce existing prejudices, is the anecdote – the good speech items (Green and MacColl 1987). The second is the so-called objective fact, with an emphasis on systematic data collection, with more nuanced, qualitative accounts commonly receiving short shift, no matter how well collected – as some researchers have made clear (e.g. Brody 1971; J. Burgess 1996; Cloke *et al.* 1997).[22] Such evidence can be collated through mechanisms like questionnaire surveys. It does not have to come from published reports. But if published (statistical) reports are to be used, then they must be approached creatively; not simply taking the material in existing tabular form, but seeing how the data can be used to 'get behind the façade' (for which data sources like the LS and the SAR are potentially extremely valuable). In the longer term the research community (indeed

22 This commentary is based on the assumption that politicians and civil servants want to respond to social 'problems' or potential for social improvement. It is not aimed at information gathering by political parties to package policies to 'sell' to the public.

society as a whole) needs to make clear that (existing) published statistics offer partial accounts. Official statistics should be approached with recognition of their silences, with bias towards particular aspects of social life and away from others. Even within the themes such figures explore they play down (and play up) particular strands of social action. These failings do not mean published sources cannot be extremely useful sources for identifying (some) general trends in societal differentiation and change. Yet over-reliance on these sources will lead to lop-sided reporting. Published sources are best utilized if their potential contribution is evaluated critically.

This stipulation is manifest in cross-national differences in reporting. As Tooze (1998: 214) reminds us: 'treating statistics simply as a mirror of reality underestimates what is in fact involved in their production … The gathering of statistics is, in fact, part of the active process through which an abstract conception of the "national economy" is turned into objective reality.' Official statistics are political in so far as they are used for purposes of nation building and as a statement of national values (hence some of the disparities in what information is and is not collected). Linked to this, the manner in which published official accounts are made available to the public reflects something of the 'national character' (at least as the already powerful define it). Of note in this regard is Hamshere's curiosity (1987: 46) that 'the British mania for keeping records is matched by an equal penchant for restricting access to them'. This position can be compared with that in the USA, where notions of free information have roots back to the founding fathers, whose driving force in part came from a desire to destroy the traditional theory of hereditary sovereignty, partly by rational and empirical evaluation (Sharpe 1978).

National biases in statistical provision once again bring out the merits of multi-method approaches to social investigation. The general point is that different theories use different analytical categories, as well as identifying different analytical problems or at least affording them dissimilar weight. This means that the usefulness of different data sources, including official statistical accounts, depends upon the theoretical problematic in which they are used. Despite silences, biases in definition and unevenness in reporting, published data can be used creatively to test theoretical propositions. We have identified how variables can be used in new combinations to enhance the theoretical value of research insights. This is made easier by new modes of access to official sources, as seen in the LS and SAR. In addition researchers are more able to draw on raw data to produce special tabulations. One example is Hanson and Pratt's use (1995) of journey-to-work data to explore gender distinctions. Another is Simon Duncan's use (1991) of building society tabulations to explore the housing market experiences of single women in London. But it is not necessary to draw on special tabulations to generate socially critical and theoretically informed research. Just because the underlying principles on which official statistics tend to be collected and presented rest substantially on positivist ideals does not mean that analytical insights using official accounts are limited to positivist epistemologies. This point is very evident in criticisms positivists make about research on non-decision-making (e.g. Polsby 1980). Adopting a traditional empiricist line, these criticisms have argued that it is unacceptable to hold that we can explain things that do not happen. Yet investigators using official statistics have been able to show that the non-occurrence of social actions is in line with established social science theory. Examples include black registrations in the US South after the 1965 Voting Rights Act being lower where the black population was more job dependent on white employers (Salamon and van Evera 1973), municipal by-laws on pollution being less evident where local economies depend on highly polluting industries (Crenson 1971; Friedman 1977) and zones of higher socio-economic standing receiving disproportionately fewer applications for new house building (Buller and Hoggart 1986). Certainly, official statistical sources have biases, but they also open the door to original research insights. They require critical engagement, not blind acceptance, nor myopic rejection.

Although his reference is to a specific data stream, Maier (1995: 99) provides a more general point to note when reminding us that, 'as with other social statistics, crime statistics are only as meaningful as the skill of the researcher who uses them'. It is beholden on us as researchers not to take published (or other) figures at face value. This means that before undertaking analysis we need to take time to understand the structure of the data and the definitions a data set uses to represent variables (Glover 1996). The five questions Hawtin and associates (1994) ask us to pose when faced with a statistic provide a good framework from which to start a critical engagement: (1) Who says so? (2) How do they know? (3) What is missing? (4) Does the result answer the question? (5) Do the assumptions of the question make sense?

4
Behind the scenes: archives and documentary records

The success or failure of historical reconstruction will largely depend on the sophistication and thoroughness of the indexing.

(MacFarlane 1977: 83)

Court 73 in London's Royal Courts of Justice saw an extraordinary libel trial in the year 2000. This culminated on 11 April with the historian David Irving losing his case against fellow historian Deborah Lipstadt. Irving had sued Lipstadt and Penguin Books on the right to dissent from mainstream views about the Holocaust. The ruling against him is important. His claims concerned Nazi leadership awareness of the slaughter of Jews, whether the executions were systematic and whether gas chambers existed. As newspaper reports reveal, the judge made clear it was 'no part of my function to attempt to make findings as to what actually happened during the Nazi regime' (S. Moss 2000: 5), rather the issue was about Irving's use and interpretation of documents. As Richard Evans, Professor of Modern History at Cambridge University, is reported to have put it, this 'was a trial about Irving's methodology' (S. Moss 2000: 5) – a trial for which Evans spent two years exploring Irving's research, after which he concluded that: 'Penetrating beneath the confident surface of his prose quickly revealed a mass of distortion and manipulation ... so tangled that detailing it sometimes took up more words than Irving's original account' (V. Dodd 2000: 4). The court ruled Irving had misrepresented evidence deliberately. The public presentation of this judgment should remind us of the centrality of history for self and communal identity. Readings of past and present are intensely political, as seen in revisionist writings that give voice to disadvantaged and less powerful people in the unfolding of history. This stress on neglected populations owes much to postmodernist ideas, with their emphasis on multiple meanings and a variety of truths (Kellner 1992). Obvious examples of such reinterpretations include work on colonialism (Chakravarty 1989), the history of native Americans (e.g. Cook-Lynn 1996; Deleria 1997) and the contributions of women to social progress (Katz and Monk 1993; Blunt and Rose 1994). This work brings a needed rebalancing in historical (and contemporary) accounts. Yet we are drawn to mark a note of caution. While many accounts of the past essentialized the role of white, upper-class males from Europe and/or North America, some would appear to counteract past lapses by making claims for which evidence is wanting (e.g. Lefkowitz 1996). This is a particular problem for (historical) documentary analysis, since more limited opportunities for data triangulation increase pressure on investigators to interpret data 'holes'. This makes for a peculiar character to research that is based on documentary sources alone.

Let us start by asking what is meant by documentary sources. As indicated in the previous chapter, drawing distinctions between documentary and published material is not straightforward. We will not make it a cause for conceptual debate. In the previous chapter we explored published materials that record contemporary or past conditions. We brought into that discussion raw data for such publications. The emphasis in Chapter 3 was on numerical records. In this chapter we explore discursive material. Some of this material might have been published at some time. Visit the UK's national archive, the Public Record Office (PRO) in Kew, and you find many files with items that were once published – newspaper cuttings, Acts of Parliament, circulars from government departments, pamphlets, reports, even books. Some unpublished information in archives is primarily numerical, with discursive commentaries used to enhance their interpretation (see e.g. L. Ward 1988). Other data sources are largely discursive, with some numerical information, as many files in the PRO demonstrate. Rather than dwelling on these seeming inconsistencies, in this chapter we interpret documentary sources as those that were not seen as 'data' when they were produced. If we restrict ourselves to written contributions for the moment, this includes official material that reported or even changed the social world (for example, legal provisions). It includes records of organizational operations, such as police reports, judicial records or the minutes of political party meetings. Personal items also fit – diaries, wills, letters and photographs are common documentary sources (J. Scott 1990; Macdonald and Tipton 1993; with C. Huff 1985 offering an example of an index on such sources, in this case for nineteenth-century British women's diaries). The classic multi-volume output of Thomas and Znaniecki (e.g. 1918) is an early study that used this kind of data. This investigation examined letters written by Polish peasants to investigate immigrant lives before and after migration to the USA. Dorst's use (1989) of postcards and museum displays offers a further taster on how documents can be used to explore the ethnography of a community. Commercial material is also valuable. We acknowledge the compilation of Webb and associates (1966/2000) here, who list materials used in early investigations to compare attitudes in Nazi Germany with those in other countries. This work included examination of youth organization handbooks (H. S. Lewin 1947), newspapers (Lasswell 1941), plays (McGranahan and Wayne 1948), songs (Sebald 1962) and speeches (R. K. White 1949). Across the various documents that have been used, some were immediately subject to public gaze (for example, newspapers), others were (originally) not intended for public scrutiny.

If we start with this thought, it is worth extending the ideas of Johnson and Joslyn (1995) on the advantages (unpublished) written records have for social scientists. First, such records allow access to subjects that may be difficult/impossible to research through direct, personal contact, perhaps because they relate to the past or to a geographically distant place. Our acceptance of this view is touched with a degree of caution. It is the case that written material is critical for investigations of the past. Even though there have been powerful investigations using oral history (e.g. Greele 1991; Woodeson 1993), these are capable of exploring only a limited period. Moreover, despite the enrichment oral history can offer, as Box 4.1 illustrates, the passage of time decreases accuracy of recall (P. G. Gray 1955; Gittins 1979; Davies and Dale 1994). In this regard, documentary evidence is essential. Our reservation is simple. This is that documentary investigation should not be associated too closely with historical accounts. Documentary analysis is an analytical approach that is pertinent for all investigations, with benefits for all to grasp and weaknesses requiring caution from all. In particular, as argued in Chapter 2, research is strengthened by analytical triangulation, with documentary accounts providing one perspective from which to investigate research questions. Added to which, there are occasions when a lack of documentary evidence opens an

Box 4.1 Distortions in recalling the past

It has long been recognized in the research literature that individual forgetfulness is a problem for the accurate data collection through interviews, with greater inaccuracy as time progresses from an event (see e.g. P. G. Gray 1955; Gittins 1979). Yet the issue of memory 'adjustment' over time, which is often associated in commentators' reports with selective recall, is more complex than is often imagined. This is seen in the distortion of recollections in ideological (and communal) ways. This point is well illustrated by Cappelletto (1998), who examined massacres in two villages following the killing of SS soldiers during the Second World War. With more than 100 persons killed in each place, this is the kind of event about which memory is unlikely to fade. But Cappelletto reveals that memories of these events display a patterning that is related to social identity. In Civitella, whose residents regard themselves as 'urbane' and civilized, with a conservative political outlook, memories saw partisans dangerously exposing the local population through killing the soldiers, without offering the villagers 'compensatory' protection. Here the massacre was seen mostly as a family or emotional issue. By contrast, in the more isolated Vallucciole, the massacre was seen more as a political event, for which the Nazi-Fascists, not the partisans, were responsible, with some blame attached to villagers for not defending themselves adequately against the Nazis. As Cappelletto makes clear here, there is a 'memory of the group' as countless retellings of a tale result in an evolution of a particular viewpoint on the event. The particularities of the perspective that evolves cannot be disentangled from communal political perspectives, nor from a 'a clearly perceivable element of social control over the way the events happen to be recollected and narrated' (Cappalletto 1998: 79), with even emigrants continuing to project the dominant vision of their home village.

Viewed in this manner, 'the dynamics of the divided memory are to be viewed as strictly intertwined with victimhood and with the more general issue of how these communities reacted against the brutality and violence of the established regime' (Cappalletto 1998: 84).

investigation to particular forms of contamination. Howard (1994) comments on this, when drawing out how the colonial heritage of Nicaragua means that interviews (intensive or not) are suspect sources of data if the interviewer comes from a position of power. The tendency, Howard reports, is for respondents to reply to questions in the manner they think the researcher wants to hear or to portray the situation as worse than it is if they believe aid might be forthcoming (see also Buzzard 1990). Providing more cautions, Derman (1990: 117) draws on Briggs (1986) to offer the following list on why interviews with the rural poor can be flawed: (a) fear of retaliation if change is opposed; (b) fear of holding different positions from officials (or assistants) who accompany the researcher; (c) a belief that it is not worth making a communication given the researcher's lack of commitment to an area; (d) a desire to avoid public discussion about personal poverty; (e) a willingness to say what the interviewer wants to hear to speed departure; (f) a sense that it is inappropriate to speak on behalf of a family or community; and (g) lack of public communication skills restricting willingness to speak.

Johnson and Joslyn (1995) see the second key advantage of documentary sources in the fact that they are non-reactive. Put another way, as they are written, the information they contain does not alter because it is used in a research project (Webb *et al.* 1966/2000). We have partial sympathy with this view, much as Johnson and Joslyn do. It is not that documentary records are exposed to the same degree of reactivity as interviews, focus groups or participant observation, but the information a researcher can obtain from them is still conditioned by the reactions of the 'informant'. On this, think about the messages you can glean from a quality newspaper. Thus, according to the *Guardian*, the UK's MI5 has started shredding large numbers of files:

> *The policy shift was prompted by embarrassing disclosures by David Shayler, a former MI5 officer who revealed that the agency kept files on a number of prominent politicians – including the Home Secretary, Jack Straw, the Social Services Secretary, Harriet Harman, and the Minister Without Portfolio, Peter Mandelson. There were also files on … John Lennon. (Norton-Taylor 1998: 7)*

If MI5 is not adverse to shredding what it once regarded as necessary files (even if this organization has revealed some paranoia over who it spies on (see e.g. Wright 1987)), then other organizations are likely to be more prone to dispense with 'embarrassing' or operational records. Johnson and Joslyn (1995: 252) recognize this in noting that 'today many record-keeping agencies employ paper shredders to ensure that a portion of the written record does *not* endure'. In so far as the survival of resources is concerned, it might seem that documents cannot be considered non-reactive if there is selective shredding. Yet selectivity in record keeping is inevitable in large organizations. Hence criteria in selection are important.

Despite shredding and other mechanisms of loss, a third potential advantage of documentary resources arises if they enable long time periods to be investigated. This does not mean that long-term trends should be sought as a matter of course. For one, as Isaac and associates (1998) point out, one criticism historians have of statistical analysis is that it can treat time as a means of analysis rather than an object of analysis. Put simply, the meaning of the same event could change over time, so clouding or even invalidating the seeming unity of temporal records. Nevertheless, under the right conditions documentary information can offer a unique database for investigators. One potentiality is environmental research, for which data recording certain events are *relatively* straightforward, with long-time data sets offering real prospects for exploring event periodicity. One example is Moodie and Catchpole's exploration (1975) of Hudson Bay Company records from 1714 to 1871 to identify temporal patterns in the break-up of ice in north Canadian estuaries. Exploration of varied sources has also been used to reconstruct social change over the centuries in rural England (MacFarlane 1977). Likewise, Floud and colleagues (1990) used 170 000 records from the Army, the Royal Marines and the Royal Military Academy Sandhurst for 1750–1980, to explore links between average height (as a measure of health) and economic growth. This is a classic example of documentary analysis, in that the preserved data were of a mundane nature, yet put to good theoretical use. Think again of the selective shredding issue. Possibly readers would not be perturbed if a company discarded all its records on paper clip purchases, whereas they might if staff records for senior personnel alone were retained. Yet it might be that paper clip purchases are a signification of broad social processes. Their disposal might thereby rob future researchers of a valuable data source. Perhaps; but to be realistic, if data were not to be lost, then every product and thought in every minute of every day would have to be retained. Researchers might wish to be involved in prioritizing what is kept and what dissipates, but the inevitability of loss has to be recognized.

For Johnson and Joslyn a fourth advantage of documentary analysis arises from the ability to use material to increase the size of a researcher's 'sample', by extending data collection beyond what is achievable through interviews or direct observation. In general this is not convincing. It is possible that a researcher is pursuing a topic others have examined, so more information can be obtained. But to view this as increasing a 'sample' size is questionable. Access can be gained to the raw material of questionnaire responses. In the UK the ESRC Data Archive provides precisely this facility (http://www.data-archive.ac.uk/), with the Inter-University Consortium for Political and Social Research (ICPSR) offering a companion facility in the USA (http://www.icpsr.umich.edu/). But the computer files that can be accessed from sources like this inevitably relate to 'the past'. They raise analytical problems of societal change, given time differences between when archived questionnaires were collected and a contemporary survey undertaken (albeit these might be slight if the time period is short). Added to which, it would be extremely fortunate if a documentary source included all variables a contemporary study wished to explore, so the material is likely to be non-comparable in this sense (albeit not necessarily for single questions). Archived questionnaire returns might offer insight on change over time but they rarely provide data that extend a 'sample'.

Yet there is a fourth advantage offered by many documentary sources, and this is an important one, even if its immediacy can be protected by confidentiality clauses restricting the release of material for some years. This comes from being able to see 'behind the scenes'. Especially in policy-making contexts, this is extremely important for theoretical evaluation. As research in Geoffrey Walford's volume (1994b) shows, those in positions of authority are well placed to deny researchers access to their decision processes in interviews. Many are skilled at deflecting questions or not answering (see also Wagstaffe and Moyser 1987; Healey and Rawlinson 1993). In these circumstances, gaining access to files that expose the mind-set of policy-makers offers evidence on what weighed in their deliberations. Consider, for example, the following Public Record Office (PRO) extract from a 4 March 1955 internal HM Treasury memo from D. M. B. Butt to Mr Strath, entitled *Local authority housing*:

> *though the overall cut in authorizations [for new public-sector dwellings] between 1954 and 1955 is about 27%, the individual cuts on a number of areas have had to be a good deal more. The process of handing out 1955 figures for local authorities is not yet complete, but there is a considerable volume of complaints and it is believed that very ferocious rows may blow up with Manchester, Leeds, London, and some other important and politically critical areas. [The Ministry of] Housing [and Local Government] freely recognize that local authority pressure derives almost entirely from the subsidy arrangements. Waiting lists have been lengthening steadily, nor is this surprising, in view of the fact that the subsidy now amounts to about £700 a house. Nevertheless they [the Ministry of Housing and Local Government] feel that if they could have authority to hand out another 10–15,000 authorizations in the course of the summer, they might be able to prevent a major political controversy. (PRO T/227/808)*[1]

The insight this passage provides on how controversy affects government willingness to adjust policies is linked to theoretical debates on state policy making (see e.g. Dunleavy and O'Leary 1987). Of course, one incidence like this is not enough to confirm or deny an explanation. Nevertheless, if such instances are regularly identified, this offers insight on decision criteria officials are unlikely to publicize at the time. This raises a conundrum,

1 We would be remiss in not mentioning that it is essential to keep good records of the location of material used in constructing arguments. Investigators who wish to challenge interpretations, or wish to follow up points, are greatly helped by clear, accurate reporting of sources. This is a key requirement for allowing others to enquire into the validity of your arguments (as with Domhoff's re-evaluation 1978 of Dahl's work of 1961). The introductory messages in this chapter on exposing David Irving's interpretations highlight the need to have confidence in your reading of documents.

for, if officials are unlikely to talk about this at the time, then it is unlikely such records will be released until after their messages are politically desensitized. In the case of national government records in the UK, for example, the great majority of materials are not released to the public until a minimum of 30 years has passed (albeit earlier access can be obtained in some county record offices). This raises the prospect that society has moved on, that processes in operation 30 years ago have been transformed. This is a real possibility, but it emphasizes the need for vigilance in linking theory to social action, rather than demeaning the value of documentary evidence. Put simply, a case must be made for the continuing theoretical relevance of documentary sources. This applies whether the investigated issue is contemporary or not.

Summary: the advantages of documentary sources

- Researchers are able to gain access to issues that are difficult or impossible to research through direct contact.
- Data collection is essentially non-reactive – the information collected is generally not influenced by the fact that sources will be used in research.
- Elongated time periods can be explored, so there is an opportunity to examine long-term trends and the periodicity of events.
- They offer a capacity to examine processes from 'behind the scenes'.

What types of documentary sources?

A distinction that can be drawn between documentary materials differentiates between episodic and running accounts. Episodic records are items such as diaries, manuscripts or correspondence (Stewart 1913/1988; King *et al.* 1995; Duncan and Gregory 1999), with autobiographies and memoirs, whether published or unpublished, often forming key records (e.g. Butterworth 1992; Blunt 1999). Such records are irregular in appearance, often one-off items. They are not part of an ongoing record-keeping programme. It might be that your great grandmother kept a diary assiduously during her life, but if this is stored in a box in the attic, possibly without the family knowing it is there, then whatever gems it contains on lifestyles, attitudes towards world events, or the changing price of sugar, will not be available. Episodic records are often 'hidden'. When they are released, finding them can be time-consuming, even frustrating. Yet they can be extremely rewarding, as they often contain detailed discursive accounts of people's lives. Moreover, centres that collect documents often concentrate on particular types of material. Some examples include the Rural History Centre Archive at the University of Reading (Box 4.2) and the Mass-Observation Archive at the University of Sussex (Box 4.3). In the UK county record offices regularly receive deposits of material from local residents, which can provide fruitful insight. Centres also set up collaborative arrangements, such as that between the Liddell Hart Archive at King's College London and the Imperial War Museum, with the former storing personal materials from military officers while the latter retains material for non-commissioned personnel. We consider access issues later, but it is worth stating that, unless you find a resource centre specializing in a subject that interests you, you are likely to devote considerable time to finding episodic material. Previous research publications or guides to archival sources are enormously helpful in such a search.

Running records are different as they offer data series, perhaps over a long period. In many cases, these have been carefully stored (for example, political speeches and

Box 4.2 The Rural History Centre Archive, University of Reading

This archive contains five main types of record:

- *Cooperative records*. These contain records of agricultural trading cooperatives.
- *Document collection*. This includes the Alec Coker and Dorothy Wright collections on crafts and rural industries, the H. J. Massingham literary papers, the J. R. Bellerby working papers on agricultural income and output (1850s–1950s), the E. S. Beaven scientific papers in breeding malt barley (1903–16), the Herbert Hunter biographical and research papers on breeding barley and oats (1898–1959), the H. B. Parry biographical and research papers on sheep scrapie (1939–82) and the R. G. Stapleton papers on grassland development (1938–early 1950s).
- *Farm records*. These include farm business records for more than 1000 farms (most held in the University Library, with application needed to the University Archivist), with additional farm material in the documents of H. D. Barley, farmer, stockbreeder and agricultural writer (1930s–1971), the records of the Earl of Selbourne's Blackmoor Estate (1867–1967), the records of the Langford Downs Farm, Gloucester, the farm records of Ruth Timbrell and papers relating to the national poultry test (1920s–1965).
- *Society records*. These include the records of the Agricultural Apprenticeship Council (permission required to view), the Council for National Parks (1970s–), the Council for the Protection of Rural England (1920s–, with the Hampshire, Penn County and Sheffield and Peak District branches of the Council also depositing some records here), the Country Landowners' Association (1919–early 1960s), the International Association of Agricultural Economists (1929–71), the Jersey Cattle Society of Great Britain (1885–1970s), the National Dairy Council (publicity and films), the National Farmers' Union (1909–1970s), the National Federation of Village Produce Associations, the National Union of Agricultural and Allied Workers (1907–early 1980s), the Open Spaces Society, the Royal Agricultural Benevolent Society (1893–1990s), the Royal Agricultural Society of England (1794–1940s), the Shorthorn Society of the United Kingdom of Great Britain and Ireland (1875–early 1960s), the Standing Conference on Countryside Sports (1977–1987) and the Wokingham and District Agricultural Association (1879–1984).
- *Trade records*. These include the business records of companies engaged in agricultural engineering, food processing, and farm and garden seed production.

Further information can be obtained at:
http://www.rdg.ac.uk/Instits/im/the_collections/the_archives.html/
Collections may be consulted by appointment with the business records officer.

Box 4.3 The Mass-Observation Archive, University of Sussex

This archive contains materials from the social organization Mass-Observation, which was active from 1937 until the early 1950s. In keeping with the spirit of the organization, this archive specializes in collecting writing by 'ordinary people' about everyday life in Britain. It also has material produced in the 1980s–1990s Mass-Observation Project. Much of the information collected consisted of reports from a sample of people about everyday life and life wishes. There was no word limit on reports and people were asked to be candid. Included amongst the mass-observation materials are reports on:

- the Fulham survey 1938;
- studies in Fulham, Stepney and Kilburn 1941;
- housing survey 1941–42 (Welwyn Garden City, Letchworth, Ipswich, Portsmouth, Fulham, Becontree and Dagenham, Roehampton, Watling, Ilford, Highgate, Worcester, etc.;
- pamphlets, leaflets, articles and cuttings from the Second World War on planning, housing and reconstruction;
- reports on living in homes (like why people chose their home, heating, storage space, neighbours and privacy, housework and cleaning, gardens, repairs and decoration);
- housing 1945 (post-war homes exhibition, temporary housing, new towns);
- housing 1946–8 (modern homes exhibition);
- post-war reconstruction (survey of views, including garden cities and town planning, extracts from mass-observation diaries on views of what people wanted after the war);
- squatting 1946 (observations on squats, overheard comments about squatters).

In addition, the archive maintains records from a 1981 (onward) follow-up, in which about 2500 people have taken part (with a 2001 mailing list of 600). This involves requests for material about three times a year, with suggested topics to comment upon. Other materials in the archive include:

- the Geoffrey Gorer papers, related to his publications *Exploring the English character, sex and marriage in Britain today*, etc.;
- BBC diaries, commissioned during the 1950s and 1960s as part of their listener research;
- Social and Community Planning Research time diaries on leisure activity in the 1980s
- One-Day for Life – a collection of 100 000 colour photographs all taken on 14 August 1988 by amateur photographers, with the 4000 selected for the competition for a charity organized into main subject themes;
- national gay and lesbian survey – undertaken in 1986, with autobiographical writing and matters of opinion;
- Checkpoint material – BBC surveys on consumer problems between 1977 and 1985, plus letters from the public; access requires approval of the head of Consumer Affairs at the BBC;
- newspaper readers' letters – from those sent to the *Daily Mirror* (1981–1987), plus responses to the *Sun*'s (1981) survey 'what's wrong with Britain';
- Victory in Europe Day memories – Television South 1985 gatherings on reminiscences of VE Day 1945.

Among the geographical publications that have used the Mass-Observation database are Thrift (1986).

Further information can be obtained at:

http://www.sussex.ac.uk/library/massobs/general.html/

Visits must be by appointment.

government documentation on policy issues). Such data are more likely to be produced by an organization rather than private citizens, or at least have the collection and storage of information coordinated by an organization. If this is not the case, as with (say) personal records of temperature and rainfall, researchers might gain access only to a few runs, often of short timescale, and commonly for different periods. Apart from the breadth of material that is offered (either within an organization, such as employee records, or across units), a general advantage running records have over episodic ones is cost. This is because the costs of bringing together material, and organizing it, have generally been borne by the record-keeper, as with company archives. This might be less of a consideration if researchers draw on government archives, such as county record offices. However, resources are far from infinite and it often takes a long time before donated materials are available to readers. Johnson and Joslyn (1995: 240) offer a sense of what can be involved in using these documentary types: 'Instead of searching packing crates, deteriorating ledgers, and musty storerooms, as users of the episodic record often do, users of the running record more often handle reference books, government publications, computer printouts, and computer tapes.'[2] But the researcher is at the mercy of collection practices by record-keeping organizations. As storage has a cost, it might be expected that in the commercial world there are different attitudes towards expenditure on preserving 'old records'. In some cases organizations have extensive collections, as with some of the older London banks (albeit access can be an issue). In other cases, funds for storage and collation are meagre. Harber (1978) brings this out for trade-union records. These tend to be scattered, with those archives that exist having too few staff and not enough space. Such records are inclined to focus on national organizations, where resources are more plentiful. As Southall and colleagues (1994: 2) note, a 'psychological' barrier exists over trade-union records, because they are not official, with local branch information more likely to be mislaid, poorly stored or even considered to be of no interest. These problems are equally apparent for episodic materials, with holders not maintaining records as consistently as researchers would like and perhaps not willing to share what they have.

Might it be made easier if materials are indexed in catalogues or printed? The printing of records is not new, for many printed series bring together key documents in public policy fields (e.g. Medlicott *et al.* 1969; Minogue 1977). Published materials of this kind often extend beyond official documents, to incorporate episodic materials from private citizens. One example is Elinore Pruitt Stewart's letters (1913/1988) from frontier

2 Resource issues can generate public debate. One example is the mid-1990s controversy over the failure of Churchill College Cambridge to allow examination of Lord Brockaway's papers, which it had received in batches in 1982, 1988 and 1991. Criticism arose from the suggestion that Lord Brockaway's papers reveal that British intelligence assisted General Franco in the Spanish Civil War. The response of the senior college archivist was reported to be that resources were insufficient to catalogue material: 'We just can't have people burrowing around here. If we let in one historian, we would have to let in everyone, and we can't have that' (*Observer*, 21 May 1995, 6).

Wyoming to a former employer in Denver, which record her efforts to show that a woman could ranch while bringing up a daughter on her own. Harris and Phillips (1984) also provide an edited collection of 1912–14 letters from Daisy Phillips to friends and relatives in England describing the creation of an English home in the Windermere Valley of British Columbia. One farmer's harrowing account of debt and drought in early 1930s Iowa (Grant and Purcell 1995) and Butterworth's more recent exposition (1992) of the pain of struggle that ends in farm failure are further examples. Such published accounts can provide vivid insight, as Anne Frank's diaries (1947/1954) portray.[3] Not unexpectedly, they have been used by geographers (e.g. Blunt's examination (1999) of middle-class colonial women's lives, as seen through the flight from Lucknow in 1857–8). Indeed, commentaries on the use of published reports within geography have tended to be favourable. Hamshere (1987: 48) offers one example, when claiming that

> *the 'purity' of the source expressed as an absolute insistence on the original rather than a printed or translation version ... has not been a particularly important issue within historical geography. Historical geographers are generally concerned with the manipulation of material derived from employment of the source rather than an analysis of the source itself.*

We are much less sanguine about the use of published material. It might be that easy access makes for quicker research and more rapid output, but, unless published accounts are seriously supplemented with wider information searches, in general they lead to inferior research and raise serious issues about the validity of results. Our use of the word 'validity' is deliberate here. It is not taken to mean there is only one view. Rather, the message is that, from whatever perspective an account is from, the 'accuracy' of the view is made shallow (possibly incorrect) by using a limited range of research material.

There are two key issues here. First, there is the question of what is published. For one thing, why are particular records published? If they are published commercially, there is the potential either that they are subsidized by someone who wants a case telling or that publishers consider them sufficiently novel that people will dispense with cash. In either event we are dealing with 'unique' cases (but see Box 4.4). Such accounts can offer slanted views of events or of how 'people' lived. Go to your local book store and look at the swathes of forests devoted to publications on politicians, actors, sports personalities or pop stars, then ask how many volumes there are on 'ordinary lives' (make sure to exclude books on any of the first list that proclaim they have had an 'ordinary life'). This might not be thought a problem, provided published accounts are

3 However, in times of great stress letters can also conceal a great deal. One illustration is Beevor's use of interviews and archives (1998) to explore the German Sixth Army's crushing defeat at Stalingrad in the Second World War. In letters from German troops, Beevor notes how innocuous commentary concealed appalling circumstances, in which food and ammunition was limited, while hunger, disease and lice were rampant.

Box 4.4 The 'unique' and the theoretically critical

There is a world of difference between investigating unique populations because records are available and persons whose 'uniqueness' has theoretical pertinence. Marilyn Monroe, David Beckham, Joe Di Maggio, Madonna, Richard Branson and Donald Trump might be 'extraordinary', such that they make good script and publishers' profits, But they provide little insight on mainstream society. This does not apply to all 'extraordinary' people or events, for there can be theoretical merit in studying the 'extraordinary'.

A pointed example is the Pullman strike and its aftermath in 1890s Chicago. This event was a defining issue in US labour relations. The Pullman Company had built its workers model houses. It provided (relatively) good working conditions and salaries. In appearance it was a model employer; one that seemingly saw mutual benefit from capitalism for workers and owners alike. The 'exceptional' event that highlighted the shortcomings of this interpretation was provoked by economic recession. The company responded by firing workers and reducing wage rates by 30 per cent or more. The labour force responded by pointing out that, as the company owned their homes and set their rents, it was reasonable to see a rent reduction. The company was shocked by the suggestion. Its aim was to maintain the company, not the workers, so revealing imbalance in the system. The company refused. Workers went on strike. The federal government signified its loyalties by sending troops to quell the workforce. The strike, and the violent disruptions that followed a few years later, have been written about extensively. A recent book on the subject is Papke (1999), but there are many more in print. The point is that this event was uncharacteristic of company–labour relations for the early history of the company (shortly after the strike the company was ordered to sell its interests in housing by the Illinois Supreme Court), but this 'unique' event is an important vehicle through which to cast light on theoretical propositions. This relevance arises from conditions that put 'the established order' under strain, such that 'real priorities' were highlighted. Seeking an event of this kind is quite different from restricting research attention to the 'unique', when the peculiarity of the source comes from the size of an ego or a (personal or publisher's) pocket. To rely on such sources alone is deliberately to limit investigative attention.

On a broader canvas, 'extraordinary' events can be critical in distinguishing eras in human life. Very evident is the impact of the Second World War on political and social values in the UK (see e.g. Thrift 1986). Addison (1994: 13–14) captures the sense of the upheaval:

> *Between the wars, politics were under the spell of the Conservatives and their ideas ... In Churchill, the descendant of Marlborough and historian of past glories, the oldest strain of ruling tradition had surfaced. Yet it was from the time Churchill came to power that the doctrines of innovation, progress and reform began to reassert themselves ... Churchill's arrival marked the real beginning of popular mobilization for total war, the era of 'blood, toil, tears and sweat'. This experience in turn bred the demand for a better society when the fighting was done, lifting the Coalition on to a new place of reforming consensus, swung opinion decisively from Conservative to Labour, dismissed Churchill from office and made Attlee Prime Minister. World War II saw the reformation of British politics for a generation to come.*

The so-called consensus politics of this time broke down at a further critical juncture, with 1970s oil-price hikes capping the disruptive effects of US economic decline and international politicking (Block 1977; R. W. Cox 1987). Keynesian welfare-state rationalities came under pressure as citizens and governments combined fiscal conservativism with social liberalism (Clark and Hoffmann-Martinot 1998). By the 1980s, the legacy of the Second World War, while leaving traces, was overlaid by new economic, political and social rationalities.

backed by clarity over their peculiarities. As Lee (1993) contends, this point applies to sources like diaries irrespective of whether they are published, for diaries that survive are likely to come from specific social fractions. Consequently, it might be more accurate to hold that these sources are valuable if other sources are used, so a more rounded picture is obtained.

Perhaps, but we would go further. Published sources are not simply problematic because they tend to cover particular types of people or events, but also because what they contain is biased, and potentially in unknown ways. A good reason for this is that the original material from which these publications are drawn is more extensive than the published volume. The material included is selective. But the selection process can be controversial (when it is known, which is often not the case for personal diaries, letters and the like). A stark example was brought into public view in the early 1990s, when newspaper articles reported on released documents on US foreign relations over the 1952–4 period. Amongst the omissions from this collection were those relating to US involvement in the coup that brought the Shah to power in Iran. This is despite the importance of this event for the late 1980s seizing of the US Embassy in Tehran, the aborted US attempt to rescue this embassy's occupants and the coming to power of Ayatollah Khomeini. This attempted massaging of US history, led to the (academic) chair of the State Department's advisory board proclaiming that the deletion of these documents meant that 'the integrity of the [book] series' was not safe. Offering a sharper critique, the historian Bruce Kuniholm of Duke University is quoted as stating: 'The misleading impression of US non-involvement constitutes a gross misrepresentation of the historical record, sufficient to deserve the label of fraud' (M. Walker 1990b: 6). This concern over published accounts highlights the need for caution in their use. There are many less controversial examples. For example, for historical research on the Philippines, analysts have understandably been drawn to a huge data source, of some 55 volumes, that provides a translation and verbatim account of original documents (Blair and Robertson 1903–9). This mammoth compilation is a major source for historical work. Yet examination of the original documents, which many researchers do not check, owing to ease of access to this multi-volume edition, reveals that the published accounts miss a great deal, for comments and information in the margins of the original materials were not transposed into the published volumes. Providing an explicit statement on limited translation from the original to the published word, La Fleur (2000: 13–17) reports that for the journal entries of the Dutch slave trader Pieter Van der Broecke:

> *The survival of Van der Broecke's handwritten draft for his published version is very unusual among early descriptions of Africa … Authors and printers routinely discarded their handwritten texts after they had been set into print … The greatest difference between the manuscript and the published editions is that approximately two-thirds of the contents of Van der Broecke's manuscript was omitted in the published version … the release of sensitive commercial intelligence and nautical information which he had recorded might have upset his benefactors.*
>
> *The published texts also added a few small items of information not seen in the manuscript … Nor is this added information necessarily historically accurate … [and] most of that third of the manuscript's contents which did get published was altered, by omitting some aspects, adding others, and in general giving new meanings to the manuscript's descriptions. When comparing the manuscript and printed texts, the most immediately striking difference lies in the changed spelling and grammar in the published editions …*
>
> *Additionally, many items were rearranged within the text …*
>
> *Van den Broecke's Dutch in the manuscript is considerably different from that of the printed text.*

Reliance on published accounts, of whatever form, implies that the researcher accepts a limited and biased data set – a data set that has often been 'tarted up' for public consumption, so precise wording cannot be relied upon. Uncertainty over what is selected for inclusion further limits what conclusions can be drawn. Here we need to recognize that what is published could be determined by anticipated reactions from potential buyers, rather than duplicating the original document.

Making this point in a different manner, we hit on a second key problem, which arises from the questionable reliability of the material included. This might seem to be less of an issue if original documents are recent, if they are already typed and if they are reported verbatim, but even here problems arise. Thus, as MacFarlane (1977: 82) noted for parish records:

> *even full transcripts are not really satisfactory for it is frequently necessary to return to the original handwriting in order to check ambiguities. [However] The need for an exact copy would seem to pose an insuperable problem … for the records for any one parish are immense.*

A pertinent illustration came from examination of Thomas Jefferson's famous 1802 letter to Baptists in Danbury, Connecticut. This letter has been cited by Supreme Court justices and policy-makers to confirm that the founding fathers of the USA wanted religion and government kept apart, with the expression 'a wall of separation between Church and State' holding the key. Readers will have to indulge us a little here, as this example is not simply about the need for examining original documents, but even about the need to 'look below the surface' of such materials. When the letter was subjected to examination by the FBI, it was found that Jefferson had originally written about an 'eternal' separation of Church and State, with various other deletions similarly revealing that the letter was not a dispassionate theoretical statement but was written to conceal Jefferson's own views while achieving a desired political end (Kettle 1998). But even if records are supposed to be straightforward accounts, difficulties appear. Thus, MacFarlane (1977: 202) has found that: 'Often the ambiguity lies in the eye of the beholder ... we do not yet know what many of the documents really mean and hence cannot use them with confidence.' Stripping passages from longer documentary accounts to place them into published volumes can create real interpretative problems. This point is a general one, but it carries more force when we are dealing with materials in a language other than our own. Here, there is little doubt that translation often does not convey original meaning, yet the ease of exploring a translation is enticing (see F. M. Smith 1996). But because manner of expression can be as important as words themselves, literal translations can 'distort' intended messages, with the translator imposing a lens between the expressions in original and translated texts. A software grammar checker offers sufficient evidence that phrases in common use can be misinterpreted or not understood by mechanistic examination. Throw into this picture errors in transcribing or translation, and the sanctity of published sources is brought further into question.[4]

Our point is not to decry the use of published material. These records can offer valuable shortcuts to what is revealed by in-depth documentary analysis.[5] They might be the only materials that are available on a particular aspect of a study. Yet, even in the latter case, as a researcher you should be dissatisfied with published accounts. They are biased in representation (in one way or another deliberately so), limited in what they represent, and give a sense of knowledge about a subject that can lull a researcher into assuming

4 Note that, for microfilms of original documents, even if in high usage – e.g. census enumerator returns from old UK censuses (Lumas, 1997), whole pages are often difficult to read, with individual names and figures commonly stretching eye capacity beyond the realms of credibility. Readers often struggle over handwritten names and numbers, only to be stumped on whether a letter is a 't' or an 'l' or the number is a '3' or an '8'. Microfilm and microfiche are convenient, as they are more readily available than raw material, but this convenience comes at the cost of reliability.

5 For example, there are collections from diverse sources on microfilm or microfiche. The publisher Chadwyck-Healey makes useful contributions here, such as the 1400–1750 records on the English village of Earls Colne, the comparable 1650–1750 collection for the Scottish village of Lasswade, news typescripts from 1939–45 Home Service broadcasts or close to 30 000 pages of commentary, maps and plans that the Royal Commission on Historical Monuments on buildings and monuments has examined since 1909 to evaluate recommendations for building preservation. Chadwyck-Healey also microfilms the publications of English (mainly county) historical record societies.

the full picture is captured. Yet it is likely to be the omitted themes and strands that need more investigation. The message is that the sources you use need to be appreciated, in terms of their purpose, whose interests they serve (if anyone's) and the angle(s) they provide on a research topic. The convenience of published accounts is no excuse for neglecting less accessible sources, although these also have problems of coverage.

Summary: what type of documentary sources?

- Episodic and running records commonly come from different sources and have dissimilar availability.
- Archives usually specialize in the type of documents they keep, making referral to indexes essential.
- Access to documents should not assumed. Gaining access to archives and documents might need to be negotiated.
- Published documentary materials should be treated with real caution. They are often biased in unknown ways, with their exclusions and manipulations making them dubious sources if precise wording needs to be explored.

Problems of coverage

It is said that the Public Record Office at Kew contains 160 kilometres of original documents, with something like 1.5 kilometres added each year (B. Hughes, 2000). Yet these records are restricted in scope. They focus on what the journalist Bettany Hughes (2000: 3) terms 'the six pillars of the establishment: the church, the military, the landlords, finance, bureaucracy and the law'. Even so, the PRO is a gigantic source, with much more on offer when we add records kept by charities, corporations, local governments, museums, universities and so on. But all documentary collections face two major problems. One is selective deposit. The other is selective survival.[6] The problems created by these 'holes' is neatly articulated by John Scott (1990: 28):

> *While all scientific research involves the 'construction' of facts, the use of documents whose representativeness is unknown involves the possibility that the 'facts' constructed from the documents may be purely functions of the bias inherent in selective survival and availability.*

6 Selective survival is affected by investment. Thus, in the archive for former Spanish colonies, the Archivo General de Indias in Seville, materials from the Philippines were prone to disintegrate, as they were written on rice paper that became brittle over time. The Archivo is converting documents to digitized records, so researchers can see documents on large computer screens (including adjusting tone and lighting, 'taking out' stains, etc.), so original materials need not be subject to handling (see Ministerio de Cultura, Fundación Ramón Areces and IBM España 1990). To add a note of caution, lest readers think that electronic media are not problematical, it is worth noting that the pace of change in computer hardware and software is such that records have to be resaved every few years, with problems of learning new systems accompanying these adjustments (see Higgs 1998).

Although there are mechanisms for mitigating this potentiality, it cannot be eradicated. But before we explore the character of any ameliorating processes, we need to understand biases in records.

In terms of selective deposit, Macdonald and Tipton (1993: 190) leave the telling message that: 'Even if there is no attempt to hide things from public gaze forever, the idiosyncracies of departmental officials can prove extremely frustrating.' This point is made for national government collections, which usually have specific policies on deposit and retention. These policies inevitably limit retention. For the UK, for example, Knightsbridge (1983) points to a reluctance to define archives as national heritage, as this would require stronger legislation, as well as raising access, cataloguing and space costs. At least the UK has national legislation, so researchers have a reasonable expectation about what they can gain access to. But there are many reasons why material is not released to the Public Record Office (E. Higgs 1996). Government files are retained if they are still 'live' (that is, needed for administrative purposes), if they are exceptionally sensitive, if the information they contain was supplied in confidence or if their release would embarrass a living person. The files on those convicted of capital punishment have a 100-year release date, as do census enumerators' books (72 years in the USA). These limits seem clear-cut, but there is a lot of uncertainty. Enormous bundles of material have to be examined, with decisions made about their inclusion or exclusion. This results in some decisions that infuriate researchers. Colin Holmes (1981), for example, reports that files on the time the Bolshevik leaders spent in England have been destroyed (including those on Lenin). Worse than destruction in many researchers' eyes is the political manipulation of records, as with the Thatcher Government closing access to material on the hunger marches of the 1930s, when massive rises in unemployment in the 1980s caused political embarrassment (C. Holmes 1981). But matters of 30 years ago causing embarrassment to contemporary governments are relatively rare. For most investigators, bigger problems arise in finding key documents in the mass of material contained within archives. Another problem is that the leads researchers follow run dry, perhaps because they cannot find documents, as they are catalogued in unexpected places,[7] or perhaps, as Macdonald and Tipton (1993) highlight, because of a ministry decision not to save certain materials. The counterpoint is that insight might be gleaned from documents held by a number of ministries or under different themes/sections within a single ministry's remit.

When we get to records beyond the national government, messages become less clear. Michael Hill's comment (1993: 8) that 'the routes by which materials come to repose in archives are neither certain nor systematic' provides one clue to why this is the

7 Two problems arise. The first is that policy issues bear on a number of departments, and the researcher might not find all combinations in which pertinent documents are stored. The second appears to be more unusual as it involves wrongly catalogued material. An example from Newson's work (1995) on sixteenth-century Ecuador was when a document on the textile industry in Ecuador was found when working on another project, as this was in a document bundle (*legajo*) dealing with Guatemala.

case. Most evidently at a local level, material is threatened by deterioration and loss, owing to poor storage.[8] In addition, since national legislation does not apply to private or sub-national records, the materials included in archives are unpredictable. There are again problems of resources, which mean that material is selectively included and excluded. Harber (1978: 149) paints a gloomy picture of this: 'I soon discovered that one of the leisure-time activities of archive hunters is swapping atrocity stories.' Included in these stories are reports on biases in the interests of archivists, such that modern history receives a relatively low priority (albeit, as with all these comments, this varies between archivists). Moreover, in common with our warnings about published materials, written works from the famous that could give prestige to a record centre tend to be accepted, with materials from ordinary people more scarce. Thus, to an uneven degree, both across time and between institutions, archives tend to reflect the power structure of society.[9] In this they are similar to buildings, which offer a stronger testament to how the wealthy lived in the past than how the poor did. In research project terms, this does not mean high-quality projects are not achievable; more that the need to piece together material from disparate sources can be forced on researchers who study the powerless. Certainly MacFarlane (1977: 202) is right to note that for local studies 'even the best documented parish will have large gaps in most sets of records'. This means that recourse to multiple sources is often inevitable, which is to be encouraged anyway, given that cross-referencing offers a validity check for interpretations, as well as access to dimensions that have been lost or were never present in single sources. It also means researchers need to be imaginative in thinking about sources. Two good examples that enabled researchers to explore the lives of 'ordinary people' are the studies of Warren (1986) and Péroz-Mallaína (1998). Warren (1986) explored reports on the lives of 'rickshaw coolies' in Singapore using coroners' reports, while Péroz-Mallaína (1998) drew on court records on claims for back pay and royal justice to explore the everyday lives of Spanish sailors.

Because documentary analysis relies on piecing partial clues together, this method raises particular process issues. In Michael Hill's words (1993, 6):

---

8 The issue of loss is a general one. Readers of history books are likely to find regular messages noting lost records. As one illustration, Andrew Ward (1996: 541) comments on how personal records from English survivors of the Cawnpore massacres of 1857 were lost in a shipwreck off Sri Lanka: 'Like the burning of the Assembly Rooms, the wreck of the *Ava* was a disaster for future historians, taking with it Martin Gubbins's journals, Thomson's notes, Teddy Vibart's letters and one of Russell's dispatches.' This brings to mind the shipwreck loss of possessions and records of the founder of Singapore, Stanford Raffles.

9 Note the early absence of the mother's name on baptismal records (Carley 1983), while citizen lists in nineteenth-century directories tended to neglect many working-class residents (G. Shaw 1982). Biases are not just about the ability to write, the production of records or the under-representation of the relatively powerless, for the willingness to make records available is also a factor, as seen in material on large corporations rarely being accessible to the public (Johnson and Joslyn 1995). Yet some researchers have put considerable effort into securing accounts from 'ordinary people', as with the 1902–04 reports found in Katzman and Tuttle (1983) or Burnett's efforts (e.g. 1982).

In archival work, what you find determines what you can analyse, and what you analyse structures what you look for in archival collections. This is blatantly circular – and points to the necessarily provisional and interactive essence of ongoing archival work. Investigations in archives cannot be predicted or neatly packaged in methodological formulas that guarantee publishable results. That for me is the attraction, but to others it may seem too indeterminate, too risky.

It follows that work is iterative, requiring checking and cross-checking, viewing ideas from divergent angles. For sure, such work can be directed by theoretical propositions, in a positivist manner.[10] However, in so far as the coverage of materials cannot be predicted easily, researchers need a degree of flexibility. This can make such work akin to grounded theory (Glaser and Strauss 1967). Although it might seem that documents lessen one's capacity to 'sample' materials, in reality researchers are commonly confronted with more material than they have time to read. In this context, the grounded theory practice of seeking theoretical rather than statistical sampling has much to commend it. Of course the conundrum for those engaged in documentary research concerns when to stop. Similar to grounded theory, there is a case for continuing exploration until 'saturation' sets in – in other words, until the messages you obtain are repeated, with no new insight gained. Most obviously if a project covers a variety of locations, time periods or human agents, the researcher needs to ensure that possible sources of variation are accounted for adequately. Unfortunately, outside a grounded theory framework, the literature offers too little guidance on how to select documents to read.

What the literature does provide is recognition that researchers can commit two forms of 'error'. The first is giving up analysing documents when more material of theoretical relevance is available. The second is to keep ploughing on and on when there is little left to find. According to Michael Hill (1993), seasoned researchers concentrate on large collections of known or highly probable relevance, in collections that are well staffed professionally, in archives with detailed finding aids. To us, this sounds like it should be a beginner's situation. If this is really what 'seasoned' researchers are up to, then we question what they are producing – seemingly more and more insight from a limited range of sources, rather than seeing if the richness of such sources can be built on.

10 This has characterized a good deal of historical geography work that focuses on census materials or long runs of demographic or economic data (e.g. Southall and Gilbert 1996).

Summary: problems of coverage

- Two critical issues are the selectivity of deposit and the selectivity of survival.
- National government collections commonly have clear rules about what is deposited, but for other collections content is less certain.
- Coverage and survival are influenced by the income available to depositories – poorer institutions (or those that allocate small resources) have greater problems maintaining and enhancing collections.
- You cannot be sure what you will find in collections, nor are there clear guidelines for when to stop collecting information. The latter should be informed by theory.

Getting down to it

For those who are new to documentary investigation, archives can seem formidable places. For one thing, there is the question of what sort of documents might be available and where they can be found. The first step in this paper chase is the existing research literature (*not* textbooks), as investigative papers provide insight on data sources, in terms of both the type of material available and its location. For some topics there are ready-made guides, some excellent in coverage and in the guidance they offer (albeit the guidance is only a first step). In this context, a useful general guide for the UK is Mortimer's publication (1999) for the Royal Commission on Historical Manuscripts, which provides information on repositories of materials. Records on more specific themes have also been produced by the Commission, such as works on Cabinet ministers, scientists, diplomats, colonial governors, church officials, politicians, business records in specific industrial sectors, and landed estate records (e.g. UK Royal Commission on Historical Manuscripts 1995). Other series on key printed documents include that edited by David Douglass on English historical documents (e.g. Handcock 1977; also Foster and Sheppard 1995). Providing detail on specific topics, a wide variety of guides exist for documents in the UK's Public Record Office (PRO). Included amongst these are source books on setting up the welfare state (Land *et al.* 1992), economic planning during the Second World War (Alford *et al.* 1992), broader activities during this war (Cantwell 1998), labour history (S. Fowler 1995), educational policy (Coulson and Crawford 1995; Morton 1997), religious non-conformity (Shorney 1996), immigration and citizenship (Kershaw and Pearsall, 2000), Foreign Office papers (Atherton 1994), and welfare policy under the 1951–63 Conservatives (Bridgen and Lowe 1998). The *PRO Guide* (UK Public Record Office, irregular), which is available in some libraries on microfiche, offers a listing of material by topic and Act of Parliament, although access to such information has been advanced since search facilities were made available online (http://www.pro.gov.uk/). A wide variety of other guides exists, but these are too numerous to list here.[11] But we can offer an eclectic list that provides a flavour of the diversity that exists, which ranges from discursive accounts to detailed listings of contents. These include reports on private documentary sources (C. Hall 1982), trade-union records (Southall *et al.* 1994), black women in Texas (Humphrey and Winegarten 1996), on tithe maps (Kain and Prince 1985), the lay subsidy of 1334 (Glasscock 1975), reconstructing historical communities in Europe (MacFarlane 1977), English towns (West 1983), London during the Second World War (Creaton 1998), investigating criminals in the past (Hawkings 1992), the archives of Rio de Janeiro (Davies 1996), guides to Caribbean and Latin American national archives (Nauman 1983), and information on materials on British residences and agencies in different parts of the globe (e.g. Tuson 1979). In case the material you read does not contain references to such guides, a good starting place for finding them is a good history library (such as London's Institute of Historical Research).

What these guides often do not tell are the 'minor' details that can infuriate researchers, like opening hours or periods of closure. It might seem trite to make this point, but as a researcher you are dependent on gaining access to a collection, on an archive's terms. Even for similar collections in the same country, such as county record offices in the UK, you find dissimilar opening hours. Many institutions have short periods of closure, as with the weekly stocktake at the PRO, but some have longer periods, as with many Spanish archives closing for much of the summer. Given numerous private collections or institutions that are not resource plentiful, opening hours may be short or only span part of the week. Web sites provide valuable information, especially for those who have to travel a long distance. As well as travel there is the issue of access. As Box 4.2 and Box 4.3 indicate, many archives require an appointment before access is granted. There is also the issue of material availability, for archivists are not unexpectedly protective of rare manuscripts or books, so there can be restrictions on reading documents. All this points to the need to approach institutions in a positive but professional manner. As Michael Hill (1993: 41) warns, introductory orientation interviews are often gate-keeper moments[12] affecting access:

11 Not all records that are passed to the PRO are retained there. About 20% pass on to other archives. UK public bodies that deposit their own and their predecessors' records include: the British Antarctic Survey, the British Geological Survey, the British Library, the British Museum, British Telecom, HM Customs and Exercise, the Family Record Centre, Girobank PLC, the Hydrographic Office, the Imperial War Museum, the Meterological Office, the National Army Museum, the National Gallery, the National Maritime Museum, the National Portrait Gallery, the Post Office, regimental headquarters and museums, the Royal Air Force Museum, the Royal Botanic Gardens (Kew), the Science Museum and the Victoria and Albert Museum (see PRO Records Information Sheet 124).

12 Seeing officials as 'gate-keepers' is also valid under Freedom of Information legislation. As US experience shows, those requesting information must know what to ask for. Interpretations of whether the 'right' material is requested (and is exempt) are left largely to administrators (Stewart and Kamins 1993).

> *It is important to frame these interactions as interviews because archivists base their decisions about access and subsequent services almost wholly on the initial conferences. The old adage about first impressions being important was never more true than here.*[13]

The importance of securing advice from archivists should not be downplayed, for collections are not organized uniformly, such that transference of knowledge from one to another is not necessarily straightforward. Commonly the organization of documents is by date rather than subject, which reflects the questions historians ask (M. R. Hill 1993). In some cases date can be date of accession, not date to which the document relates. Most certainly it is rare to find good index systems by place, as those who trawl PRO catalogues for anything other than large cities can testify. So it is essential to establish how material is organized. This extends to what heading to look under for particular topics. The norm is that researchers need to check different headings for material on the same topic – much as you might look under agriculture, countryside, farming, rural, agribusiness, etc., if you were searching a library for books on agricultural change. A key question is what categories are most fruitful and are there quirks in the system you are unaware of. While archivists can be extremely helpful in providing a map to chart you through unknown territory, it is the researcher's responsibility to go beyond this. Investigators need to bear in mind that archivists often have specific responsibilities (such as caring for the collection of photographs) and have uneven knowledge about the rest of the system. Moreover, the demands placed on archivists tend to reinforce particular types of knowledge, while leaving others latent or, through lack of use, forgotten. Visits to many county record offices in the UK bring this out with clarity, with high percentages of users seeking information about their own property or family history. Moreover, with material not classified, and documents covering a wide range of topics, archivists can have restricted knowledge of all that is in a collection (M. R. Hill 1993). This comment is unfair on some archivists. But if you find one with such knowledge, treat this as a bonus, and start by assuming that you need to be innovative in thinking about where documents might reside in a collection.

Documentary projects tend to be time-consuming. MacFarlane (1977) makes this clear, noting estimates that reconstructing the demographic history of a single parish of 1000 people over a 300-year period can take 500 hours.[14] Of course, straightforward

13 Not all collections have these interviews, which can be common in private collections. In county record offices, archivists often ask if they can be of assistance, for which Hill's points about being prepared and first impressions are valid.

14 A reminder of the need to check multiple sources and to be vigilant, and that research using documentary sources is likely to be time-consuming, is Erikson's work (1966) on 'wayward' Puritans. This reveals the Puritan casualness over spelling names. In the case of Francis Usselton, who made many appearances before the Essex County Court, this led to his name being spelt in 14 different ways in seventeenth-century records. Similarly, with names often passed from generation to generation, was the same John Brown convicted of seven offences between 1656 and 1681? An indication that he was not is the 1679 entry: 'John Brown, son of John Brown, chose John Brown, his grandfather, as his guardian' (Erikson 1966: 210). A more recent instance of name problems in official records, albeit with less variability, is found in Box 3.3.

analyses of tabulated data, as found in census enumerations or trade directories, can involve less time. Such investigations are akin to contemporary official statistics, except that they look at the past. If researchers restrict attention to such tabulations, they neglect a key advantage of documentary material, which lies in the capacity to 'look behind the scenes'.[15] Another downside is that collecting and analysing data can be tedious 'and the work includes a high amount of dross' (Webb *et al.* 1966/2000: 63). Readers should not take from this an expectation that more discursive accounts are filled with startling discoveries. Although discursive records often prove interesting to read, many records have an annoying habit of concealing more than they reveal. Take the following complete passage from the minutes of Hatfield Rural District Council (21 November 1951), which relates to the annual report of the Medical Officer of Health:

> *It was reported that copies of the Annual Report of the Medical Officer of Health had been circulated to all members of the Council. Various members asked questions arising out of the report and the Medical Officer of Health replied verbally. In reply to questions by Councillors Wenn and Mrs Campbell, the Medical Officer said he would investigate and submit a report to the Health Committee. (Hertfordshire Record Office RDC 7/1/15, 211)*

Could this passage be more informative? The key message for this section has already been stated, but is worth repeating. Investigators must piece together a picture from a variety of sources. Often the materials that provide insight are not those you expect. Often, as above, reports on an item are uninformative. Other sources must be explored (or later entries checked for any reporting back). As an example, in work on public housing, a file was uncovered in Bedfordshire Record Office (BRO) during a search for materials on builders. These records were explored to establish whether building companies saw particular problems or preferences over erecting homes for local councils in particular locations. The file provided information on a legal case involving Biggleswade Rural District Council and Messrs Alban Richards & Co. Ltd (BRO CD/957). The court case concerned building by the company under the 1919 Government Assisted Housing Scheme, in which the contractor erected 424 houses for the council. Immediately after the contract had been completed, the firm went into liquidation, with the court case focusing on receipt of the last payment for the houses. The contract for the houses, which was large for a rural district at that time, was signed on 10 February 1921. From reading just the minutes of council meetings around this time, the researcher would be oblivious to the signing of so large a contract. Indeed, for a year before and after the signing of this contract the main information on housing in the minutes concerns

15 We are thinking here about rather dry accounts of topics like social segregation in cities (Goheen 1970), or marriage distances as a measure of social segregation (B. S. Morgan 1979; P. E. White 1981), which could explore builders' decisions or diaries of social activities.

complaints about rent levels, owing to high unemployment, and the council's attempts to get national government approval for rent reductions. One uninformative entry that might relate to this contract, in the minutes for 21 July 1920, reports that the 'Ministry sanctions housing bonds under which the council is authorized to borrow under the Housing Acts' (BRO RDBwM 1/8, 32). Searching relevant files for lists of houses built by the council does not identify this contract, since the first systematic report on house building in the record office relates to the 1923 and 1924 Housing Acts, with figures for units constructed from 1928 (BRO RDBwH 2/1). This example is given by way of caution. You cannot dabble with documentary research. It requires effort, cross-checking, and seeking out a variety of sources to gain insight on one issue. More often than not, it involves piecing together disparate sources of evidence to construct a theoretically convincing case. This will always be a partial case, but it can be presented with conviction.

Questioning validity

Eugene Webb and associates (1966/2000: 86) warn us that:

> *There is no easy way of knowing the degree to which reactive measurement errors exist among running archival records. These are second hand measures, and many of them are contaminated by reactive biases, while others are not. The politician voting on a bill is well aware that his action will be noted by others ... those errors that come from awareness of being tested, from the role elicitation, from response sets, and from the act of measurement as a change agent, are all potentially working to confound comparisons.*

Lest the reader fall into the trap of associating the concept of 'validity' with positivism, it might be appropriate to replace the word with 'the consistency of evidence'. Although documentary material is often used to cross-validate insight from other data types, for research that involves documents alone, as Webb and associates imply, this limits validity options. Thus, in questionnaire surveys, issues of reliability and validity can be addressed by employing different questions to assess the same phenomenon, as with incorporating different attitude scales (see e.g. Henerson *et al.* 1978; Oppenheim 1986). For documentary research this option is less feasible, because the researcher has less control over what data are available. This can be influenced by searching more widely and innovatively for sources, but it cannot be determined (in terms of the objectives of data collection). As a consequence, a key to convincing yourself, and then others, of the case you build is the triangulation of sources to reinforce messages (Macdonald and Tipton 1993). Added to which, as a researcher you need to be confident in the documents you use.

To put this another way, what is required is a sense of surety that the records you use have not been deliberately manipulated to present a particular viewpoint (or, if they have, to be aware of that viewpoint, and take it into account in assessing records). In this

Box 4.5 The Zinoviev letter

The Zinoviev letter has often been blamed for the general election defeat of the first Labour Government in 1924. Released four days before the election, the letter exhorted the UK Communist Party to organize workers in the armed forces and munitions industries, and to mobilize support for the Anglo-Soviet trade treaty. Most newspapers used the letter to attack the government and its policies. The letter is credited with ruining the Anglo-Soviet trade pact. The Soviet Government of the day declared the letter to be a forgery and this is the conclusion Chester and associates (1967) reached some decades later after examining the available evidence. Most commonly the assumption is that the British Establishment, or those elements of it opposed to 'their party' (the Conservatives) not being in power, concocted the letter to discredit Labour. In this context, those on the political left point to how the British Establishment responded adversely to Labour holding national office (e.g. Wright 1987). However, as John Scott (1990: 43–8) indicates, no UK government has released the letter for public scrutiny. While the balance of evidence suggests that it was a forgery, there is uncertainty over the issue. The Hitler Diaries, by contrast, were (seemingly fraudulently) composed to extract money from the German media, which would pay significant sums for exclusive rights to publishing diary entries (Hamilton 1991).

regard, researchers have generally concluded that deliberate falsehoods, like the Zinoviev letter (Box 4.5) or the Hitler Diaries, are rare (J. Scott 1990). That said, there are instances when documents offer false witness. The Watergate records are reported to be one such set, as they were allegedly altered by those worried about the legality of their role in this scandal (Johnson and Joslyn 1995: 254). Platt (1981) likewise reports on suspect documentation, and suggests questions researchers should ask about the authenticity of material. These include:

- Does the document make sense or contain glaring errors?
- Are there different versions of the original document available?
- Is there consistency in literary style, handwriting, typeface within the document?
- Has the document been transcribed by many copyists?
- Has the document been circulated via someone with an interest in passing it off as correct?
- Does the version of the document derive from a reliable source?

There are no easy answers to these questions, nor simple means of assessing them. Nevertheless, it is appropriate to be cautious. Even if outright falsehood is not common, as John Scott (1990) devoted his book to elucidating, the quality of documentary evidence should be assessed by the criteria of authenticity, credibility, representativeness and meaning.

As regards credibility, some massaging of messages to reveal oneself in a better light might be anticipated. This applies especially for documents that are expected to see the light of day when they are written. Webb and associates (1966/2000: 56) offer the following message that gives a sense of this:

A rich source of continuing data is the Congressional Record, *that weighty but sometimes humorous document which records the speeches and activities of the Congress. A congressman [sic] may deliver a vituperative speech which looks, upon reflection, to be unflattering. Since proofs are submitted to the congressman, he [sic] can easily alter the speech to eliminate his peccadilloes. A naïve reader of the* Record *might be misled in an analysis of material which [s/]he thinks is spontaneous but which is in fact studied.*

More overt in this regard are mass media reports. At one level these contain deep-seated biases, which might not be overt in individual articles, but which come through clearly in systematic analysis of words and tone (see e.g. Glasgow University Media Group 1976, 1980, 1982). Most obviously these biases reflect the interests of proprietors and editors. These might arise from their political interests (Parenti 1986; Curran and Seaton 1988),[16] which can be manifest in support for particular politicians or programmes (see e.g. Stein and Fleischmann 1987), or from commercial interests, which can be apparent in support for measures that might increase newspaper circulation and profits (see e.g. Molotch 1976).

Offering one example of such distortion, Matthews (1957: 165) reports that Lord Northcliffe used the *Daily Mail* to try to get rid of then Prime Minister Asquith and replace him with Lloyd George. As part of this strategy he ordered a smiling picture of Lloyd George to be put in the paper with the caption 'Do it now', while the worst picture of Asquith was accompanied by 'Wait and see'. Malcolm X felt he was subjected to similar biases in reporting, as the following passage reveals: 'A television interviewer told Malcolm X, the late black Nationalist leader, that he was surprised at how much Malcolm smiled. The Negro leader said that the newspapers refused to print smiling pictures of him' (Webb *et al.* 1996/2000: 78). But caution is required not just because of deliberate manipulation but also because of subtle bias. Across nations coverage differences can be stark, as US readers often find when they see with horror the content of the UK's tabloid press, with many readers of the UK's broadsheet papers experiencing similar sinking feelings about the parochialism of US city papers. Providing systematic analysis of national disparities, Ann Johnson (1996) compared US and USSR newspaper coverage of the 1965–8 city riots in the USA. The conclusion was that both were penetrated by ideological coverage, with US reports passing by issues of racism, while those in the USSR suggested that riots resulted from the undemocratic and racist nature of US society. Likewise, in comparing non-German and national newspaper articles on events in East Germany in 1989, Mueller (1997) found that international papers were more ideologically biased, with the added twist of scant reporting if events were distant from a newspaper's head office, unless the incident was violent.

16 This point is not restricted to the mass media. As Gareth Shaw (1982) notes, entries in trade directories reflect the priorities of compilers, while Harley (1988) reminds us that maps are filtered records of the landscape, with presentation biases that reflect priorities of the powerful.

Such considerations should lead investigators to interrogate the sources they use (Box 4.6). Prime questions should be, what likely biases do sources contain and how might these be checked using other data? Most evidently bias is visible in letters. As Plummer (1983: 23–4) holds, letters have two parents; one is the author, the other the recipient.

Box 4.6 Tendencies towards bias in newspaper reports

One of the advantages of newspapers is the immediacy of their reports, but this is also a source of weakness, as there is little opportunity for journalists to consider the implications of their story or even adequately to check the 'facts'. If you want to grasp this by way of a very readable and, for any understanding of the pressures democracy is placed under, vital piece of journalism, see Bernstein and Woodward's *All the President's men* (1974). This recounts how two *Washington Post* journalists discovered how President Nixon was breaking the law for political gain – a process that would eventually lead to him being forced from office. What this account brings out visibly is the way in which pressure for a story led to information from informants being misinterpreted, and incorrect articles printed, so the uncovering of the conspiracy was threatened. Taking this point further, Matthews's evaluation (1957: 165) of the newspaper industry leads to the observation that: 'Journalists themselves generally have a horror of being interviewed, "written-up" or even noticed by the Press – they know all too well from their own experience how inept and cruel a distortion the result is likely to be.' Emphasizing the point, Matthews (1957: 15) reports on a memo sent by a former editor of the London *Sunday Express* to his staff, which stated: 'I do not wish to be hyper-critical, but the plain fact is – and we all know it to be true – that whenever we see a story in a newspaper concerning something we know about, it is more often wrong than right.' This comment is made long before the sensationalizing tendencies that so characterize newspapers in the UK in the late twentieth century. What Matthews stresses is that much of this misreporting is not mischievous or deliberately cruel, but arises from the time pressures under which journalists operate. Moreover, as the title of Matthews's book, *The sugar pill*, suggests, the dynamic that drives journalistic writing is market driven: 'coating the pill of hard news with the sugar of entertainment' (1957: 34).

Throw into the pot for consideration deliberate newspaper efforts to manipulate citizen views. Systematic biases in newspaper reporting have often been seen to be linked to the values of their proprietors. This point is made visible in the account of Harold Evans (1983), the former editor of *The Times*, who highlights the role of Mrs Thatcher in bringing the newspaper under the control of a right-wing owner (Rupert Murdoch), with evident impacts on biases in news reporting after the takeover. That such biases cause concern amongst public agencies is evinced in reports by the Commission of the European Communities on widespread inaccuracies in UK newspaper stories on the European Union (see e.g. CEC–London 1995).

The message from all this is that newspaper reports should be treated with a high degree of caution. They can provide indicative insight, but investigators are advised to seek confirmation from other sources rather than assuming the accuracy of reports. This applies just as much to the newspaper items cited in this book as other writings. Perhaps readers should approach newspaper reports with the caveat that 'if the report can be trusted', then the message it contains highlights an issue that merits following up.

Both affect inclusions and exclusions. It takes only a second to realize this. Contrast the kind of material you do (or would) include in letters (or e-mails) to your mother and to a best friend from school days. Letters contain different material depending on who they are sent to, as well as varying in the manner issues are raised or described. Put simply, letters do not reveal the way the writer thought about X but the way he or she wished to portray feelings about X to the recipient of the letter. All this should be ringing bells about post-structuralism in the reader's mind (Chapter 1). The issue of understanding context in making interpretations is considered in greater length in Chapter 6. Revealing that letters can have a notable political message, Webb and associates (1966/2000: 108) summarize a *Chicago Sun-Times* report on an investigation by the Xerox Corporation in 1965. Having received more negative than positive letters after sponsoring a TV series on the United Nations, the Corporation submitted the letters to a handwriting expert. This analysis concluded that the 51 279 letters of criticism were written by 12 785 persons, which was about the number of positive responses received. Quite apart from such manipulations, examination of letters and personal accounts need to be 'read' with a view to the implicit assumptions that guide the writer. One only has to recall reading Jane Austen to bring this point out. With all that visiting of neighbours some distance away, it is clear that the working classes who must have lived nearer the heroine's home did not qualify as 'people' (so they could not count as 'neighbours'). In similar vein, geographical interest in nineteenth-century travel writing has brought out that European voyeurs in exotic lands gave scant attention to local people and romanticized the places they visited (J. S. Duncan 1999), even if that 'idealized' vision was 'real' to them (Gregory 1999).

The interaction effect of writer and receiver can raise problems for researchers who interrogate letters and reports. Often correspondence is not that plentiful and information on the writer and the written is sparse (except from material in the letter). This can place strain on the researcher, for it can be tempting to interpret when there is insufficient evidence to convince yourself an interpretation is convincing. Calling on multiple sources can build confidence, by offering insight from different angles. Here the ease with which material is brought together depends on the organization of sources. In the case of Bedfordshire Record Office, for example, correspondence and parish council materials are in files relating to specific themes. Thus, in parish file PC Arlesey 9/29 we find documents signifying concern in this parish about dog control. First, there is the decision of the Parish Council to join the League for the Introduction of Canine Control in 1984. Then, on 30 October 1985 the Parish Clerk wrote to Mr Hemmings of Mid-Bedfordshire District Council (MBDC) that: 'The Town Council have received numerous complaints regarding dog fouling in the Village. Would it be possible for MBDC to erect the usual notices regarding dog fouling on lamp posts throughout the village.' This is followed by a 16 July 1985 letter requesting 100 copies of 'Train your dog to "go" at home' leaflets from the League for the Introduction of Canine Control.[17] That this correspondence is indicative of (some) citizens' concerns is suggested by parishioners' letters complaining about 'a yellow van' that brings 'nine Alsatian guard dogs' to roam the common, of dogs fouling footpaths, and so on. Insights from written documentation about individual lives and values can provide one sense of the pulse of a community, as studies of distant centuries show (e.g. Le Roy Ladurie 1978). However, the questions of which views are projected, and how widespread they are, linger in the background. Are the problems created by dogs in Arlesey felt by members of one social class, by recent arrivals and not longer term residents, by parents more than others, or generally?[18] To answer this question, correspondence alone often proves insufficient. Other sources have to be consulted. The implication this offers provides a consistent message for documentary sources – namely, that any single source needs backing, with the task of searching for additional sources often taking considerable time, albeit the end product is justified.

Summary: getting down to it

- A multitude of excellent indexes to documentary sources exist – use them.
- Be sure to seek the archivists' help in exploring collections. This can be a time-saver. Recognize that 'interviews' with archivists are gate-keeper moments.
- Be conscious that documents can conceal as much as they reveal.
- Validity requirements necessitate that sources are cross-checked using other types of document and/or other types of data.

Extracting messages

For Macdonald and Tipton (1993: 188, 193),

> *the exercise of interpretation which is so obviously part of the reading of the 'meaning' of public statutes or a genre of movies is not so very different from the assessment that must be undertaken of any archive or set of records … With paintings, sculpture and architecture, however, it is unlikely that the artist is trying to deceive the beholder, but it is often necessary to have some knowledge of the circumstances in which an object was produced before it can be interpreted as a social document.*

17 These dates are within the 30-year rule for the release of national government documents. Yet the announcement in 2000 that the UK Government was to release information on 1960s–1970s negotiations to enter the European Union indicates that some national material can be accessed in under 30 years. This demonstrates the political nature of document availability, for it is difficult to disentangle this decision from the Labour Government's support for the euro as its national currency. Such early releases are not common at the national level, even if some country record offices release material sooner.

18 Rob Burgess (1984), for instance, casts doubt on the reliability of the classic study of Polish immigrants of Thomas and Znaniecki (1918), because the material they used came from responses to advertisements (as with Pooley and Turnbull 1998). This is a reasonable point, in so far as readership and response to adverts (in particular magazines or newspapers) are selective. Set against this it suggests that, if research is not based on 'random' samples, it is questionable. This stipulation would eliminate any prospect of undertaking many research projects (Chapter 2, Chapter 6 and Chapter 7).

Hardly surprisingly, given this view, analysts have approached written documents, as with texts such as advertisements, buildings, films, landscapes, maps, novels, paintings or speeches, as having surface and deeper meaning (see also Geertz 1973/1975; Duncan and Duncan 1988). In positivist social science, it is surface messages that generate interest. Literary critics argue that these surface impressions do not do justice to a creator, but some social scientists counter-claim that social science is not primarily concerned with the artistic, so this criticism carries less force (Macdonald and Tipton 1993). We reject this view. It carries within it the message that there is one right way, or at least one way is more correct than another. Epistemologically lauding the surface level implies that positivism is superior to frameworks that seek deeper understanding, like those rooted in interpretative traditions. We accept that there might be better ways of analysis, if by this we mean more coherent identification and elucidation of themes. There might be more rigorous ways of undertaking research. There might be more valid research designs (for particular research contexts/problems). But all these are set within the confines of an underlying philosophy and the objectives of a research project. Searches for deeper meanings in texts offer one perspective, just as surface examinations offer another. To undertake a surface-level investigation and assume deeper messages would be misguided, but this does not discount potential contributions from more surface-level surveys. In a similar light, to ignore potential gains from investigating multiple layers of meaning would be to neglect a powerful addition to the social scientist's armoury. Taking this as our standpoint, let us begin by examining how surface-level messages can be extracted from documentary (or textual) sources.

Extracting surface level messages

Indexing information

A primary feature of analytical methods that explore surface-level understandings of documents (in-depth interviews or field notes from participant observation) arises from indexing information by topic. As we explore in Chapter 7, when discussing computer programs that assist in this procedure, the process of indexing observations by theoretical categories is fundamental to analysing data that are not predefined into categories. For this kind of analysis there are well-established procedures of investigation, although, as Sanjek (1990: 391) makes clear, far too little has been written about 'how others do it' (a partial help is Fielding 1993). In a nutshell, what the analyst is seeking in indexing is to develop a coding system to enable the investigator to draw together material on the same topic or explore similar themes from a variety of sources. Offering one insight on what is involved using interviews, William Foote Whyte (1984: 118) gives the following account:

In writing a report we can work directly from the index to the outline of the paper. A few minutes spent in re-reading the whole index gives a systematic idea of the material to be drawn on. Then, for each topic covered in the report, we can write into the outline the numbers of the interviews and the page numbers of the relevant material. For example, in writing a section on relations between hostesses and waitresses, we write in the outline some general heading referring to the supervision of waitresses. Then we note in the outline all the interviews where we find in the index 'waitresses–hostess' – plus the page number of those particular interview sections. This may refer us to a dozen or more interviews. Perusal of the index will refresh our memory on these interviews ... some of them merely duplicate each other. We pull out of the file perhaps a half dozen interviews, then turn to the section where 'waitresses–hostess' is marked in the margin, re-read these sections, and finally use the material from three or four.

While this kind of examination is 'surface-level', this does not mean the analysts simply report what people or documents tell them. Almost certainly you find there are silences, uncertainties and inconsistencies in messages provided (here we mean beyond matters like changes in the role of occupations like clerks and nurses over time, so that analysts cannot take single words at face value). The researcher thereby needs to investigate messages in a theoretically informed manner. Procedures for this are available in works like Strauss and Corbin (1990). As John Scott (1990: 1) puts it, what is involved is a systematic and disciplined search for knowledge. Ideas are cross-validated by seeking confirmatory evidence, inconsistencies are evaluated through critical assessment of current theorization (current for that stage of the research analysis), with the researchers seeking to challenge the understanding they have developed to assess its veracity. This is a foremost concern. Inconsistencies or silences are common. The researcher needs to explore what distinguishes those giving different accounts and what separates those who mentioned an item or issue from those who did not. At its most simple level, the analyst is trying to rebuild a picture of last week's local derby, drawing on accounts offered by supporters of Manchester City and Manchester United. The accounts initially seem wildly different, with some reports emphasizing how unlucky one side was to lose, while some stress how the winning team thrashed the other. By categorizing informants according to the team they support, some sense of the inconsistency is possible. Similarly through categorization, we might be able to understand why many reports charged that the 'unfair' sending-off of a key defender 'turned the match', whereas other accounts never mentioned this dismissal. To clarify here, we are not saying that there is 'one view' of the game. We know a number of people who live with the delusion that either Manchester City or Manchester United is the best team in the world. There would appear to be no 'objective' account capable of bringing these mind-sets together. Yet, even if we are seeking to present an account from one perspective, we need to challenge the explanation we are constructing by considering an alternative vision. This can reveal issues we have ignored or downplayed that, even within the logic of the account that is emerging, reveal a shallowness of interpretation, requiring more complex or simply different conceptualizing.

Of course, the assumption that accounts will break down neatly into two groups, as allied to the football team supported, is almost always too simplistic. Other life experiences, or positionalities, are likely to distinguish accounts. As a consequence, what the researcher must engage with is a process of contrasting, cataloguing, classifying and re-evaluating assumptions and theorizations. For sure there are problems of false accounts. For sure failure to recognize (or find documents covering) a key dimension of events can lead to 'partial explanation'. However, there are procedures to mitigate these potentialities (see e.g. Strauss and Corbin 1990).

Content analysis

Amongst those who look for surface understandings of documentary material, the most commonly used method after 'scoring' a document is content analysis.[19] At its most basic, content analysis gives a quantitative measure of the frequency of occurrence (or salience) of an item or of its co-variation with other items. As such, content analysis is used to identify key themes or associations in documents. As one example, van Meter (1994) reports that open-ended questionnaire responses on proposals to reform French language spelling showed that, when informants from the political left and from the political right described reform proposals, the frequency they used words revealed different values. Concerns about reform from the political right questioned the impact on cultural values, national heritage and national identity. Those from the left more commonly mentioned lack of debate, government interference and the absence of public consultation. Research by the Glasgow University Media Group (1976, 1980, 1982) offers a long list of examples, as with reports that newspaper coverage of a 1974 strike by British Waterways workers led to 1800 column centimetres on the effects of the strike, its progress and the progress of negotiations, compared with about 60 for the settlement and under 40 for the background to the claim – for example, the reasons for the strike (1976: 223).

The above description might imply that content analysis is straightforward, since it primarily offers a quantitative indication of the presence or co-variation of items. Given the quantitative nature of content analysis, it might seem ideal for computer-based procedures. These can be helpful, but a great deal of manual work is always required as well. Most obviously this is seen in the difficulties computer packages have in interpreting the meaning of words. Think through the complications of words like 'mine' or note Hodson's recognition (1999) of seven definitions for the word 'state'. The starting point

19 To clarify our meaning, there are projects involving little more than recording data in documents. For example, when examining applications to build new homes, the investigator might primarily assess whether successful applications are associated with the size of developments, their location, their per unit cost and so on. Content analysis examines more 'subliminal' messages than this.

of any analysis is manual, for this involves defining precisely what the subject matter of the study is. Data are recorded for relevant topics rather than everything being 'scored'. To represent key concepts there should be explicit selection rules, which should be predicated on theory. However, even with a strong theoretical base, there can be complications over identifying critical concepts, since these can be represented in a multitude of ways. As one example, Moodie and Catchpole (1975) recorded 149 root words and phrases to describe the break-up of ice in their examination of records for the Hudson Bay area. There must be explicit ground rules for accepting that passages of text represent a particular concept, otherwise different researchers will reach different conclusions from the same document. As Robert Weber (1985) outlines, one aid in this direction is the existence of general dictionaries (such as the Lasswell value dictionary). Box 4.7 provides one illustration by recording Laswell words on wealth-related issues, with Stryker (1996) providing detailed norms and standards for coding content on social movements. After coding text into such categories, researchers are most commonly concerned with identifying associations. When did ice break-ups occur in the annual calendar? Were words like 'extremist' or 'radical' more commonly linked to those with one set of political views but not another? Is there gender bias in fiction stories (see Box 4.8). Offering one insight, Slater (1998: 236) outlines how we might explore whether soap opera images of women have changed over time:

We might categorize these images in terms of roles: what changes have there been in the kinds of social roles that women characters have been allocated in the soap opera over this period? We need to draw out of this research question a set of concepts by which to categorize the data in a relevant manner. For example, we might look at the women represented in each soap opera in our sample and ask whether they are presented as mothers, girlfriends, housewives, workers, career women, and so on. The aim would be to be able to assign each represented woman to those categories and see how the frequencies change: for example, is it the case that career women have become more common and housewives less common in soap operas?

In this case, the primary item used to assess content is female role. In terms of the jargon, this is one recording unit in content analysis. Others include words, themes, sentences, paragraphs and so on. The recording unit, which must be decided beforehand, should be selected by the research question. Following Slater (1998: 236), the categories selected should be:

- exhaustive (we must be able to assign every representation);
- mutually exclusive (we should not be able to code an item into more than one category);
- enlightening (analytically relevant and coherent).

Whichever coding unit is used, all raise analytical problems (R. Weber 1985). Yet choice of unit is not made without difficulty: 'The selection of the appropriate recording unit is often a matter of trial and error, adjustment, and compromise in the pursuit of measures that capture the content of the material being coded' (Johnson and Joslyn 1995: 246). Most evidently, as the recording unit becomes larger, the reliability of recording items wanes. This is brought to light in investigations that have two or more investigators coding documents, then comparing results. Much of the literature suggests that a good rule of thumb is to have at least 10 per cent of cases coded by a second reviewer. Without such reliability checks (known as inter-coder reliability) researchers have little insight on data quality. In this regard, it is worth noting Tilly's comparison (1981: 74) of coding into two sets of categories, with 59 per cent agreement for more abstract categories but 94 per cent consistency for more concrete ones. These figures are in line with those in Thomas Cook and associates' volume (1992: 304), which were in the 56–100 per cent range. (For more detail on reliability and bias checks, see Hodson 1999.) With this level of disagreement, it is clear that some recording units are not well suited to (particular) research questions. This at least is the initial situation, for one of the characteristics of content analysis is that investigators go back and forth, progressively refining categories. Reliability can be questioned in this way, so an explanation of procedures followed and of content categories used needs to be provided when writing up such work.

In terms of what content analysis can achieve, loss of richness from coding in a numerical manner means that the data analysed have limited sensitivity and precision regarding the meaning of human actions. Depth is sacrificed to increase the generalizability of conclusions. As Slater (1998: 237) points out:

> *because methodological controls have been used all along the way, and because a significantly sized sample has been drawn to represent a coherently defined universe, comparisons and generalizations can be made across a social field and represented in meaningful numerical terms.*

As content analysis converts qualitative material into numbers, it neglects the symbolic form of the text in favour of highlighting incidences (or co-variations). Its strength comes from its (relative) simplicity. Using documentary sources, this capacity is added to by being able to explore materials that cover long periods of time or different places.[20] For all that, it is not a subtle method of analysis. Moreover, in the selection of categories for

20 For example, through meta-analysis, content analysis has been used to compare materials from studies of different places – with codes for time and locational attributes (T. D. Cook *et al.* 1992). A review of comparative analyses using case studies is provided by Hodson (1999).

Box 4.7 Alphabetical list of wealth nouns in Lasswell's value dictionary

abundance
account
acre
affluence
agriculture
allowance
annuity
appropriation
article
auto
automobile
backwardness
bale
bankruptcy
bargain
belong
benefit
bill
bonus
bookkeeping
bounty
branch
brass
bread
budget
business
capital
car
cartel
cash
cattle
cent
charge
check
cheque
clear
coal
coffee
coin
collateral
commerce
commodity
copper
copyright
corn
cotton
crop
currency
custom
debt
deficit
department
depreciation
depression
dollar
earn
economics
economist
economy
electricity
employment
end
endowment
energy
engine
enterprise
equity
estate
expenditure
expense
export
factory
farm
fertilizer
finance
forest
forestry
fortune
freight
frugality
fund
fur
garden
gift
gold
goods
grain
grant
herd
hide
highway
hold
horticulture
household
incentive
income
indemnity
industrialization
industry
inflation
input
interest
inventory
investment
iron
irrigation
ledger
liability
livestock
loan
lot
low-cost
luxury
manufacture
manufacturer
market
mine
mineral
mint
money
mortgage
oil
ore
output
ownership
parity
pay
payroll
penny
pension
piece
plant
plantation
poor
population
port
poultry
pound
poverty
present
price
proceed
produce
producer
productivity
property
prosperity
ranch
rancher
rate
real
receipt
reclamation
redevelopment
refund
remuneration
rent
rental
resource
retail
retirement
return
rich
road
royalty
rubber
salary
salesmanship
save
scarcity
security
sell
shop
silk
steel
sterling
stock
store
supplier
supply
surplus
tariff
tax
taxation
textile
timber
tin
train
transport
transportation
treasure
treasury
trust
unemployment
wage
wealth
wheat
wholesale
win
wood
wool
worth

[to which we might today add Yen, Deutschmark and Euro, amongst others]

Source: R. Weber (1985: 29).

Box 4.8 Checking for bias in whole accounts using content analysis

Content analysis is commonly used to explore single words or phrases, but it can also be employed to investigate bias in whole reports. As one example, consider the following table, which is taken from Bouma (1993). This is designed to be used for whole articles; you might employ it to compare different magazines, to see if they are characterized by a higher or lower level of gender bias in the stories they present. The list of questions here is for illustration. A much longer list would be appropriate in many research situations. The key to this approach is to develop a theoretically informed and comprehensive list of questions (that is, sufficient to the purpose), and to be sure that the categories along which magazines will be evaluated are appropriate. To clarify this last point, it is not satisfactory to divide categories into quartiles (as shown below), if the critical distinctions occur within 10 per cent either side of the median.

Questions on the content of stories	Place tick in appropriate column				
	Always	In over half the stories	Half and half	In less than half the stories	Never
1. Are boys shown to dominate girls?					
2. Are girls shown to win against boys?					
3. Is unisex clothing used?					
4. Are women shown in traditionally male roles (e.g. a female physician or female priest)?					
5. Does a male ask a female for help, directions or information?					
6. Are females shown to be helpless?					

Source: Bouma (1993: 79).

study, content analysis depends on interpretative judgement on the part of the researcher. In this it shares analytical elements with interpretative approaches. Yet by emphasizing quantitative representations, content analysis lays stress on generalization rather than understanding. It neglects the contextualization of meaning that interpretative approaches seek, while its emphasis on surface impressions distinguishes it from the search for abstraction that characterizes critical realism. This does not mean that the insight gained might not inform studies in either of these genres, but this would be background insight, not core revelation. In this book more weight is placed on interpreting the significance of events or images, rather than totting up the incidence of them. This is because we recognize that:

> *documentary reality does not consist of descriptions of the social world that can be used directly as evidence about it. ... Rather they construct their own kinds of reality. It is therefore important to approach them as texts. Texts are constructed according to conventions that are themselves part of a documentary reality. Hence rather than ask whether an account is true, or whether it can be used as 'valid' evidence about a research setting, it is more fruitful to ask ourselves questions about the form and function of the texts themselves. (Atkinson and Coffey 1997: 60–1)*

Extracting deeper meanings

Textual analysis

What is meant by textual analysis is explained by Ron Johnston and associates (2000: 825) as being

> *associated with the hermeneutic method of Wilhelm Dilthey and others. According to this method, an interpretation is produced which results from the interaction between the text being studied and the intellectual framework of the interpreter. Throughout the twentieth century, competing methods of textual analysis have been put forward such as structuralism, which would include semiotics, and post-structuralism, which includes discourse analysis and deconstruction.*

Captured in the expression 'interpretative understanding', the textual analysis has long been used to explore social and cultural meaning. In the early part of the twentieth century it was an important element in research on human societies, especially in Germany (Dilthey 1923/1988). More recent interest in interpretative understanding emerged as analysts sought to appreciate how those they were studying 'constructed' their world view (e.g. Schutz 1960). The emphasis in this approach, which was referred to as *verstehen*, was empathizing with informants (hermeneutics). In this guise, as captured in the introduction to Cooper's book (1967), the approach provided a significant critique of positivist methodologies. More recently, the work of Geertz (1973/1975: 1983), amongst others, has made this approach more widely popular, in promoting understanding of culture through textual analysis (Chapter 1).[21] This approach has been widely, if perhaps rather uncritically, adopted by human geographers. Its essential conceptualization emphasizes that there are layers of meaning in a text. The research goal is to get at the

21 Textual analysis can embody hermeneutic and semiotic approaches, although Geertz (1988: 12) is dismissive of their shared features, seeing overt rationalism in the view that semiotics is the science of signs as inappropriate. In this book our use of the expression 'textual analysis' is more closely linked to that used by Geertz. In passing we should note Geertz's rejection of *verstehen*. He saw attempts to see through the eyes of the investigated as infeasible, as researchers can never 'get inside' their mind-set.

underlying message of the text, not by highlighting key words or phrases but by identifying the system of rules that structure the construction of the text. Following Max Weber, Geertz (1973/1975: 5) captures the essence of this understanding by noting that people are 'suspended in webs of signification' (namely, culture). From this perspective, the aim of researchers should not be to search for laws but should be 'an interpretive one in search[ing] for meaning' (Geertz 1973/1975: 5). Drawing on Ryle, Geertz proposes that this can be achieved through 'thick description' (see above Chapter 1) aimed at 'sorting out the structures of signification' (Geertz 1973/1975: 9). Although the concept of a 'text' involves more than written material (see e.g. Duncan and Duncan 1988), the link with documentary analysis is obvious. Yet Geertz's comments relate to ethnography as a whole, including participant observation (Chapter 7). For Geertz (1973/1975: 10):

> *Doing ethnography is like trying to read (in the sense of 'construct a reading of') a manuscript – foreign, faded, full of ellipses, incoherencies, suspicious emendations, and tendentious commentaries, but written not in conventional graphs of sound but in transient examples of shaped behaviour.*

This linkage between what some see as the divergent methods of documentary analysis and participant observation has been recognized by John Scott (1990). He holds that, save for some specific features, the general principles of documentary analysis are no different from other social methods. This is an argument we have sympathy with. However, it raises peculiar problems for writing a book like this. If we briefly return to Geertz (1973/1975: 19), he provides a guide to the underlying dilemma. This is achieved when he asks and answers the question: 'What does the ethnographer do? – [s/]he writes.' The research process embodies conceptualization, theorization, data collection, data analysis and writing-up. One can add 're-' in front of all of these and then add some 're-re-' entries, most likely locating them in a different order. Research is not conducted in a series of discrete steps. It is an iterative process, where clean lines and neat packages are something poor textbooks imply, but which honest research accounts disabuse you of. This is readily appreciated if you read contributions to collections like that of Bell and Newby (1977), with reminiscences from the quantitative revolution also revealing (e.g. Billinge *et al.* 1984). Some personal accounts by young geographers are likewise instructive (e.g. Robson and Willis 1994; A. Hughes *et al.*, 2000). The essential message is that the researcher must think through questions of analysis and writing-up from the start of an investigation, with many considerations varying little with the manner of data collection or type of data examined.

This understanding might help answer those who question why we introduce ideas of content analysis *and* interpretative understanding in this chapter. As readers will find in later chapters, both these and other analytical procedures (with associated implications for writing-up) are relevant across a range of methods. Our dilemma is how to examine similar analytical issues in several places without being repetitive. Our response is to raise issues where they are pertinent to the discussion under way, giving weight to basic questions about an approach – as we have just done for content analysis.

In the case of interpretative understanding we are left in a quandary. This comes from providing a precise understanding of what the 'method' embodies. Let us start with what is involved. Here we wish to draw a clear distinction between two approaches. One tends not to have a formal name, since it is more commonly associated with a particular theoretical position. The second is textual analysis. For the first of these positions, whether the aim is theoretical abstraction, in a Marxist sense, or empirical theorization, in a grounded theory sense (Glaser and Strauss 1967), the insights from a document tend to derive primarily from the surface layer; as with Fogel and Engerman's research (1974/1989) on slavery (see Box 3.5). The essential distinction between these studies and textual analysis arise from the focus of the study. As Feldman (1995) has put it, the focus in textual analysis is *how* sense is made of social life, not *what* sense is made. In this regard, research that seeks theoretical abstraction or empirical theorization more commonly assesses material conditions or the creation of those conditions. Textual analysis by contrast examines the meaning of social acts. As such, textual analysis is seeking a deeper understanding of (particular) aspects of society. In doing so, as Geertz articulates, analysts have been inclined to separate their interpretations from materiality (see Lees 1997, 2001, for attempts to *re*materialize cultural analysis). As Geertz (1973/1975: 30) recognizes, there is a danger that in searching for an 'all-too-deep-lying' signification, cultural analysis will 'lose touch with the hard surfaces of life – with the political, economic, stratificatory realities within which men [and women] are everywhere contained – and with the biological and physical necessities on which those surfaces rest …'. Those concerned with *what* sense is made of social life tend to focus more on the 'hard' surface, whereas textual analysis seeks meanings below that surface. Drawing rather too sharp a divide for illustration, those who focus on material conditions are likely to examine nutritional levels amongst slaves, how disputes over legislation affecting slavery were resolved, and why particular forms of slave transhipment or auction were changed. Those adopting Geertz's textual analysis perspective would be more likely to explore how the practices of slave transhipment and auction inform us about slave-owner values or how readings of 'evidence' were used by politicians to justify the continuance (or not) of slavery.

In textual analysis, procedures for understanding meaning are commonly less than crystal clear. As Aitken (1997: 204–5) expressed it:

> *The adoption of textual analysis and the reading metaphor has given rise to a methodology which is often thought of as being largely implicit and derived from years of apprenticeship and practice. This makes things a little difficult for the beginning researcher who has little training and experience in interpreting texts.*

A good illustration of this is Aitken's chapter itself, for, despite its inclusion in a methods book, this contribution tells us more about research conclusions and epistemological underpinnings than how to read a text. For Miles and Huberman (1984), part of the problem is that researchers still see the analysis of texts as 'art', in which intuition is critical. Rycroft (1996: 427) offers a good exposition of the almost mystical vision some proponents present. He does so not simply in telling how 'there is no formula with which to interpret novels' (or any texts) but in emphasizing how the meanings of novels are fluid (cf. post-structuralist interpretations in Chapter 1). A superficial reading of Geertz (1973/1975: 20) might suggest he bolsters this view, when he notes that: 'Cultural analysis is (or should be) guessing at meanings, assessing the guesses, and drawing explanatory conclusions from the better guesses ...'. But this common-world language is not saying much that is too different from inductive approaches to positivism (J. U. Marshall 1985: 117), realism (A. C. Pratt 1994: 51) or grounded theory (Glaser and Strauss 1967: 4). As Geertz (1973/1975: 27) goes on to stress, these 'guesses' are informed by prior knowledge and theorization:

> *Although one starts any effort at thick description, beyond the obvious and superficial, from a state of general bewilderment as to what the devil is going on – trying to find one's feet – one does not start (or ought not) intellectually empty-handed. Theoretical ideas are not created wholly anew in each study.*

When researchers go beyond this recognition, textual analysis methods are not described in accessible ways to the uninitiated in the literature. Examine Geertz (1973/1975), or take geographical commentaries like Duncan and Duncan (1988) or Winchester (1992), and what you may come away with is a sense of intent and end-description, but not much on the method that links the two.

Textual analysis, then, is a craft skill that is difficult to render. However, we offer an analytical framework that should give the researcher some direction:

1. Look at the language used in the text(s). This might involve an analysis of rhetorical features – that is, how the text persuades its readers/hearers. Official texts use a different language from a personal diary. Examine the characteristics of the language used – styles, conventions and so on.
2. Think about the text(s) in relation to their authorship and readership. How are they produced and consumed? Remember that

> *in textual terms [the authorship and readership] are not coterminous with the particular individual social actors who write and read. We need to pay close attention to the implied readers, and to the implied claims of authorship. This becomes particularly*

> *interesting when we are examining how a text implicitly claims a special kind of status – as factual, authoritative, objective or scientific. Linking this perspective with that of rhetoric, then, we can ask ourselves what 'claims' a text seems to inscribe, and what devices are brought to bear in order to enter that implied claim. The same would be true – though the rhetorical devices would differ – if the document in question had a different function (such as constructing a complaint, a confession or a personal reminiscence). (Atkinson and Coffey 1997: 61)*

Remember texts are often aimed at a particular reader, and it may take a particular highly socialized member of a subculture to make any sense of a text.

3. Emphasize intertextuality. Texts often refer to other texts. This is especially the case with, for example, government documents and legal documents.

> *Intertextuality thus alerts us to the fact that organisational and official documents are part of wider systems of distribution and exchange. Official documents in particular circulate (though often in restricted social spheres) through social networks which in turn help to identify and delineate divisions of labour and official positions. (Atkinson and Coffey 1997: 57)*

4. Capture the point of view from which a document acquires its relevance. However, 'grasping this frame of meaning is no easy task, for no researcher can escape the concepts and assumptions of his or her own frame of reference' (J. Scott 1990: 31). Here we are indebted to Macdonald and Tipton (1993: 198), who remind us of how, in *Foucault's pendulum*, Umberto Eco (1989) draws out how preconceptions can trap interpretations (Box 4.9). In order to avoid this, mediation should occur between the researcher's understanding and that of the author(s):

> *The researcher must seek to discover as much as possible about the condition under which the text was produced, and must relate the use of individual concepts to this context. The ultimate interpretation of the meaning of the text will derive from the researcher's judgement that this interpretation 'makes sense', given his or her understanding of the author's situation and intentions. (J. Scott 1990: 31)*

The idea that the interpretation 'makes sense' is a powerful one for textual analysis, but it is an idea that is imbued with potential problems.

Box 4.9 Foucault's pendulum

In *Foucault's pendulum* (1989), Umberto Eco, a Professor of Semiotica at the University of Bologna, provides a long account of a documentary mystery whose 'solution' revolves around interpreting the meaning of documents. Although the story takes more than 600 pages to unfold, in essence it centres on the discovery of a manuscript, which is soon taken to be a fourteenth-century secret plan for a 600-year project undertaken by the Templars (Eco 1989: 134–40 offers the first glimmerings of this). This rather grand explanation is built on over the year, but takes a couple of days to be undone by one of the main character's girlfriend, with what comes across as a much more plausible interpretation, that the manuscript is a laundry list (Eco 1989: 534–41). In a story that seeps with intrigue and plotting, with murder thrown in, the main character is found fleeing from those who believe in the original interpretation as the book ends. Expecting capture from those who pursue him, the book comes to a close with the following message:

In a little while, They'll be here. I would have liked to write down everything I thought today. But if They were to read it, They would only derive another dark theory and spend another eternity trying to decipher the secret message hidden behind my words. It's impossible, They would say; he can't only have been making fun of us. No. Perhaps without his realizing it, Being was sending us a message through its oblivion.

It makes no difference whether I write or not. They will look for other meanings, even in my silence. That's how They are. Blind to revelation.

At one level this is a nonsense story. But in an environment in which advocates of a method are not particularly open about how they extract conclusions, it is powerful, as John Scott (1990) indicates. The fact that analysts have been shy about telling others how they reached their conclusions, or even what support there is for them, leads other analysts to question the veracity of such accounts. Holding that there is no single reading of a text is not good enough. What needs to follow this statement is an account of why attention should be given to the view that is presented from a particular perspective; for Geertz (1983: 148), convincing others as to the veracity of an account must be a priority. As the academic profession in most of the advanced economies is predominantly male, middle class, from the dominant ethnic group and heterosexual, one has to question whether the readings of texts that offer little evidence to support them, alongside little insight on methodology, are little else but pontificating by dominant groups. The merits of this view are strengthened by some severe questioning of theorizing by art historians, who have been at the forefront of textual interpretation. One example that brought this to the fore was a re-examination, through x-ray analysis, of the work of the fifteenth-century Flemish painter van Eyck. What this revealed was that the finished version of van Eyck's Arnolfini painting was a product of trial and error, as drawn items were painted over in a manner suggesting he could not make up his mind. Rather than the painting being embedded with well planned, crystalline and decisive symbolism, as art historians had claimed, x-ray analysis suggests that it was simply a portrait and not a symbolic edifice – the newsworthy aspect of this story being made evident by its appearance in broadsheet newspapers (see e.g. Checkland 1995).

Semiotic analysis

The foundations of semiotics are in structuralism, but the ideas were developed by French post-structuralists, who applied linguistic analysis to the study of culture (Gottdiener 1995: 1–25). Semiotics is concerned with the process of interpretation – namely, how a text gets a message across.[22] For Slater (1998: 237–8):

> *semiotics represents the exact opposite to content analysis along every dimension. It is closer to interpretive methodologies than to quantitative and survey methods and is utterly open-ended rather than closed in its questions and investigations. It is strong on rich interpretations of single texts or codes but offers almost no basis for rigorous generalization outwards to a population. It argues that elements of a text derive their meaning from their interrelation with a code rather than looking at them as discrete entities to be counted. Where content analysis is all method and no theory, hoping that theory will emerge from observation, semiotics is all theory and very little method ... Above all semiotics is essentially preoccupied with precisely that cultural feature which content analysis treats as a barrier to objectivity and seeks to avoid: the process of interpretation.*

The semiotic approach used by geographers has developed from the work of French structural linguist de Saussure (1974). His argument emphasized relations between words – namely, the system of differences. He argued that language, not objects, produce conceptual worlds:

> *de Saussure's argument implies that descriptive language of any kind cannot be understood through a consideration of just the words that have been uttered, or written. Semiology is primarily concerned with understanding what is present through understanding what is not present. You need to understand the underlying system that gives the words their full sense, and this system is only realized through the whole set of possible utterances, it is never apparent in any one utterance. For de Saussure, then, the aim of semiology is to elucidate the underlying system of differences that gives sense to any domain of meaning, whether it is a language, fashion, architecture or road signs. (Potter 1996: 70)*

The method first involves breaking the text into fragments, which can be words or sentences or signs. This is followed by demonstrating how each fragment draws on cultural codes. Thus, to say that someone is wearing baggy jeans might not just provide a reference to what the person is wearing but can be interpreted to indicate age and social class. The central argument in semiotics is that the sense of a text is produced

22 See the website 'Semiotics for Beginners', at http://www.aber.ac.uk/~dgc/semiotic.html. For other helpful introductions, read Eagleton (1983) and Hawkes (1992).

through the codes of connotation they project, not raw referents (for example, baggy trousers).

In examining film, Fiske (1987) divided such codes into three levels. In semiotic analysis meaning is explored at all these levels:

1. *Reality.* What is meant here is what is encoded through social codes, such as dress, posture and so on. These codes generate particular impressions.
2. *Representation.* In film/video there are codes of camera angle and position through which, much as in language, different meanings are formed. As one illustration, a close-up shot is often used in filming to create a sense of intimacy. Lighting, editing and sound all construct representations and meanings. These are part of the language of the audio-visual. When a narrative is produced, these codes are deployed.
3. *Ideology.* At this level the individual story takes place within higher conceptual levels – namely, within broader systems of meaning and belief concerning the nature of individuals, of individualism, the nation, power and so on.

In semiotic analysis the researcher is prompted to ask a variety of questions that help elucidate meaning. The following are illustrative (L. Fuller 1992):

- To whom is the text addressed (audience)?
- What kinds of social relationships are constructed or reproduced in the text and in audiences viewing the text?
- What definable discourses (such as masculinity, individualism, capitalist economics and so on) does the narrative affirm as natural, preferred ways of viewing the world?
- Whose interests are best served by this view?
- What and how are social roles defined?
- Does the text allow the viewer to construct meaningful alternatives or even opposing readings that question the dominant ideologies (re)presented?

Natter and Jones (1993) offer a demonstration of Fiske's approach in an analysis of Michael Moore's film *Roger and me*. This film offers an account of the demise of Fordism in the USA, focusing on Flint, Michigan. For Natter and Jones (1993: 142):

> *The narrative principle organizing Moore's film is montage. It juxtaposes past and present, class perspectives of the rich and poor, views of capital as both a private and social power, and the seemingly impersonal laws of the economy against their personalized effects. The trope of irony which organizes the film's stylistic juxtapositions does not permit a reconciliation of these differences, but instead exposes them as unresolved contradictions. It is between the positions thus revealed that the reviewer is forced to choose.*

Natter and Jones demonstrate that similarities and differences between two forms of ideology (community and post-Fordist competitive individualism) come across as a key feature of *Roger and me*. Moore's narrative strategy is to juxtapose a 1950s vision of community with a contemporary one. This manœuvre provides the basis for Moore's critique of the decline of Flint in class, family and individual terms. The implications in terms of public and private space are made obvious. The private enclaves of the golf club and General Motors corporate building stand in contrast to the 'private' homes of those evicted by the deputy sheriff. 'Moore never includes scenes of the evicted refusing access to his voyeuristic camera, thus highlighting the class character of what constitutes the private' (Natter and Jones 1993: 146). For Natter and Jones (1993: 153) Moore's use of a Flint woman selling rabbits (the central metaphor for labour) as 'pets or meat' is a privileged scene that weaves the film together:

> *her managerial flexibility bespeaks the emerging post-Fordist economy; the force of her on-camera butchering a rabbit renders impossible any distanced viewing of the act committed; and finally, the encounter symbolizes the failure of GM workers to offer resistance to the company. It is in the 'Pets or meat' sequences that Moore is most artful, and in which the ironic structure that he employs becomes evident.*

If you view this film carefully and then read Natter and Jones's analysis, you should gain a clear picture of Fiske's codes of meaning. Gottdiener (1995) also provides case studies adopting a socio-semiotic approach to analyse material cultures or texts like malls, Disney, architecture, real estate signs and so on.

But, as revealed by the limited nature of our commentary (of step-by-step procedures), semiotic analysis can be critiqued for a lack of analytical clarity. Outputs can come across as interpretative flares of genius rather than emerging from a research method capable of general application. This need not discourage researchers from investigating *and* practising this stimulating approach.

Discourse analysis

For this post-structuralist method, the focus is on talk and texts as part of social practices. Discourse analysis approaches text and language as forms of discourse* that help create and reproduce social meaning. The focus is on the use of language and strategies of argument. Discourse analysts view language as one domain in which knowledge of the social world is shaped (Chapter 1). As Tonkiss (1998b: 246) puts it: 'Discourse analysis involves a perspective on language which sees this not as *reflecting* reality in a transparent or straightforward way, but as *constructing* and organizing that social reality for us.' Tonkiss provides the example of disability. Because there is a greater sensitivity today towards the *power* of discourse in shaping attitudes and identities, the term 'mentally handicapped' has been replaced by 'learning difficulties'. Discourse can refer to a conversation, a piece of writing or a type of language. One example would be academic discourse – the academic language used by social scientists. Academic discourse is a *system* of language that *socializes* us into a profession. It consequently has parallels with

Kuhn's paradigms. This discourse authorizes and professionalizes us. Not surprisingly in this context, discourse analysis has been used to investigate how scientific knowledge is socially constructed (Latour and Woolgar 1979; Demeritt 1996).

Following Tonkiss (1998b: 249–50), in undertaking discourse analysis the researcher seeks to highlight two themes. The first is the *interpretative context*, which is the social setting in which discourse is located. Here the researcher looks beyond the text to social relations external to it. Secondly, the researcher examines the *rhetorical organization* of discourse – that is, the argumentative schemes that organize a text and establish its authority. The work of prosecution and defence lawyers provide an example:

> *Whenever a speaker aims to use language persuasively, to dismiss alternative claims, and to produce certain outcomes (forgiveness, agreement, apology, a purchase, and so on), attention to the rhetorical or argumentative organization of their account will be fruitful. (Tonkiss 1998b: 250).*

Added to which, factual accounts have a double orientation. On the one hand, they embody an action orientation, with people using descriptions to perform acts. On the other hand, there is an epistemological orientation, in which description builds its own status as a factual version, making a description credible (Potter 1996: 108–18).

As for how discourse analysis is undertaken, here Potter and Wetherell (1994: 55) offer five analytical considerations, but warn that, as the analysts becomes more skilled with discourse analysis, their separation becomes less clear-cut:

- *Use variation as a lever.* Potter and Wetherell consider this to be the most important analytical principle, with attention focusing on differences in a text. Here even minor variation in discourse is important, since this is seen to be oriented towards action. In practice it is variation between texts that is likely to be more common than unevenness within texts. Disparate usage of words and phrases is taken to reveal dissimilar emphases and meanings.
- *Read the detail.* Analysts must be attentive to the fine detail of discourse. The kind of detail that is lost in quantitative approaches is crucial for discourse analysis, especially as what is a 'big issue' and what is trivial cannot be identified from the outset.
- *Look for rhetorical organization.* This is a central feature of the approach, which 'is best thought of as an orientation built into the analytic mentality through practice rather than the basis for any specific procedure' (Potter and Wetherell 1994: 59). The focus on rhetoric is intended to draw out how a discourse relates to competing alternatives rather than to some putative 'reality'.
- *Look for accountability.* This concerns making claims and actions accountable, which is linked to making them hard to rebut or undermine. Discourse is explored to identify justifications or excuses.

- *Cross-refer*. This involves using other discourse studies as an analytic resource. This is not designed to generate general laws, which discourse analysts tend to eschew, but to examine whether features of discourse construction in other investigations can inform analysis.

Readers will appreciate that there is no simple way of describing discourse analysis, with aficionados of the approach emphasizing the 'craft' nature of the approach. Potter and Wetherell (1994: 53–5) also found difficulties:

> *One of the difficulties in writing about the process of discourse analysis is that the very category 'analysis' comes from a discourse developed for quantitative, positivist methodologies ... Analysis in those settings consists in a distinct set of procedures; aggregating scores, categorising instances ... [for discourse analysis] To see things in this way would be very misleading, although, given the authority which accrues to these procedures, it is tempting to try ... this chapter is peppered with disclaimers about what we are not doing; this is because, to some extent, we are writing against prevalent expectations about analysis ... [in discourse analysis] How you arrive at some view about what is going on in a piece of text may be quite different from how you justify that interpretation.*

Discourse analysis is a craft skill, something like bike riding or chicken sexing, which is not easy to render or describe in an explicit manner. Indeed, as the analyst becomes more practised, it becomes harder and harder to identify explicit procedures that could be called analysis.

Of course, many research methods are improved with practice. Discourse analysis fits this schema. As for the quality of its results, these are seen in 'how well they [the analytical interpretations] account for the detail in material, how well potential alternatives can be discounted, how plausible the overall account seems, whether it meshes with other studies, and so on' (Potter and Wetherell 1994: 63).

Deconstruction

We mentioned this post-structuralist method briefly in Chapter 1 in our discussion of Derrida. Deconstruction has been used in geography for some time, with Olsson (1980) an early exponent and Doel (1999) a recent follower. The deconstructive method involves reading texts 'against the grain' to destabilize truth claims by challenging the authority on which they are (claimed to be) based. Often this involves focusing on dualisms that structure and ground meaning. This involves reversing 'oppositions', like female/male, black/white or culture/nature, showing how the meaning of one term (male) is grounded by denial of what it is not (female). A careful deconstructive reading of a text can reverse this binary logic, so the prior term (male) can be shown to be secondary to the other term (female). The aim is to show how the (multiple) position(s) of an author influence the writing of a text, so the range of meanings of a text is considerable, not singular.

As pointed out in the *Dictionary of human geography* (R. J. Johnston *et al.* 2000), the emphasis on multiplicity laces deconstructivist approaches with overt relativism, with accompanying question marks about how researchers can represent others when representations by themselves are so contaminated. Yet it is possible to provide multiple interpretations and visions within single works, as investigations by a number of geographers have revealed. In this regard, deconstructive criticism and method have been popular among feminist and 'new' cultural geographers. Specific applications include Harley's employment (1989) of such methods to deconstruct mapping and the logic of cartography (although he has been criticized for not being faithful to Derrida's conceptions of deconstruction), Barnes's pursuit (1996) of a deconstructive critique of economic geography and Lees's deconstruction (1996) of texts on gentrification. Lees (1996), for example, questions the way gentrification is represented as binary oppositions, like inner city versus suburb, marginal versus mainstream and so on. She reveals how these binaries are becoming or can be displaced through the analysis of four gentrification texts: academic, journalist, realtor and gentrifier. Following Derrida, she asserts that the meaning of gentrification is continually mobile.

As with the other textual analysis methods, researchers using deconstruction techniques have no step-by-step guide for the novice. As Barnett (1999: 279) argues: 'There appears to be an unacknowledged investment in the idea that deconstruction is too difficult, or too precious, to be opened up and made accessible.'

Summary: extracting messages

- Content analysis can provide a means of exploring consistent associations at the surface level. The method appears easy but determining the right 'scale' for units of analysis is not necessarily straightforward.
- Understanding issues of materiality can be undertaken by 'scoring' documents to identify the basis of relationships. Having something in common with content analysis, for its surface-level analysis, this approach tends to focus on explaining behaviour using qualitative data. This approach explores *what* sense is made in social life.
- Exploring deeper meaning, textual analysis focuses more on *how* sense is made of social life. Textual analyses have the potential to offer powerful messages but as yet there is little clarity over method.

Messages and sources

One problem for approaches to textual analysis is how to assess the end product? Certainly, as researchers see a text through the lens of divergent social experience, their interpretations differ. As Winchester (1992), amongst others, states, the fact that there are different 'stories' does not mean one is right and others wrong. At one level we have no problem with this. At another the argument is flawed, as it implies that everyone is equally capable of interpretative understanding. This is not so. Take any serious evaluation of textual analysis, look at Geertz (1973/1975), look at John Scott (1990), you find the message it is very difficult. It is not something researchers are equally capable of. On this basis alone interpretations will be uneven in quality. Moreover, as Gutting (1996) draws out, concepts like identity, that are the food for much textual analysis, are difficult to elucidate. Not only is the method difficult but so are concepts associated with it. Hardly surprisingly, many researchers seek reassurance about the quality and validity of interpretations from such work, most especially as many interpretations leave the impression that a text is coherent, when cultures are comprised of conflicting discourses (Clifford 1990). Moreover, commentary on validity in textual analysis is sparse, perhaps even problematical (J. Scott 1990: 32–3). With few analysts revealing much about their methods, this is worrying (Sanjek 1990).

Contrast this with content analysis. Here the approach is atomistic, in that it breaks texts down to their individual elements and counts their incidence. There are good grounds for criticizing this approach, not simply because meaning is qualitative rather than quantitative but also because an atomistic approach cannot provide holistic understanding (see e.g. Riessman 1993: 4). But content analysis provides a criterion, in coding reliability (which links to validity if meaning is seen to inhere), by which competing assessments can be judged. Textual analysis rejects this approach but its advocates are too quiet over presenting criteria for convincing others about their own methods. This means it is difficult to gain an appreciation of the quality of conclusions.

What are the implications for documentary analysis? In a sense the answer is none. The problematics identified in the last section arise from the analytical approach, not whether material is documentary or not. Despite the problems identified above, textual analysis should not be associated with poor research. At least if we take the criterion of whether explanations are convincing, in that they help those who adopt other perspectives to interpret society, then textual analysis has much to commend it. The problem for sympathetic readers who wish to interpret the work of others has more to do with the silences that complicate their reading. In particular, if we use as our criterion for 'success' the ideal of an argument being persuasive, then we fall neatly into the Kuhnian (1970) 'trap' of allowing 'ideology' (or accepted interpretations) to dominate our vision of the world. Rather than being 'critical', textual analysis becomes deeply entrenched in conformity. It does not require the oppression we associate with Orwell's *1984*, Stalin's USSR or Hitler's Germany for this effect to occur. The point of Kuhn's articulation is to draw out how scientific progress is dominated by conformity in central understandings. If we do not challenge research conclusions through critically engaging with their methodology, but accept that a convincing message is sufficient, we are likely to accept poor research simply because it reaches conclusions that (scientifically or otherwise) we want to hear. The power relations in all this are stark.

5 Superficial encounters: social survey methods

There is no substitute in survey-methodology training for apprenticing with a survey-research firm or agency devoted exclusively to a daily routine of survey design, collection and analysis.

(Goyder 1987: 8)

The history of the social survey is connected to the history of the social fact (Poovey 1998). One of the earliest social surveys was the Doomsday Book of 1085. By the late seventeenth century the collection of demographic data, known as political arithmetic, was commonplace. Yet eighteenth-century classical political economists, such as Adam Smith (1776/1991) and Thomas Malthus (1798/1998), based their ideas and theories more on thought experiments. As with Malthus's influential *Essays on population*, speculative argument dominated, rather than the investigation of 'facts'. One consequence of Enlightenment reactions to this 'idealism' was the emergence of 'statistical science'. This arose partly from political fears about economic and social conditions, with nineteenth-century economic and political elites hoping to respond to socio-economic problems through social engineering. To act, a first need was information on populations to be 'engineered'. This led to the collection of social statistics. Initial efforts at systematic information gathering, while praiseworthy in their comprehensiveness, had little appreciation of the complexities and epistemologies that underwrite data collection. Instead, data were taken as obvious and unproblematic. Empiricism was the foundation on which claims were made. Even for the first UK population census (1801),[1] data collection was undertaken with only a scanty theoretical background. As Tonkiss (1998a: 59; see also Hacking 1990) argued:

> *The population in this sense came to be thought about not just as a mass of people but also as a* datum *– a quantitative entity which might be measured and monitored in respect of its size, distribution and growth, and increasingly in terms of a plethora of local rates of disease, marriage, age, employment, wealth, birth, death, and so on.*

1 Early social statistics (state-istics) were closely tied to government and state rationalities.

The rationale for collecting statistics was that if enough facts were available explanation would follow. Science was seen to be objective, with subjectivity denied (Poovey 1998). Partly this was driven by Baconian empiricism in which theories were built inductively from 'facts'. Is this a familiar model? Think of the Victorian (colonial/imperial) geographer, in *his* gumboots, in the field, doing *active* fieldwork (Driver 2001), collecting data, sorting data (think of naturalist Charles Darwin's trips on the *Beagle*), inducing from data (Livingstone 1984, 1985; Seale 1998: ch.2). This attitude is understandable for the time. After all, significant findings came through induction. The discovery of anaesthesia, Charles Darwin's theorizations (1861), and the link between impure water and cholera are examples (Kearns 1985). Such 'discoveries' enhanced the desire to collect 'facts'. As Steckel (1991) notes for the USA, by the 1830s the desire for knowledge about society was such that statistical almanacs were produced for profit. Yet such data sources left much to be desired. Indeed the very looseness of the UK's 1801 Census galvanized a commitment to 'producing sound numerical information about the state of the nation' (Tonkiss 1998a: 60). Statistical societies were formed, such as the American Statistical Association in 1839, with the tabulation of conditions under which the working classes lived being a notable interest on both sides of the Atlantic (see Kent 1981). But as these early efforts assumed that statistical facts spoke for themselves, many reports read more like morality statements than scientific accounts (see Kent 1981: 23).

For our purposes the failings of such data gathering are less fundamental than the inner drive they represented. What was grounded in these beginnings was an impetus towards more systematic data collection. As we move from the early nineteenth century towards the present a series of key studies contributed to and transformed our understanding of data collection. What distinguishes these early investigations is that they collected data through extensive surveys of people or institutions. There were also qualitative social surveys, or systematic data collection combining qualitative and quantitative material, with this work often undertaken by philanthropists, novelists and journalists. Charles Booth's classic series (1891) on the life and labour of the poor in London powerfully revealed the capacity of surveys to change the public psyche by elucidating the circumstances of populations, in this case through a poverty map. Set alongside the powerful imagery such surveys provided on the 'hidden realities' of society, highly publicized surveys also exposed their flaws. Perhaps most spectacular was the *Literary Digest*'s survey of electoral intentions in the 1936 US Presidential Election. This was the largest poll in history at that time, with 10 million people sent questionnaires. However, the sample was drawn largely from telephone directories, when poorer people often had no phone. The *Digest* predicted victory for the Republican candidate, Alfred Landon (by 57 per cent to 43 per cent), when Roosevelt secured a landslide (by 62 per cent to 38 per cent) (Frankfort-Nachmias and Nachmias 1992: 174). Such 'errors' prompted deeper exploration of the mechanisms and potentialities of social surveys.

Captured within the above two paragraphs are central ideas for this chapter. First, there is the question of what surveys can show. This question arises whether the survey is elucidating information from a whole population or from a sample. Textbook accounts of research design stress the necessity of a theoretical base for data collection, but social surveys can still be trawling exercises, in which the theoretical base of the study is weak. Hold onto this thought. Add to it the message that this may not matter, provided the intention is clearly understood and interpretations of findings are not extended beyond their capacities. If the reader is beginning to suspect we are losing our marbles, then think through what a survey is capable of. A key advantage of a survey is that the researcher identifies broad trends. As the title of this chapter highlights, what characterizes this search is the 'superficial' nature of the data collected. While we will discuss this particular aspect of surveys below, for the moment the point to grasp is that surveys are aimed towards establishing general trends. Certainly, investigators could undertake lengthy in-depth interviews with large numbers of informants. However, this strategy is not common, for the cost and time implications of an extensive, in-depth procedure are beyond the capacity of virtually all researchers. More commonly, leaving aside epistemological predilections for the moment, researchers are confronted with deciding whether to engage in detailed investigations of a relatively small number of informants or seek evidence from a larger group in less depth (the latter is the survey approach). If we reintroduce epistemological considerations, then, as Chapter 2 pinpointed, particular researchers favour one approach above the other. Our concern throughout this book is not to suggest one is better than another. Rather the expression 'horses for courses' is appropriate. As later chapters show, we see great strengths, alongside weaknesses, in in-depth investigations. The same applies for surveys. The critical consideration centres on whether the information that surveys elicit validly match the stated aims of the study.

What we are suggesting may strike readers of some social science philosophizing as unusual. Direct your thoughts towards the issue of explanation. Recall a passage from Chapter 2, in which we quote Dey's words (1993: 40): 'To explain is to account for action, not just or necessarily through reference to actors' intentions. It requires the development of conceptual tools through which to apprehend the significance of social action and how actions interrelate.' The message is that we need to identify not just intentions but also the significance of actions and connections between acts (with theory to link them). Questionnaires are not equipped to reveal either 'significance' or 'connections'. Think about asking a person why he or she had bought a house. You might receive an array of responses, ranging from why he or she wanted to buy rather than rent (security, monetary savings or whatever), alongside the rationale for purchasing a specific home (ease of transport access, price, number of rooms, original features, inconvenient for the in-laws or whatever). These represent intentions. But how do you get at the significance of the action and its links with other social acts? Is a questionnaire capable of probing to establish why the person has a need to feel secure, why he or she believes the propaganda mortgage companies spread about monetary savings or why he or she wants to be sure the in-laws find it difficult to access their home? Do we get to underlying rationalities and signification by asking simple questions, with limited time for responses? We think not. This requires exploring decision criteria and contexts in greater depth than survey questionnaires can attain. As such, social surveys provide a weak base for explanation. What they offer is a tabulation of tendencies. But to know that something happens often does not explain why it happens. If university lecturers are to be believed, there is an empirical association between the presence of a large number of books (that is, libraries) and the regularity with which undergraduates are either induced to sleep or provoked to talk incessantly. Do the books cause these responses? Surveys record regularities, and superficial indications of rationalities, but lack the depth for real explanation.

This has important implications. Recall the power of surveys in revealing what people believe – watch politicians scurrying around for excuses and squirming under media gaze after unfavourable public opinion polls. Recall that if surveys get it badly wrong this can dent public confidence in them. This raises peculiar demands on the survey method. They do not involve in-depth exploration of people's rationalities, with multiple opportunities to tease out reasons, interpretations, meanings and the significance of acts. They are built around snapshot insights. Cost and time pressures mean you have one crack at getting it right. It is rare for researchers to be able to go back and collect more information. Moreover, if you go back, people might be unwilling to impart further information. This might seem to place a peculiar necessity on ensuring a survey is well grounded in a theoretical framework. For some studies this is true. However, surveys are also used when little is known about a social phenomenon, or it is poorly tabulated. Booth's maps (1891) are an obvious example. Buller and Hoggart's investigation (1994) of British home-owners in rural France is another. Here the 'trawling exercise' analogy is appropriate, for researchers in these contexts can find they have asked questions that were not important and omitted questions that would have given more insight. But, whether surveys are theoretically informed (deductive) or more of a trawl (inductive), their main advantage arises from eliciting information on incidences. Surveys can derive qualitative information, but they lack the capacity of in-depth approaches to reveal meaning and significance. By their nature they only achieve this at a superficial level. At this level their potential to highlight incidences or trends is powerful.

Summary: the uses of social surveys

- Social surveys have a long history, with many feeding into government policy.
- The key advantage of survey methods arises from securing broad coverage of populations.
- In securing this coverage, they address surface level aspects of human behaviour and values. The demands of seeking (relative) uniformities in response limit the capacity to probe meaning and intention.

But are surveys valid?

Despite pronouncements that surveys can record the incidence of social phenomena or even public values, there are potent criticisms of this method. These go beyond obvious failures, such as the 1936 US Presidential Election prediction, which can be interpreted as an error of technique rather than principle. A most telling commentary is that of Briggs (1986). His critique focuses on the interactional and communicative norms that underlie interviews or other communications that embody personal meaning. Briggs is concerned that so much literature on social surveys is concerned with technical issues, like assessing the impact of interviewer characteristics on respondent answers or how the wording of questions affects responses. For Briggs (1986: 13) these issues obscure the real problematic, which is the dialogic, contextualized nature of discourse, whether this be formal questionnaire surveys, in-depth interviews or oral history (also Cooper 1967). At heart this critique applies to all research that does not encompass long enduring interaction with study participants, for Briggs (1986: 43) insists that the design, implementation and analysis of interviews must be grounded in an awareness of the communicative 'competence' of an informant (and on communication conventions (Barley 1983: 67)). This is not just about linguistic abilities but extends to researcher understanding of the information imparted. A key distinction for Briggs is between 'referential meaning', by which people indicate humans, objects, events or processes, with some (perceived) correspondence between the 'content' of what they say and 'the real world', and 'indexical meaning', which is defined by the context in which an expression is uttered. For Briggs (1986: 43): 'If the ability to communicate effectively in interviews and to interpret the data is identified with the ability to encode and decode referential meaning alone, inaccuracy, distortion and misunderstanding are inevitable.' This is because the context of the interview impacts on every response an interviewee gives. If an interviewee does not accept the interviewer's definition of their 'shared' situation, responses cannot be interpreted 'accurately' by an interviewer. Briggs provides various examples of respondent answers bearing no seeming relationship to questions, yet responses were rational for interviewees. As one instance, when asking questions in a Mexican community, Briggs's (young) age and the fact he was not married meant older people took it as their task to 'teach' him. As a result, what Briggs heard was what older people thought would be advantageous for his 'education', not the information he requested. The message Briggs provides is that statements fit into a communicative web, the understanding of which is essential for interviews to be interpreted. Viewed in this way, the whole must be grasped before individual parts can be interpreted: 'Interviewees do not draw, even ideally, on a fixed idea or feeling in answering a question, but connect questions with some element or elements of a vast and dynamic range of responses' (Briggs 1986: 22).

The hefty tone of this critique should remind us at one level of Buzzard's stipulation (1990) that researchers should be sure what data they need before undertaking a survey, since they might be able to secure this from other sources (such as records kept by public agencies). However, before we let this thought carry us too far, it is worth reminding ourselves that many of these sources are based on surveys. As such this solution might not help circumvent Briggs's central criticism. But what weight should be put on this criticism is unclear. Taken to an extreme position, the implications of Briggs's comments raise doubts over the validity of any data. Figures on university student attributes are presumably suspect, because students could have filled in registration forms in a different frame of mind from that intended by the authorities. Being realistic, if you ask someone if he or she has any sisters, while there might be reluctance to answer, is it reasonable to assume that the question is interpreted so differently that we should assume answers are grounded in a different world view? In a few cases perhaps. Certainly we need to be conscious that the 'distance' between researcher and respondents can affect the advisability of undertaking questionnaire surveys. Offering a reminder of this at a cross-national scale, Gow (1990: 144) notes that using questionnaires in stratified, closed societies like Afghanistan is foolhardy. Closer to home, Goyder (1987) notes that surveys are poor mechanisms for gaining information about groups at the margins of conventional life, such as vagrants (e.g. Brody 1971) and elites (e.g. Zuckerman 1972). More commonly, when questions require 'factual' answers, it seems reasonable to assume some consistency in understanding. If there is not, governments have been basing policies on haphazard data for centuries. But as questions get more complex, as researchers try to understand causation, or to identify meaning and event significance, interpretative difficulties mushroom. In this context, more in-depth investigations offer greater potential for capturing valid information, even if the problems Briggs highlights do not go away.

But the essence of what Briggs is saying offers a lot to note. For one thing, as Foddy (1993) draws out, there has been a tendency amongst survey researchers to approach their work from a stimulus-response framework, where with the right stimulus informants respond uniformly (that is, they understand the same thing by the question asked). This has led to more effort to improve the reliability of question responses than their validity. This lies at the core of the Briggs criticism. Expressed more in list form, Vaus (1991) offers the following philosophical critiques of questionnaire surveys: (1) they cannot establish causal connections between variables, merely co-variation; (2) they are incapable of eliciting the meaning of social actions; (3) they gain information on people's beliefs and/or actions but cannot contextualize them; (4) too many assume human action is determined by external forces and neglect consciousness, goals, intentions and values; (5) there is too close an association with a sterile, ritualistic and rigid model of science; (6) many are essentially empiricist; and (7) many key elements in explaining behaviour, such as power relationships, cannot be captured. Confronted with this list, we might assume that the answer to the question 'Do surveys yield valid data' is no. This is too simplistic. The real answer depends upon the questions asked and the use to which surveys are put.

What can be boldly stated is that planning and conducting successful social surveys require a great deal of time and energy. Illustrative of the complexity, *The survey kit* (Fink 1995) contains nine volumes to help the researcher who has not undertaken a social survey before (as well as a needed refresher and thought-sharpener for those that have). The kit provides information from initial planning stages through to the practicalities of carrying them out, analysing data and making reports. As well as printed guides, computer software is available to help design, administer, process and analyse questionnaires. One example is *Sphinx survey*, which offers an array of procedures for analysing data (visit http://www.scolari.co.uk/ for information on related software). Given the ready availability of high-quality aids to the design, implementation and analysis of social surveys (e.g. Institute for Social Research Survey Research Center 1976; F. J. Fowler 1984; Converse and Presser 1986; Oppenheim 1986; Nichols 1991; Vaus 1991; Foddy 1993; Fink and Kosecoff 1998), our aim is not to summarize key factors in survey design and implementation. Rather we weave a picture of useful tips to help undertake social surveys, while drawing out inconsistencies in the social survey approach that lead to rather dogmatic statements about its strengths and weaknesses.

Summary: are surveys valid?

- Questionnaire information is limited by respondents placing responses in different (personal and social) contexts.
- Surveys are better suited to uncomplicated, 'factual' data gathering, where the chances of uneven reaction should be lessened.

Types of surveys

Before we move on to the execution and design of surveys, it is useful to outline different types of survey used by geographers. There are five main types: the *personal survey*, the *intercept survey*, the *mail or postal survey*, the *telephone survey* and, finally, the most recent type of survey, the *online survey* (Gray and Guppy 1994). Different procedures for surveys create differential distances between researcher and respondent. The personal survey, for example, involves less distance than the postal survey. While questionnaire surveys can be either oral or written, in oral (personal or telephone) surveys the respondent is interviewed verbally according to a pre-designed and structured set of questions. The written survey (like the postal survey) may ask the same questions but it is impersonal, as there is no dialogue between the researcher and the respondent. This differential distancing is associated with dissimilar problems. But in deciding on which type of survey to undertake, researchers also have to consider important cost, time and social interaction implications.

The *personal survey* and the *intercept survey* involve face-to-face interaction between interviewer and respondent. The personal survey usually takes place in a relatively comfortable setting, such as the respondent's home. The intercept survey has people approached whilst, for example, shopping, with the survey completed at the point of interception (the doorstep, a shop or the street). Although the personal survey will usually gain a higher response rate than a telephone or postal survey, it is more time-consuming and involves more legwork. Moreover, because of its interactional nature, the personal survey introduces 'interviewer effects' into the equation, as respondents express different views according to their reaction to the interviewer (Fowler and Mangione 1990).

The *postal survey* involves mailing out questionnaires to selected respondents who complete and return them. This survey type can be effective but it is coming under threat from the volume of junk mail people receive, which has a tendency to cool people's willingness to respond (especially if completion of commercial questionnaires results in unsolicited follow-up calls from sales personnel). Hence, a concerted effort is needed to persuade people to complete a mail questionnaire. Even the appearance of the envelope in which the questionnaire arrives can influence whether people respond. Implementing a postal survey is often cheaper than personal interviews owing to reduced travel costs, but it can be expensive and time-consuming relative to the proportion of questionnaires sent out that are returned (that is, the response rate). As many commentators note, a key to raising response rates from mail surveys is convincing people that the information they provide is important (Gray and Guppy 1994). This process can begin with sending an initial letter advising people the questionnaire is due to arrive, followed by a letter of introduction, the questionnaire and a stamped addressed envelope for its return. If no response has been received after a few weeks, it is common to send a follow-up letter (questionnaire and stamped addressed envelope) or at least a reminder postcard. Commentators regularly state that at least one follow-up questionnaire should be sent to non-respondents, others favour at least two, with returns seeing a sharp percentage fall at each round of re-circulation. Significantly, more than for face-to-face interviews, mail questionnaires are notorious for low responses rates, with returns biased towards those of higher income or more formal education. On this basis, it is often recommended that special attention is given to non-respondents to try to counter this effect (see e.g. Goyder 1987); although this suggestion could be made for all social survey types, given their key benefit is securing broad coverage of populations.

Non-coverage used to be a major problem with *telephone surveys*, but as access to telephones has increased this is now less of a problem (albeit uneven coverage has not gone away, with Tortora (1994) reporting that 20 per cent of rural households in the US South had no phone in 1986). This kind of survey is conducted with conversation taped or notes taken. Those who use this approach increasingly see it as a computer-based strategy (as with CATI (computer assisted telephone interviewing)), with random digit dialling and on-screen questionnaires becoming more standard. Geographers have so far made limited use of such technological treats. One reason is that a telephone survey can be expensive owing to the cost of the calls. Moreover, the public has become less willing to engage in telephone surveys, especially given the increased uptake of call-screening and answering machines. The skills required for this approach are additional concerns (especially given that little training is available in the university sector), with 'interviewer effects' on participation and responses heightened (R. M. Groves *et al.* 1988).

This problem is avoided in the *online survey*, which is when a survey is posted online and answers are received electronically. Although not commonplace in geography, the online survey is of growing importance in market research. For some, these surveys have a 'novelty' value, which can encourage participation. Believers in online research say that it is relatively cheap and can deliver in a short period of time. Moreover, such surveys might reach hard-to-find groups (for example, paedophiles or computer hackers) and they allow respondents to address the survey at their leisure. Researchers also report that online respondents often take care in replying, with lengthy commentaries on open-ended questions. Added to which, online research opens up technological opportunities; for example, images can be loaded onto the web and reactions collected.

The disadvantage is that online surveys are unlikely to attract a random sample of a population. Whereas sample selection for other survey approaches is deliberative (and partly handicapped by this), except for targeted e-mail surveys (which are little different from mail questionnaires), online surveys are dependent on who finds the questionnaire and decides to respond. This commonly means young wealthy males. Attempts to get around this problem in market research include applications that route every *n*th web-site visitor to a survey section. This does not provide a random sample, but it does target those who visit a web site, and so (presumably) have particular interests (Edmondson 1997). But this does not verify who completes the questionnaire. Stories abound of men pretending to be women online, of kids pretending to be adults and so on. Some argue that technological developments, like the ability of computers to read fingerprints, will solve verification issues in the future. This will be a development to watch. There is too little case law for firm conclusions, but it is worth noting that telephone and mail surveys (especially) tend to draw self-selecting populations. Perhaps this effect is heightened in online surveys, but publication of texts like *Internet communication and qualitative research* (Mann and Stewart 2000) reveal a trend toward online research that could change this pattern. We have found these surveys useful for tapping particular audiences. Thus, in examining the marginalization of youth from public space in Portland, Maine, Lees posted an online questionnaire on Portland's *teengo* web site to gain data on experiences of eviction or being moved on from public spaces. This questionnaire sought to establish whether respondents felt they had been harassed by the police, with open-ended questions about feelings towards this marginalization (Box 5.1). As it was recognized the survey would be biased towards those who accessed the *teengo* web site, all Portland high schools were faxed asking teachers to inform students of the survey and persuade them to take part. This yielded useful, cost-effective data, although there was a bias towards those who normally accessed the *teengo* site and the responses were superficial, as the researcher was distant from respondents (Lees in the UK, the respondents in the USA). Nevertheless, this was an instructive data source to accompany other research procedures.

Box 5.1 The teen Go web site questionnaire

Reprinted by permission. Copyright Maine Today.com and Blethen Maine Newspapers

Examples of questions asked of visitors to the Teen GO web site

Are you male or female

- ◯ Male
- ◯ Female

Vote!

To which age group do you belong?

- ◯ 12–13
- ◯ 14–15
- ◯ 16–17
- ◯ 18–19
- ◯ 20–21

Vote!

Do you feel excluded from public space in downtown Portland?

- ◯ Yes
- ◯ No

Vote!

With whom do you hang out downtown?

- ◯ School friends
- ◯ Other friends
- ◯ Gangs
- ◯ Street Kids

Vote!

Have you ever been the victim of aggression in downtown public space from the police?

- ◯ Yes
- ◯ No

Vote!

Thank-you for your time and help – it is much appreciated.

I am interested in hearing about:

- Your experiences of and in downtown public space in Portland
- Who you hang out with? When? Where?
- Your feelings about adult control(s) of youth, youth rights, etc.
- Exclusions, inclusions, conflicts in and over downtown public space
- How might you design downtown public space so that it is a democratic space, ie. inclusive of both youth and adults?

When choosing between different survey types, sample limitations, response rates, cost, subject matter, speed and timing all inform choices. Some researchers opt to mix types, combining, for example, a postal questionnaire with a more limited face-to-face survey. What is crucial in reaching decisions is to take stock of what a survey method can and cannot do, as well as thinking through advantages and disadvantages for your project. One means of reaching such decisions is to undertake pilot investigations that try out investigative styles.

Summary: types of survey

- These vary significantly in likely response rates, in the type of people who are likely to reply, in cost of execution and in speed of response.

What is a pilot study?

In what McCracken (1988: p. ix) calls 'the winter of positivism that prevailed in the social sciences in the 1960s and 1970s', in-depth qualitative investigations were commonly portrayed as suitable for exploratory studies before social surveys were undertaken. The implication was that in-depth qualitative work is not suited to theory testing, even if it prompts hypotheses to 'verify' through survey analysis. This contrived misrepresentation should be somewhat obvious. Surveys do not establish causality, even if they highlight coincidences that theory predicts. To provide explanation we need to understand the context in which behaviour occurs, as well as the mechanisms through which it occurs (Sayer 1984). Surveys are not capable of this. What they offer is thin description. This distinguishes their insights from the thick description symbolic anthropologists favour (Geertz 1973/1975), with description (or at best narrowly contrived positivist theoretical evaluation) the order of the day.[2] Moreover, casting surveys simply as deductive evaluator of theory neglects a powerful element in the survey armoury. This arises from inductive insight. In-depth investigations of a small number of people might provide a bounty of information on intentions and meanings (on the home relocation decisions of 35 households, see e.g. Forrest and Murie 1990) but cannot yield the relative importance of trends (on tenure, employment, family location and housing histories for 1408 householders, see e.g. Forrest and Murie 1994). Both in-depth *and* broad social survey data can act as 'pilot studies', depending on the question the researcher wishes to answer. In-depth qualitative work might suggest behaviour patterns whose generality can (or perhaps cannot) be assessed through a survey, just as a survey can paint a picture

2 Read this sentence with caution. We are not claiming that the information surveys provide is not critical to theory. What we charge is that on their own surveys cannot elicit data for 'real' explanation (Dey 1993). Surveys remove descriptions of actions from social contexts. Even if careful manipulation of data can provide powerful insight on behaviour (Hellevik 1984), by dissociating action from context, the danger of lauding spurious associations is ever present.

requiring in-depth identification of rationalities for behaviour. Analysts can derive benefits from undertaking both in-depth and survey work, with the relative strength of each recognized in the research design, so as to exploit mutual benefits. The search for validity requires that we explore multiple data sources. So, while analysts have long held that questionnaires should contain a series of items that tap the same issue, so as to cross-validate responses on single items (Henerson *et al.* 1978; Oppenheim 1986), these steps are not of much value if the questionnaire is itself invalid, in that it cannot tap the concept of interest. Yet construct validity – which relates to how well a measure or representation conforms with what it is designed to measure or represent (P. E. Spector 1993) – is not easy to confirm empirically. The fact that two data sources do not provide the same answer to a question might reflect not the invalidity of one source but the fact that these sources capture different aspects of a process. Moreover, we cannot assume that, if messages from multiple sources converge, then we have valid results. The issue of validity must be addressed by constructive argument, with evidence to persuade a reader to have confidence in research results (see Baxter and Eyles 1997). One element in this process is to evaluate the potential of a data source to provide an accurate representation of concepts and issues.

In survey work, a first step in checking the credibility of an instrument is a pilot survey. To distinguish what is meant here from the naive vision of qualitative analysis being only appropriate for exploratory analysis, let us be clear that a pilot study is a mechanism for questioning whether a proposed instrument is suited to the purpose at hand. Although the manner in which a pilot investigation is conducted might be slightly different, perhaps owing to the relative difficulty or ease of identifying participants, the need for a pilot enquiry applies whether the underlying thrust of a survey is inductive or deductive. As Caunce (1994) states, it is always sensible to run a trial to see how things work out before tackling a large-scale enterprise, even when there seems to be a clear path to follow. But what are the characteristics of a pilot study?

As the intention is to check the operational effectiveness of survey instruments (most evidently the questionnaire, but also how informants will be identified and how data will be collected), what a pilot study involves is a scaled-down version of the (at this stage proposed) survey procedure. The aim is to try out, not to secure the first survey responses. The intention is to challenge the proposed instrument, so that it can be adjusted for better effect, not to hope that all goes well and that pilot responses can be used in the final analysis. In this context, Fink and Kosecoff (1985, 1998) provide valuable guidelines. First, the researcher should try to anticipate the circumstances under which the survey will be conducted and make plans to handle them. This means that, if you intend to use letters of introduction to encourage involvement in the survey, these should be tested in the pilot and not left to the final project. Second, select pilot informants who will be similar to those in the full survey. In so far as possible, it helps if participants are of similar age, socio-economic background and so on to those targeted in the final survey. This does not mean an exact fit. It does mean that, if you are studying young people, do not use your parents to pilot your questionnaire. It means that, if you are drawing on a broad spectrum of the population for the eventual survey, then seek a broad spectrum for the pilot. Third, try to enlist as many participants as possible without wasting resources. This is a difficult stipulation, as numbers cannot be given with confidence. This does not mean some have not tried, but their efforts reveal the arbitrary nature of their recommendations. Thus, Nichols (1991) suggests that to get a feel for the problem investigated a pilot group of 30–50 respondents is usually enough. Buzzard (1990), by contrast, suggests that 5–10 who are not in the survey might be sufficient. Rather than offering some inevitably false numerical security, we are more inclined to accept Fink and Kosecoff's suggestion (1985, 1998) that when the pilot process is not yielding information that helps improve the study, then stop. Our caveat is an obvious one – namely, that this stipulation assumes the pilot does not bunch informant types. If you are studying young people at school, then even if all seems well amongst the 14 year olds, if you do not pilot beyond this group you might find your procedures and questions are not much use when you get to the 11 or 17 year olds.

The next two pointers Fink and Kosecoff provide are more to do with content and aims. Critical here is to try to establish whether the questions you propose 'work'. This is difficult to assess. Fink and Kosecoff's fourth consideration, for example, is to focus on the clarity of questions, to see if they are understood. Some indications of what might be looked for are offered by Converse and Presser (1986), who write about using a number of different interviewers in a study. They suggest that group discussions are held with interviewers after pilot surveys. The objective is the same as for conducting surveys on your own: seek to identify whether questions made respondents feel uncomfortable, were awkward to read, led respondents to ask for them to be repeated, did not permit informants to say what they wanted in response or elicited little response.[3] Lee (1993) adds the further suggestion that informants be asked if any questions are too personal or discomforting to discuss. More difficult to answer is Foddy's indication (1993) that, if respondents typically search for contextual clues to help them interpret a question, this is likely to mean they will use different cues to answer the question, so responses will be non-comparable. To clarify this point, think back to the question about why people bought their house. If some answer by telling you a series of horror stories about renting and the security of being a home-owner, while others drool over garden gnomes, bathroom tiles, regularity of bus services and so on, then these responses are not really comparable. It might be better to ask two separate questions, one about buying rather than renting, the other about the selection of this house in preference to others. Identifying the potential advantages from such question clarification lies at the heart of pilot studies. But, as Foddy (1993) notes, piloting is not that helpful for identifying whether questions are interpreted as intended. For one thing, if a question is felt to be

3 Converse and Presser (1986) and Fink and Kosecoff (1985) suggest pilot study interviewers make comments in the margins of the questionnaire about reactions to questions.

difficult, respondents might modify it to provide an answer (Belson 1981: 370). This is something you might establish from dissimilar reactions to the same question (as with the house purchase example), but, even if you ask people directly if they understand a question, they might not want to tell you if they do not. A more direct approach that has been used is to ask pilot study informants to restate questions in their own words. The intention here is to identify inconsistent interpretations. It takes little imagination to recognize that this approach can make respondents uncomfortable, unless they were aware this was a 'trial run' of the questionnaire. But if people are asked to trial run a questionnaire, this raises questions about them agreeing to participate. At the very least those running the survey should ask if it is likely people will participate who are ideal for trialling the questions (that is, similar to eventual respondents, rather than friends or family).

This is likely to be less of a problem if the survey proposed is national in scope and supported by a large organization, like the government or a major corporation. The 'selling' of the need for a trial run is easier in such circumstances, with the 'authority' of a major organization encouraging involvement. However, for smaller-scale projects, possibilities of intensive interrogation over survey questions are more restricted. Recognizing this, Fink and Kosecoff (1985: 48) offer the advice that: 'One way to make sure that you have a reliable and valid survey is to use one that someone else has prepared and demonstrated to be reliable and valid through careful testing.' This does not absolve the researcher of the responsibility of piloting a survey, for new questionnaires commonly combine questions from other surveys so they are unique. Moreover, just because a questionnaire has been employed in one context does not mean it can be transferred trouble-free to another context. Linked to this, the literature provides regular reminders that researchers should consult community leaders and local researchers on research design and data collection approaches (Buzzard 1990). Potentially add to this list government officials, as getting permission for a research project may be a legal requirement, not just a courtesy or a possible aid in research design (Barley 1983; Howard 1994).

Summary: pilot studies

- These are critical for evaluating sample frames, approaches to informants to secure involvement, survey method, questions and recording responses.
- In setting up a pilot project, the conditions under which the final survey will be undertaken should be approximated (or built towards).

The informants

The literature has a number of helpful contributions that focus on sampling populations (e.g. Dixon and Leach 1977; Bulmer 1983). For most projects sampling is undertaken because resources limit capacities to study whole populations. But how should samples be chosen? There is no easy answer. For one thing, there is limited information on individuals within populations. Added to which, when sources are available they often give limited information to help in sample selection. But what are these sources? The technical jargon is that they are sample frames. Such frames often come in the form of a population list. In the UK the most widely recognized sample frame for the adult population is the Electoral Register. This is compiled annually, using information gathered in October, although the register does not come into force until the following February. This seemingly universal source inevitably contains errors.[4] Estimates have suggested that it becomes at least 0.67 per cent more inaccurate as each month progresses, owing to deaths and migration (Gray and Gee 1967). Inaccuracy creeps in from the start, as some people are missed, with the introduction of the Poll Tax in 1990 resulting in many avoiding registration in order to avoid the tax (see e.g. Dorling *et al.* 1996). Pre-Poll tax, the Electoral Register was estimated to be 96 per cent accurate at the time information was collected in October, with this figure falling to 85 per cent by the time the register was replaced 16 months later (Gray and Gee 1967). Omissions are also uneven, with recent movers and those under 30 being especially prone to be missed (even pre-Poll Tax); one estimate put a 20 per cent shortfall for these groups in Inner London (Todd and Eldridge 1987). It might seem this is a far from perfect source for selecting a random sample, but at least a register exists, which is not the case for many countries. Even then, the Electoral Register gives only names and addresses. When we seek to identify people with particular demographic, economic or social characteristics, problems multiply.

Searching for specific populations requires a specialized sample frame. Such frames have their own problems of accuracy, with the merits of specific sources requiring evaluation. Take the telephone directory as an example. Some researchers have explored whether this is a useful device for identifying occupational groups like farmers, who might be expected to be in such a directory for commercial reasons. In general this source is valuable for this purpose, with Errington (1985) holding that there is little difference between farmers who are and are not listed in Yellow Pages. Adding a note of caution, Burton and Wilson (1999) warn that lifestyle farmers may be missed, as a fee has to be paid for a Yellow Page entry. More generally, despite the wider availability of telephones than in the past, Tortora (1994) warns that the Asian and African-American population is under-represented in US telephone directories. This symbolizes an underlying trend in population lists. Fundamentally they are compiled with underlying commercial, political or social rationales. As such they tend to highlight groups with specific characteristics, as well as reflecting the purpose of the compiler. Thus Tortora (1994) notes that lists of US agricultural producers contain data on landlords who do not farm land, which limits their usefulness as a source on farmers. Commercial directories have similar troubles, with entries biased towards particular economic activities (e.g. Shaw 1982). Organizational lists, which provide information to members about one

4 The Electoral Register is available in public reference libraries for public perusal.

another, can be valuable (e.g. London Federation of Housing Associations 1996), although gaining access to them might not be straightforward for some organizations. Indeed, many groups are 'hidden'. Newby (1977) offers one illustration, when noting that access to farm labourers in the UK necessitates contact through farmers, which offers a clear message on power structure. The more specialized the population, the more difficult it might be to identify it. Yet, if the activities of this population have commercial value, if the population has organized an association for itself (whether for cultural, educational, hobby, professional, social or sporting reasons), or if governments believe information on this population is of general interest, then the chances of a listing increase.

Sampling

When such lists are available, a (sufficiently large) survey can reflect the population if a sample taken is 'random'. In simple terms this means inclusion has the same probability as exclusion. In practice, researchers rarely have the resources to draw samples from whole populations, with a variety of devices used to narrow the range of potential survey members. Most evidently this is seen in imposing geographical restrictions on the areas survey members come from. This strategy is used for many national surveys, such as the US national opinion poll survey (Davis and Smith 1992). Here, probability sampling is used to select successively smaller geographical areas within which informants are selected. The main advantages of this strategy are that it reduces interviewer travel time and cost. Set against these, the areal zones in which interviews are drawn are not neutral in their impact on population diversity. For relatively small geographical zones, populations are likely to be relatively homogeneous. If a broad selection of the population is required, the project should not be restricted to a few zones. Larger geographical areas are inclined to be more heterogeneous, so a smaller number of zones might draw in a more diverse population. That noted, it is worth breathing caution into clustering strategies. As Nichols (1991: 63) argued: 'Always use cluster sampling with caution. Because you are concentrating your sample in small areas within the target population it is easy to miss out some important sub-groups completely.'

In a high percentage of geographical studies, the research question relates either to differences between specific population groups or to the behaviour of a particular group. Commonly lists of members of such groups are not available. When a group is relatively large, such as an occupational or a large ethnic group, sources like the census can identify population concentrations. Compared with random selection procedures, this increases the chance of finding relevant informants (relative to effort expended). Identifying potential informants in this way is often desirable owing to the lack of sample frames that 'stratify' (or categorize) by key research attributes. An inability to identify population strata commonly means compromises are made. Especially when the population of interest is small, a common identification method is snowballing, in which each informant is asked to identify others with the target characteristic. A variant of this approach is to use key informants to identify relevant people, as Huw Jones and associates (1986) employed to identify recent in-migrants into villages. More commonly it has been employed to identify people who share the same characteristic, as with Byron's identification (1994) of people from the island of Nevis living in Leicester (see also Batterbury 1994).

What seems like a disadvantage of this approach is that it is not based on a random sample, so there is no base for undertaking (inferential) statistical tests. There are certainly problems with such selection devices, as Buzzard (1990) notes, for informants might move in narrow social circles, so snowballing leads to (unknown and unwanted) social concentrations. But, if steps are taken to guard against this, by seeking varied entry channels into groups, this approach brings one major advantage. This is that informants are theoretically central to a study. The selection procedure thereby prioritizes theoretical sampling over statistical sampling (on the difference, see Glaser and Strauss 1967). This is a critical distinction, as it is linked to the nature of the research problem posed. As Bernard (1994: 96) reminds us: 'All samples are representative of something. The trick is to make them representative of what you want them to be.' Statistical sampling is highly desirable when the population is large and the objective is to gather a representative picture of it. In such circumstances, the absence of accurate sampling frames is not necessarily terminal. Nichols (1991) offers signals in this direction, drawing attention to maps, lists of standpipe licences and so on, as potential sources for Third World investigations, with postcodes a possibility in some nations (Raper *et al.* 1992). When populations are small or scattered, difficulties in gaining a statistically representative sample increase. In such circumstances, the chances of conducting informative pilot studies are more limited. But for small populations, it is a mute point whether it is advisable to use a standardized questionnaire, since the 'peculiarity' of the population means more intensive interviews might be appropriate (Chapter 6).

Readers can draw from this one further message, which relates to sample size. A regular question for research projects is the number of informants that are needed for the project to be (numerically) 'adequate'. The answer is simple and complex (Bouma 1993). The simple response is that there is no answer, hence the complexity. A desirable size for a sample is affected by the realities of a research project. This means there must be an interaction of considerations related to the time that is available to complete the work (you cannot expect an undergraduate dissertation to be equivalent to a Ph.D. in workload terms), alongside the cost of different strategies. The answer is also determined by the nature of the project – whether the study seeks a picture of a general population, elucidates dimensions of human decisions, explores a unique event or whatever. Even for questionnaire surveys designed to identify broad trends, an answer is not easy to provide. What we can say is that if the population is heterogeneous then a larger sample is desirable (Buzzard 1990). Otherwise the chances of picking out subgroups is diminished. Drawing on statistical theory for a moment, Hedrick and associates (1993) note that sample sizes in social research are often too small to identify 'real' statistical effects. Indicating this, they highlight that if 10 per cent of variation in people's behaviour is accounted for by a particular factor (say 10 per cent of decisions to buy a house are accounted for by access to public transport), then, if the researcher wishes to be 95 per cent certain of picking up this effect, a sample size of 2600 informants is needed. For the same level of confidence, if the factor under scrutiny accounts for 30 per cent of behaviour differences, a sample of 290 will suffice. The message is, for random samples, if sample size is small, you are only sure to pick dominant elements in human behaviour. The rich diversity of human action is less likely to be identified; you might get hints but these are unlikely to distinguish between the unique and a rich, minority theme. For in-depth qualitative analyses the reverse problem is faced. Rich diversity is identified but uncertainty abounds over whether this covers the variability that exists. The question is, is what I/we found unique or part of a general trend?

Getting responses

It is a reality of social surveys that people have become less willing to complete questionnaires. While Caunce (1994) might be right that oral history projects find people remarkably willing to participate (even if this should not be presumed), this is in a context in which they are generally asked to relate what they want to say within a broad subject area. In questionnaire surveys the issues raised, and type of response called for, are more restricted. As Goyder (1987: 21) noted, non-response to social surveys has risen since the 1960s as public antipathy has increased. Indicating this sense of disquiet, Buzzard (1990) reports that in parts of Indonesia family planning researchers have made themselves such a nuisance that families post the number and sex of their children, alongside the family planning method they use, on a sign in their front yard. In the 1980s, many undergraduate and school projects were attracted to London's Docklands, owing to the peculiar transformations occurring in the area (Brownill 1990). Here residents were more blunt, with posters, like 'no surveys here', making regular appearances in homes' windows. If a reminder was needed, these posters brought home the message that interviewing and questionnaire completion are an imposition on people's time. Undertaking either in a state of unpreparedness can be unethical, not simply because it wastes the time of informants but because it can lessen the willingness of informants to engage with future research.

Add to this a feeling some people have, that what they have to say is not worth listening to. Palriwala (1991) noted this for her Rajasthan village study. Informants could not understand the value of sampling, nor the need to ask so many questions. They would ask, 'Why not just talk to one or two "knowledgeable" men and women and then relax and gossip with us' (Palriwala 1991: 32). Experience conducting social surveys in Europe and North America reveals similar experiences. Often potential informants cannot believe the information they can impart is helpful. The exclamation that they are ordinary and not worth getting information from, as they do nothing that is interesting or unusual, is common. The benefit of securing a random sample to elucidate what is 'ordinary', the fact we often do not know what 'ordinary' is (if it exists), is often not recognized. 'I am no different from the others around here' is a common sentiment. The Achilles heel of questionnaire surveys in this regard is that they are superficial encounters with informants. Surveys do not hold the key advantage of in-depth qualitative interviews (albeit these can be intimidating to some informants, who foresee having to 'reveal all'). In more in-depth investigations the researcher can be up front in proclaiming the centrality of an individual's history, lifestyle or values to a study. In this research approach the single informant has a major part to play. In social surveys, by contrast, the interviewee is one in a long list, whose importance arises from getting a sense of general trends and distinctions. The general is the priority not the individual, and the potential informant is likely to be sensitized to this by the manner in which he or she is approached and asked to take part in the study. Being asked to participate in a 30-minute survey to identify general trends can be much less appealing than being told you are personally important and a researcher will devote a lot of time to you (albeit some will prefer the former owing to time constraints).

For the social survey, the objective is generally to secure a representative sample of a population. To achieve this, researchers engage in a variety of measures to encourage participation. Where literacy levels are not high, some researchers use local radio announcements or local meetings to introduce a project and encourage participation (Buzzard 1990). More broadly, introductory letters explaining the purpose of a study, followed by phone calls a few days later to arrange meetings, are common (see e.g. Healey and Rawlinson 1993).[5] Of course there are differences between studies in the reaction researchers receive from potential informants. It is generally recognized that the interest people have in a topic increases participation, just as more controversial topics can intensify problems of securing cooperation (G. Walford 1994a). There is also unevenness in people's willingness to engage with questionnaires. Those with more formal education generally feel more comfortable with the questionnaire format, with a result that surveys tend to be biased in responses. If the aim is to secure a representative picture, such biases are problematical.

If response rates to surveys are high, this might appear to be less of a problem. In reality, response rates are often comparatively low. Thus, Morris and Potter (1995) note that their interview response rate of 43.5 per cent was not uncommon for farm surveys. Commenting on mail surveys of rural housing needs, Mullins (1993) likewise reports a response rate of about 45 per cent but notes variation from 25 per cent in large villages to 87 per cent in small villages. There are countervailing forces at work here, for mail questionnaires generally have lower response rates than interview-based surveys, while surveys on behalf of government agencies that suggest a potential policy input receive superior return levels.

5 Extending this idea, Tim Butler (1997: 76) followed his questionnaire survey by circulating a report on its results to participants, inviting them to participate in in-depth interviews. This invitation received a 'very positive response'.

That less than half of initially targeted respondents commonly complete standardized questionnaires is a weakness in this approach. Substituting non-respondents with others often raises the response rate but with the potential to distort survey conclusions: 'In the United States, at least, interviewing only people who are at home during the day produces data that represent women with small children, shut-ins, and the elderly – and little else' (Bernard 1994: 92). For this reason, in national surveys like the US General Social Survey, interviewers generally seek to contact people after 16:00 hours on weekdays and at weekends. If they obtain no response, they ask neighbours when people are usually home, leave notes and call back. With the population at home exhibiting temporal fluxes during a week, survey researchers should make at least two (preferably three) attempts to make contact rather than assuming surveys are one-stop operations.

It might seem that one way around this problem is a mail questionnaire or a telephone survey. Both raise peculiar problems (see e.g. Groves *et al.* 1988). For one thing, both deny the possibility of encouraging respondents to answer questions through developing trust between interviewer and interviewee. As Healey and Rawlinson (1993) note, it is difficult to build a trusting relationship on the phone. As for mail questionnaires, here Goyder (1987: 13) offers a stark reminder of their character:

> *The mailed questionnaire is so mechanised a technique it is sometimes hard to remember that living people receive the survey material through their letter boxes. Because the mailed questionnaire is inexpensive, and because response is often low, the scale of the mailings tend to be large, further abstracting from the human element.*

Set against this, interviewers have very evident impacts on the information gathered.

Summary: the informants

- Potential informants can be difficult to identify, as sample frames are not plentiful, nor particularly accurate.
- The sampling method chosen will depend on its target population, cost considerations and the subject of the survey.
- Investigators commonly focus samples on particular geographical zones, with the character and number of zones impacting on study conclusions.
- An Achilles heel of survey methods is securing high levels of response – this is a serious concern for a method whose main advantage is securing information on general trends.

The information

Most commentators agree that obtaining information in interviews is aided by respondents having confidence in an interviewer. This is one reason why an association with a prestigious or official agency can encourage engagement with a project (even a letter headed from your university department can help). This also helps explain why it is a good idea to contact and inform potential informants about the project before arranging a time for an interview. The interviewee may be more likely to participate if he or she knows the kind of questions you will ask. Adopting this approach enables you to give interviewees more control over the interview. Knocking on front doors with a request for an immediate interview leads to many rejections because people are occupied with other activities (Goyder 1987). If interviews are an imposition on people's time, seeking interviews at times that suit you, rather than them, is an even greater imposition. The same applies for the location of the interview. We would not advocate using noisy locations as venues for interviews as a matter of course. But by arranging an interview beforehand the interviewee can select the location for a meeting. It might be the interviewee is uncomfortable letting a stranger into his or her home, even if this is by pre-arrangement; a neutral venue might be preferred, and might be necessary if an interview is agreed to. As such, interviewers need to be sensitive to asking about a preferred location, even if most people want the interviewer to come to their home or office.

Starting the interview on the right note is important. This is not simply a matter of the order in which questions are asked. It also relates to the tone put across in reminding (or if cold-calling telling) the informant who is conducting the interview, why it is being conducted, why the respondent has been chosen, the importance of the answers, who is funding the work, how long the interview is likely to take, how the information will be used (including confidentiality) and so on. As regards the questions asked, one practice that can make interviewees more comfortable is to give them a copy of the questionnaire so they can follow as questions are asked. This allows time for complex questions to be read by the interviewees and enables them to check ahead that there are no discomforting questions. There are regularly questions in social surveys that some people are unwilling to answer. One example is age, but many questionnaires exclude income questions because administers do not believe they will get a good response. If there is uncertainty over whether informants will answer a question, place the question toward the end of the questionnaire (albeit this does not help with mail or online questionnaires). When the informant has a copy of the questionnaire, the phrasing of the question and the categories into which an answer is placed are important. Presenting the questionnaire so that its structure and logical progression are evident is consequently critical (see Gray and Guppy 1994; Fink and Kosecoff 1998; Gillham 2000, for good discussions on drafting and designing questionnaires).

Despite efforts to encourage respondents to provide frank and open responses, interview results are impacted upon by the interview as a social event. A key factor in the quality of responses is the motivation of the respondent (Cannell and Fowler 1968). Agreeing to complete a questionnaire is not equivalent to putting equal effort into responses (see e.g. Box 5.2). Here interviews can score over self-enumerated

Box 5.2 Uneven responses to mail questionnaire items

These are verbatim extracts from the first two responses from the département of Lot in Buller and Hoggart's survey (1994) of British home-owners in France. When the study was conducted, the absence of research on this topic meant that reasons why British people bought homes, the changes made to properties and the use made of them were little known. Hence, most questions allowed for open-ended responses.

Respondent A

Question 11c: *What were the main reasons for buying [your home] in France?*
Love of France.

Question 12c: *What were the main reasons for buying in this area?*
Knew the area.

Question 13: *What were the most important features of this property and the immediate area that attracted you?*
Value for money. Ability to build a pool. Attractive location.

Respondent B

Question 11c: *What were the main reasons for buying [your home] in France?*
I spoke French quite well, was familiar with its culture, cuisine, etc., yet familiarity had not dulled the edge of excitement for me. Its size appealed – the vast rolling forests, medieval villages, only just becoming acquainted with the media society. It didn't seem like a big step to me: moving to Scotland or Ireland would have been a radical step, a real exploration of the unknown.

Question 12c: *What were the main reasons for buying in this area?*
I wanted to live in an upland rural unspoilt département, with forests, hills, sheep and cattle, rather than arable agriculture. For me there's no mystery in a 'patchwork quilt' panorama of cereals, etc. But here in the Lot, it's wild, romantic walking country. Also I needed to be within 30 km of a prefecture for work and I liked Cahors – it's the meeting of the short black-haired Mediterranean people and the taller fair-skinned Franks and Celts.

Question 13: *What were the most important features of this property and the immediate area that attracted you?*
Property – on the edge of tiny village with south-west aspect, totally private. Great architectural interest – stone built with arches, dormer windows, huge cave area which could be incorporated into house, delightful (wild!) garden with view over valley, bearing vines, figs, nut trees, hazelnuts. So features were peace, beauty and sense of Nature's values.

questionnaires (for example, mail questionnaires), as the interviewer can encourage the interviewee. This is well recognized in the US General Social Survey, for which especially persuasive interviewers (known as converters) are used to encourage those who are reluctant to take part (Davis and Smith 1992). The fact that these converters are 35–45 per cent successful is enough to provoke a scream of 'beware'. This fact is double-edged. If interviewers can gain converts, they can lose them. Assessments of the impact of interviewers bring this out clearly. To contextualize this point, most of the literature emphasizes the need to train interviewers, so they understand the meaning of the questions, appreciate the type of responses that might emerge and know how to mark up responses (see e.g. Buzzard 1990). Apart from encouraging interviewers to travel in twos, so there is mutual support and an ability to talk over problems, it is generally stressed that interviewers need to follow instructions in the training sessions and their performance needs to be monitored.

Training and monitoring are not foolproof means of ensuring even interviewee responses. Interviewers might be equally polite, might ask questions in the same order and the same way, but the impact of underlying values or impressions is difficult to assess. One study that points to this is Leal and Hess (1999). They explored skilled interviewer evaluations of both how informed interviewees were about political issues and of interviewees' intelligence. They found that after taking account of the accuracy of responses (that is, compared to 'correct' answers to the knowledge questions posed), interviewers were more likely to see those of lower socio-economic status as less informed and less intelligent. They also graded black and young respondents more negatively, even if their responses were equivalent to others. How these biases filter into the recording of survey data is difficult to assess. But for McDowell (1992b: 405; see also Sparke 1996) such biases are significant:

> *Conventional research methods in human geography especially those involving interviews have also been criticised on the grounds of gender blindness of those administering the survey or undertaking the interviews. For example, male researchers may privilege male respondents without considering whether the information so obtained is systematically biased.*

Can one presume women researchers are equally prone to prioritize women respondents? We are not aware of research evidence on this, although Robson (1994) hints as such by noting that she found it easier to interview women (for commentary on this, see Lee 1993). But Leal and Hess (1999) suggest such biases exist on a broader front, as Derman (1990) reports for interviewers questioning different ethnic groups in Nigeria. Offering a detailed exposition of this effect, Gow (1990) reports on the 1976 Nepal Fertility Study (NFS) on awareness of family planning practices. He compared NFS data with information gathered by experienced interviewers who were strangers to the three study villages, alongside researcher cross-checking using key informants and unstructured interviews. Consistent differences were found. When asked if they had heard of family planning, 22 per cent said yes in the NFS, 88 per cent to the experienced interviewers, while the researcher recorded a figure of 97 per cent. When people were asked if they had heard of a condom, the figures were 5, 45 and 95 per cent. These startling divergences owe much to the temporal and social context in which information was gathered, so they should not be generalized. Nonetheless, they signify the potential for questionnaire surveys to reach wildly differing conclusions. The information collected is not 'fact' but a particular response to a specific social situation.

Is it the question?

Awareness of the problem of interviewer bias is not new, with research in this topic popular half a century ago (see e.g. Ferber and Wells 1952). But there is more to this problem than interviewer effects. Both interviews and the self-completion of questionnaires are socially contrived situations. Either on paper or in person the researcher is asking interviewees to give up part of themselves to someone they do not know, for a purpose they might little understand, with outcomes they might little comprehend. Some are happy to expound effusively on topics raised but others are cautious. The topic of the questionnaire, its sensitivity and its interest to the informant will bear on its outcomes, and the impact will not be even. Moreover, with a primary consideration being to obtain a broad sense of the population, issues of cost and time loom large in operational procedures. In part they do so of necessity, for the literature regularly warns us that those who agree to respond to questionnaires are less likely to do so if the effort takes much time. Thus, according to Nichols (1991: 42), questioning should be kept to within 45 minutes, as you cannot expect people to give up more than an hour, including the introduction to the study. But if this is the case, then the researcher has a real problem in constructing the questionnaire. If a shotgun approach is used, with questions scattered across a broad range of issues, in case they are relevant, this only leaves space for short answers if a 45-minute deadline is met. Being wedded to 45 minutes might seem too constraining, but it provides a pertinent frame to focus your mind on. It reinforces the need to ask questions concisely. The aim is not elucidating what issues mean precisely for informants but securing responses that are comparable.

Hence, the literature is replete with a litany of edicts on constructing questionnaires. Do not over-complicate questions, avoid double-barrelled questions,[6] do not identify a policy with a public figure,[7] do not use double negatives in questions, mix question order so positive and negative responses say the same thing, do not use leading questions,[8] ask cross-cutting questions as a validity check,[9] do not ask prestige questions,[10] and on and on (see Converse and Presser 1986; Foddy 1993; Gray and Guppy 1994). If you think all this seems fussy, the literature provides remarkable examples of responses being substantively influenced by slight wording variations (see Dijkstra and van der Zouwen 1982). An example

---

6 A double-barrelled question asks about two issues, so you cannot be sure what the respondent replied to – something like 'Do you think Arsenal is the most overrated soccer team in England and Europe?'.

7 If you ask about the government's policy on public transport, responses can embody reactions to the policy with feelings about the government.

8 A leading question invites (dis)agreement with a question. 'Do you like school?' might be less loaded than 'You do like school?' but it is less neutral than 'Do you like school or not?'.

9 Do not just ask if the person was ill last month, but also ask whether he or she took time off work, whether he or she was hospitalized and/or whether he or she needed to take medicine (Foddy 1993: 96).

10 Offering early evidence on this, Hyman (1944) reported on respondents interviewed within a week of redeeming War Bonds, without them knowing the interviewer knew this. In all, 17% denied redeeming bonds, with the chance of denial greater as one moved up the social ladder.

cited by Foddy (1993) came from the work of Butler and Kitzinger on the 1975 UK referendum on the European Community. When respondents were asked, 'Do you accept the government's recommendation that the United Kingdom should come out of the Common Market?', there was a 0.2 per cent difference in the magnitude of pro and anti votes. By contrast, there was an 18.2 per cent disparity when the question was phrased, 'Do you accept the government's recommendation that the United Kingdom should stay in the Common Market?' Offering a further taster, Belson (1981) reports on responses to a 1944 question that asked: 'After the war is over, do you think people will have to work harder, about the same, or not so hard as before?' When respondents were asked how they had answered the question, this revealed that the expression 'as before' was interpreted as before the war started by about half, but as after the war started by the other half. Revealing a similar pattern of confusion, about half took 'people' to mean everybody. One-third read this to mean one social class. This example illustrates a feature that is often not grasped sufficiently in questionnaire design, which is the notable unevenness in meanings that informants attach to words. Thus, Foddy (1993: 4) reports on studies that asked interviewees (what analysts regarded as) the same question, yet where close to 20 per cent provided different responses. Forcefully cautioning on wording, Payne (1951) provides a list of 428 of the 1000 most commonly used words in the English language that he claims raise interpretative problems for (interviews or) questionnaires (Box 5.3).

But the problems of question wording relate not just to words, but to the nature of a question. Even seemingly innocuous questions can elicit 'biased' responses. Thus, Briggs (1986: 97) reports that asking about views on neighbourhood facilities yielded a lower response from Navajo informants than others, not because interest in such facilities was less, but on account of the Navajo belief that it is inappropriate to speculate about the beliefs of others. We might think the chances of coming across value systems that distort responses in this way is low, or, in line with the underlying assumption of so much survey work, that biases in one direction are compensated by those in another direction – that these can be treated as random 'noise' (a large sample size reducing the chance of bias in one direction). However, the direction, magnitude, likelihood and source of such biases are unknown before surveys are undertaken (and most usually after). The assumption that such biases compensate for one another is convenient, but we do not know what impact such 'distortions' really have. Generally, owing to limitations on resources, alongside the vast array of influences that could impact on question interpretation, analysts are not in a position to evaluate possibilities. In this context, it is no surprise that many analysts recommend drawing on well-established questionnaire items that have been rigorously tested and used by others.[11]

11 Foddy (1993: 5) reports on Peterson's evidence on asking people their age. This is a notoriously difficult question. Ways of getting around sensitivity include asking people to place themselves in a broad age brackets (like 20–39, 40–59, etc., years). What Peterson found for similar random samples is that, when people were asked 'What is your age?', 3.2% refused to answer. When the question was phrased 'How old are you?', the refusal rate was 9.7%.

Box 5.3 Problem words for questionnaires

These 428 words were taken by Payne (1951: 151–7) from the 1000 most frequently used in the English language. Payne holds that they all cause problems of interpretation in questionnaires. Those marked * were seen to have multiple meanings.

about*	act	after	all*	always*	America*
American*	and*	any*	anybody*	anyone*	anything*
anyway*	around	art*	as	at	back
bad*	ball	bank	bear	bent	bed
behind	believe*	best*	bill	bit	black
blind	board	body	book	box	break
broke	broken	brush	bum	business*	by
call	came	camp	can	carry	case
catch	centre	change	charge	check	class
clean	clear	close	colour	come	cool
comer	could*	count	country*	course	court
cover	cream	cross	cry	cup	cut
daily*	dance	dark	dead	deal	deep
did	dinner*	direct	do	does	done
down	draw	dress	drew	drive	drop
drove	dry*	each*	easy	even	ever*
every*	everybody*	everything*	eye	face	fair*
fall	fast	fat	father	feel	fell
fellow	felt	few*	field	figure	find
fine	fire	fit	flat	fly	follow
foot	for	form	found	free	fresh
front	full	gather	general	get*	give*
given	go*	goes	gone	good	got
govemment*	green	ground	guard	had	hand
hard	has	have*	hear*	heard*	heart
heat	heavy	held	help	high	hit
hold	home	horse	house	hung	ice
in	iron	it*	its*	job	jump
just*	keep	kept	kill	knew*	know*
laid	land	last	law*	lay	lead
leave	led	less*	let	letter	life
lift	light	like*	line	list	live
lose	lost	lot	love	low	made
make*	march	mark	market	match	matter
mean	meet	met	might*	mind	mine
more*	most*	mother	move	much	name
myself*	near*	new	niece	nobody*	none*
nose	note	nothing*	now*	number	nurse
of	off	old	on	one	only*
open	or*	order	out	outside	over
own*	pack	paid	paper	part	pass

past	pay	people*	pick	piece	place
plain	plane	plant	play	pocket	point
poor*	possible*	post	pound	power	present
print	public*	pull	push	put*	quickly
quite*	race	raise	ran	reach	read*
receive	record	repeat	report	rest	return
ride	right	ring	rise	rock	roll
rose	round	rule	run	rush	salt
sat	save	saw*	school	season	seat
second	see*	seen*	serve	service*	set
settle	several*	shade	shadow	shape	ship
shook	short	shot	should*	show	sick
side	sight	sign*	simple	sing	single
sit	skin	slip	slow	small	smart
smoke	so	soft	some	sometimes*	sound
space	speak	spirit	spoke	sport	spread
spring	square	stage	stand	star	start
state	stay	step	stick	still	stir
stock	stood	stop	straight	stretch	strong
study	stuff	subject	such*	suit	supper*
sure*	table	take*	taken	talk	taste
tear	tell	term	that*	the*	there*
thick	thin	this*	those*	tie	time
to	today*	told	tone	too*	took
top	touch	town*	train	tree	tried
trip*	true	trust	try	turn	under
up	upon	use	voice	walk	warm
wash	watch	water	wave	way	wear
well	went	where*	while	who*	wide
wild	will	wind	with	wood	word
work	you*				

This approach is not foolproof. For one thing, new issues cannot be covered. For another, as the potential array of biases is large, and target populations vary socially and culturally, the problems a particular target poses might not have been covered in the past. Even so, there is much to be said for taking already tested questionnaire items. Researchers do not have to invent new ways of asking questions. They should draw on the experience of others as much as possible. Questionnaire design is difficult, especially for those unfamiliar with the process. Calling on others' experience is important for securing more valid survey results. That said, do not accept questionnaire items just because someone else has used them. As these items might not have been tested effectively, draw on work by well-established survey research units (amongst which British Social Attitudes reports, Eurobarometer and MORI are noteworthy opinion surveys). Drawing on the advice of local researchers or even government officials, as Howard (1994) did in Nicaragua, can also help.

Taking questions off previous surveys is not enough on its own. The wording of a question is just one problematic. Also important is the manner in which answers are requested. It is well established, for example, that if questions are left open for respondents to provide answers, people name less items than if a checklist of possible responses is provided. Converse and Presser (1986) cite the work of Belson and Duncan as a good illustration. They asked informants about their reading the previous day. When an open-ended question was used, 7 per cent of informants mentioned the *Radio Times*. When the same question was posed with a checklist of popular reading material, 38 per cent claimed to have looked at the *Radio Times*. Quite feasibly, a quick glance at this magazine might not have been recalled except for the prompt of the checklist. But fewer mentions for items from open-ended questions occur across a broad range of question types. Add to this, the fact that for many commentators checklists exaggerate, as well as downplaying items that are not on the checklist (the 'other' category at the end). The implication is stark: for many items, questionnaires cannot be relied on to give accurate, absolute values. They might provide an index of the relative importance of items, but absolute values depend on how the question is posed (see Box 5.4 on how this could distort interpretations). Without belabouring the point, we should note that this issue is pertinent even for the category 'do not know/no opinion'. Analysts have shown that the placing of this option, whether in the middle or at the end of a scale ranging from strong agreement to strong disagreement (say strongly agree = 5, strongly disagree = 1), has an effect (if in the middle this category appears to be used to represent 'neutral' views, not don't knows). Even having a 'don't know' category has an effect, since more people use it than say they do not know if no such category exists. Put simply, for many items in questionnaires, people do not have strong views, or they have a degree of uncertainty over 'facts', especially when events are not recent (see e.g. Cherry and Rodgers 1979). In such contexts, differences in response can owe a great deal to the availability of prompts.

The structure of a questionnaire is not neutral in prompting informant answers to questions. The order of questions is important. Again there is an abundance of research on this, so we will not provide an over-weighty commentary. Suffice it to note that anticipating such effects can be difficult. Thus, Schuman (1992) reports on research in Detroit that varied question order in split sample surveys for which there was little difference in response for the great majority of questions. Yet Schuman notes that a 1980 survey that asked whether communist reporters should be allowed into the USA received 75 per cent agreement if the question was posed after informants were asked if US reporters should be allowed into the USSR but achieved only 55 per cent agreement if it was posed before. Context effects exist for some questionnaire items, and these might interact with other factors affecting responses (such as interest in a subject, formal education and so on).

Box 5.4 Differences between open-ended and checklist answers to survey questions

Checklist question	%	Open-ended question	%
Please look at this card and tell me which thing on this list you would prefer in a job?		People look for different things in a job. What would you most prefer in a job?	
1. High income	12.4	1. Pay	11.5
2. No danger of being fired	7.2	2. Secure employment	6.7
3. Working hours are short – lots of free time	3.0	3. Short hours, time for other things	0.9
4. Chances of advancement	17.2	4. Opportunity for promotion	1.8
5. The work is important and gives a feeling of accomplishment	59.1	5. Stimulating work – challenging and gives a sense of accomplishment	21.3
		6. Pleasant – enjoyable work	15.4
		7. Conditions include control of work and physical conditions	14.9
		8. Like job, not codable under 5 or 6	17.0
		9. Responses specific to one job	3.0
		10. More than one codable response	1.4
		11. Other reasons	2.1
6. Don't know, no opinion	1.1	12. Don't know, no opinion	4.1

Source: Foddy (1993: 145; based on work by Schuman and Presser)

Summary: the information

- Questionnaires and survey implementation need to be interrogated vigorously, as the potential for uneven informant response to the same stimulus is high.
- Responses to questionnaires can vary with question wording, question order, the structure of the questionnaire and interviewer characteristics.

Limitations on strengths

In this chapter we have not provided a blow-by-blow account of 'how to construct' or code a questionnaire (for suggestions on coding, see Bourque and Clarke 1992). We are not seeking to present a long list of do and don't stipulations. Many good texts on questionnaire design do this already (e.g. Oppenheim 1986; Foddy 1993; Fink and Kosecoff 1998; Gillham 2000). They should be referred to if a questionnaire survey is contemplated. Our interest is to draw out how different methods offer dissimilar opportunities and constraints for research. Underlying our understanding is the message that all methods are imbued with epistemological biases, as well as dissimilar potentialities. As the title of this chapter implies, larger-scale social surveys offer a relatively superficial encounter with informants. This social situation generally offers informants little opportunity for reflection on responses. It allows little time, or because of its superficial nature perhaps encourages little effort, for respondents to check the accuracy of answers (for example, checking credit card statements to see when you were last in Singapore or phone bills for what you actually pay on overseas calls). On this score, the demands placed on ensuring that informants understand questions in the way intended, as well as replying as accurately as possible, are high. The strength of social surveys arises in their capacity to offer messages about general trends (or distinctions) in society. This main strength is also a key failing. This arises because of the fundamental holes in the fabric of social surveys. For one thing, without high response rates, they are biased reports. More so for some styles of delivery than others, but all survey methods are handicapped by non-response (Goyder 1987). It is not just the percentage of non-responses that is critical but self-selecting biases in non-response. Commonly this method produces returns that sing with a particular social tune. All too often reports on the experiences and values of certain population groups are thin. With bias in completions, interviewer effects and the uneven impact of questionnaire structure and question wording, the key advantage of social surveys in identifying general trends is brought into doubt. Such effects are not equally potent for all surveys, or even all questions. In this regard we should be more suspect of attitudinal questions, which tear responses from their social context, with interviewer bias long ago reported to be more notable for attitudinal questions (Ferber and Wells 1952). With careful planning and execution, alongside sufficient resources, these limitations can be mitigated, but this does not eradicate the common problems of questions being misunderstood or questions not allowing informants to express their views adequately.

We do not want to misdirect readers, for it is unclear how important misunderstandings are. There is a bounty of evidence that problems arise in informant understanding of questions (see e.g. Belson 1981). Yet evidence points to surveys securing accurate views from the citizenry. Examine the results of opinion poll surveys shortly before an election. Bar the technical problems the US presidential elections in 1936 and 1948 identified, there is often a high degree of congruence between survey results and election outcomes (once intention to vote is taken into account). When we add the fact that private companies use survey-based market research to plot corporate strategy, perhaps we should not get too carried away with the shortcomings of the method. If corporations are able to rely on the method, perhaps proof is consuming the pudding. Thus, despite imperfections and biases, the professional polls for the 2000 US presidential election for the most part got it right, even down to predicting the swing towards Al Gore in the final days. However, as mass media network efforts revealed for Florida, doing a rushed job and not taking adequate precautions produce suspect results. Our caution then is that, while surveys can be very powerful methods of data collection, when the aim is to gain insight on general trends, they are not easy vehicles through which to gather valid data. Rather than being a 'soft option', they place demands on investigators that require very serious engagement. In general, researchers do not have the resources nor the expertise of professional organizations. As a result, as researchers we have to be extra cautious, making good use of pilot investigations, in order to design and implement surveys to the best of our abilities.

There is no doubt that geographers find the social survey useful (see e.g. Dixon and Leach 1984; Sheskin 1985). However, comfort with the survey method was higher when positivism held a stronger grip on the discipline. With the rise of new perspectives and new analytical questions (Chapter 1), doubts over the value of the insights they can provide have increased. Questionnaires are capable of yielding large amounts of data on incidences, but are weak in addressing questions such as 'what does this mean?', 'why?', and so on:

> *Survey research will do well enough to find out how many locks people have on their doors, or whether they have installed security lights, and other easily measurable factors. But, with something as complex (and hence unquantifiable) as fear, survey research has not been able to answer the 'what', never mind the 'why', of a given person's, or community's, fear of crime. To use another example that interests us, namely, sexuality, survey research might be able to find out how many sexual partners a person has had in a given period (although actually it might signally fail to do so), but it cannot find out what this means, nor why sexual behaviour of whatever kind is engaged in. (Hollway and Jefferson 2000: 2)*

Surveys can elicit detailed responses, but do not hold your breath expecting them. Interviewees generally agree to complete a questionnaire, not engage in an in-depth interview. Even with a bounty of open-ended questions, you cannot guarantee informants will pass on more than a minimal amount. Ask about why a particular place was selected to buy a second home and some reply 'cheap and accessible to home', while others provide reams of script (see Box 5.2). However, as we discuss in the following chapters, it does not follow that more intensive research methods have ready answers to such questions. We cannot assume participants tell it like it is. The data gleaned in any method are not 'fact' statements but filtered reports, depending on who is telling, who is listening and who might listen.

6
Close encounters: interviews and focus groups

[The] choice of research practices depends upon the questions that are asked, and the questions depend on their context.

(Nelson et al. 1992, 2)

As Chapters 1 and 2 illustrated, there have been numerous attempts to delineate philosophical moments in academic research. These are useful hermeneutic devices for students stepping into the complex arena of research, but they are limiting in four respects: first, they can isolate theory and method, apparently denying the epistemological decisions that infuse every research choice; second, they can create a paradigmatic illusion of discrete historical phases, when research approaches are used simultaneously (Denzin and Lincoln 1998); third, they can essentialize philosophical traditions and deflect from their internal differentiation; and, finally, they can artificially isolate 'phases' (design, implementation and analysis) in research, which negates their interlinkages. The intention here is to reunite theory and method through an acknowledgement of complexity in research acts. Research methods are not owned by a single discipline, nor are they dominated by one conceptual framework. It should not be disciplinary background that inclines researchers to favour methods, but research problems. Human geography's concern with the interactions of people in places makes the range of approaches considered in this chapter relevant for its research questions.

This chapter deals with a range of techniques, including focus groups, in-depth interviews and oral histories. The common strand in these methods is that they are located towards the heightened end of the intensity continuum of research approaches, where intensity refers to level of direct interaction with research participants (Fig. 6.1). Commonalties also arise from these methods generally being used to understand social phenomena in terms of their complexity and/or the meanings people bring to them. In the realm of 'closer encounters', the dimensions drawn out in this chapter are concerned with communication, interaction and mutual understanding between researchers and the researched.

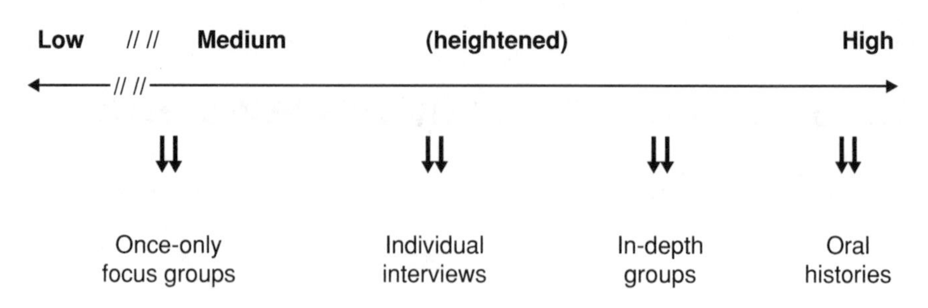

Figure 6.1 *The spectrum of intensity*

The attractions of intensive methods come from 'getting closer' to lived experiences; to exploring beliefs and actions in terms used by those under investigation. The depth of insight that can emerge through intensive research can make the research process exciting. While this heady combination should caution us towards a dispassionate consideration of the capacities of intensive methods, too often it has provoked an essentialization of method:

> *To a person the qualitative researchers are scathing in their denunciation of the survey researcher's failure to attend seriously to the commonplace observation that different respondents often give quite different interpretations to the same question, and that the answer given by a particular respondent to a question in one social situation is often quite different to the answer given by the same respondent in another situation. (Foddy 1993, 15)*

Through intensive analysis, the researcher should derive a more nuanced understanding of the meanings of social acts, as well as a greater appreciation of interacting and contextualized rationalities that impact on behaviour. Yet critics commonly raise issues of researcher bias, contamination, subjectivity, reliability, validity and the generalizability of findings. Many of these criticisms emerge from an epistemology affiliated with notions of detached, objective and value-free research. In these attacks, failure to appreciate the aims and claims of intensive research is common. As Greele (1991: 131) noted, when critics charge that:

> *interviewees are not representative of the population at large or any particular segment of it, they raise a false issue and thereby obscure a much deeper problem. Interviewees are selected not because they represent some abstract statistical norm, but because they typify historical processes.*

An appropriate objective for intensive methods is not to seek representative information (Glaser and Strauss 1967), but 'to gain access to the cultural categories and assumptions according to which one culture construes the world ... In other words, qualitative research does not survey the terrain, it mines it' (McCracken 1988: 17).

To use the somewhat archaic language of critics, this means intensive research principally uses case studies. For us, this does not hold true. As the Ragin and Becker (1992) volume shows, what is a 'case' is open to dispute. A quantitative investigation of 1982–90 socio-economic change in cities in France is still a case study – France is not the world and the study is of one time period. What is an appropriate 'case' for investigation differs (Wieviorka 1992). In medicine it is rare and diagnosable, in social research it is usually the theoretical significance of the phenomenon or process (as argued by Walton 1992), while in history it is impact on subsequent events. All studies are case studies in some regard, irrespective of method. This should encourage us to have an open mind towards dissimilar approaches – much as some cultural geographers have shown towards quantitative material in recent years (e.g. Philip 1998; Philo *et al.* 1998). Certainly, as the Foddy (1993) citation suggests above, turning a blind eye towards the advantages of less intensive approaches is not helpful (Chapter 5). In this chapter, therefore, the reader should not be looking for a reification of intensive methods, nor for messages suggesting that a single research method holds the explanatory key for human geographers. Rather the purpose is to discuss the implications of conducting intensive research and making knowledge claims from associated analysis.

It is unfortunate the literature contains instances in which intensive methods are criticized as 'soft' while extensive methods using statistical strategies are valorized as 'hard' (Cantley 1992; Openshaw 1998). This is misleading. It traces its roots to the rise of positivistic social science, in which qualitative approaches were often delegated to the generation of hypotheses (Dey 1993). Indeed, such was the dominance of quantitative approaches in the heyday of positivism that McCracken (1988) holds that only the pioneering work of Glaser and Strauss (1967) saved intensive, qualitative approaches from going under. This period fostered a defensive posture. Although intensive approaches have risen in favour since, defensiveness is still seen occasionally, as when criticisms of the limitations of quantitative analysis are exaggerated. We do not wish the reader to fall into this trap. As indicated in Chapter 5, achieving rigour, reliability and validity in numerically based questionnaire surveys is not straightforward. Investigators must be assiduous in invoking critical, reflective self-examination in such projects to attain (self and communally) convincing outcomes, much as for analyses that rely on government statistics (Chapter 3) or documentary sources (Chapter 4). The same applies for intensive methods. As Strauss and Corbin (1990: 18) put it: 'The requisite skills for qualitative research are these: to step back and critically analyse situations, to recognize and avoid bias, to obtain valid and reliable data and to think abstractly.'

Far from being 'soft', in the sense of simplistic, idiosyncratic or easy, methods of heightened intensity are 'hard', because they embrace rather than deny the dynamic complexity of society. Moreover, in close embraces with research subjects, only the most myopic can fail to grasp the value-laden nature of research enquiry – the essential 'soft' humanness of the research process. If anything merits associating intensive methods with 'softness', then perhaps this is justified by encouraging researcher humility. To be clear, intensive methods are not appropriate for tabulating the incidence of events but are appropriate for exploring rationalities, implications and meanings. The thrust is not on counting how many homes are demolished under urban renewal schemes, where former residents are relocated to, what are the socio-economic characteristics of those relocated or other scorings. These are important, but we also need to appreciate the subtleties of what home relocations mean. What the loss of a home, and of a neighbourhood, means to people. About how relocation, with loss of neighbours and nearby friends, dislocates social and psychological well-being (Young and Willmott 1957; Fried 1966). When the depth of personal feelings and experiences are engaged with, only the most thick-skinned can avoid being humbled. Moreover, confrontation with rationalities that are different from a researcher's can be powerful provocations for self-critical reflection and awareness of personal biases. Such self-realization can be a catalyst to increasing researcher maturity, which is critical for high-quality research.

A further key to this process is self-awareness of epistemological groundings. Too often the epistemological frameworks that shape research practice go unacknowledged, whether consciously or unconsciously. This chapter is predicated on the significance of the situated positioning of both researcher and researched. Both are visible and creative agents who interact with the institutions of academic convention and inherited methodological traditions. It follows that the research act should be a deliberative and interpretative process that is renegotiated and reflected on before, during and after data 'production'. To reiterate the message of Nelson and associates (1992), methodological approaches not only shape the questions asked, but affect the content of answers and conclusions drawn. From the position of heightened intensity, each research situation demands a unique approach. Methods should not be merely procedural, but also epistemological.

Intensive interviews

According to Ann Oakley (1981: 31), 'Interviewing is rather like marriage: everybody knows what it is, an awful lot of people do it, and yet behind each closed front door there is a world of secrets.' It is easy to concur with Oakley that interviewing is a strangely familiar, yet mystical, social process. There are many forms of research interview and many contingent variables that affect and influence the research act. As noted by Silverman (1997), we live in an 'interview society', where interviews are 'windows on the world'. They are filtering mechanisms used to gain employment or enter educational courses. They improve profit margins or develop new products through market research. They entertain on TV and radio by communicating debate and life histories. Different interview types have dissimilar purposes.[1] They are all special forms of discussion involving interaction between participants. It is this interaction that is of interest here, for it is crucial and problematic.

It is appropriate to emphasize the generic characteristic of intensive research interviews that distinguishes them from other research methods. The intensive interview has a range of possibilities, from the in-depth semi-structured interview (Valentine 1997; A. R. Davies 1999), through the more fluid long interview (Spradley 1979; McCracken 1988), to the narratives of oral histories (Caunce 1994; Perks and Thomson 1998). What is common to all is the level of contact between the researcher and the participant, conceptualized as a 'close encounter' in this chapter. Often the attractiveness of such close encounters lies in their familiarity and thus apparent simplicity. As Brenner and associates (1985: 2) noted: 'if you want to know something about people's activities, the best way of finding out is to ask them' through the 'everyday activity of talk' (Brenner *et al.* 1985: 7). This enables opinions, networks of relationships and ideas to be presented and qualified. In intensive interviews it is possible to search, clarify and probe, effectively to ask why a story was told 'that' way (Riessman 1993: 2). Yet at the heart of interviewing practice is the assumption that people are willing, and able, to comment on their experiences and articulate their feelings and values, thus allowing culture to 'speak itself' through individuals' stories.

Why use this approach?

A suitable topic for intensive interviewing requires in-depth understanding that is best communicated through detailed examples and rich narratives. Not 'how many people are homeless', but 'how do people exist without conventional accommodation structures'. Completed projects in human geography using intensive interviews are too numerous to recount. The range of applications is illustrated in diverse projects like a study of women in corporate banks (McDowell and Court 1994), the migration decisions and experiences of immigrants (Western 1992), the lives of homeless women (Doyle 1999) and new geographies created by genetic engineering (Parry 1998). In essence intensive interviewing is appropriate when research seeks to unravel complicated relationships or slowly evolving events. This

1 Interviews are usually conceptualized along a structured–open-ended continuum. This means interviews are more or less directed by the intervention of the researcher, ranging from an unvarying, fixed order of questions, to a more free-ranging interview schedule, so that issues can be explored in depth. This chapter concentrates on interviews that require an intensified relationship between researcher and researched, falling towards the open-ended extremity of the continuum.

approach is warranted whenever depth is required. Conducted sensitively, intensive interviews can facilitate the explanation of events and experiences in their complexity, including their potential contradictions (Bryman 1988). This can lead to insight far beyond the initial imagination of the researcher (Silverman 1993). But, if the richness this approach can bring is so desirable, why do investigators not go the whole hog and engage in participant observation (see Chapter 7)? A good reason is given by McCracken (1988: 10):

> *if qualitative methods are important, their use in the study of modern societies is not by any means straightforward. The difficulty is that respondents lead hectic, deeply segmented and privacy-centred lives. Even the most willing of them have only limited time and attention to give the investigator. Qualitative methods may have the power to take the investigator into the minds and lives of the respondents, to capture them warts and all. But few respondents are willing to sit for all the hours it takes to complete the portrait.*

This point applies especially to those who have little time to 'spare', whether these be parents with sole responsibility for a lot of offspring, corporate executives or those whose economic needs require significant over-time employment. Add to this time demands on researchers, which often limit research options. Apart from time, there is the matter of appropriateness. For instance, in her study of science Nobel laureates, Zuckerman (1972: 167) found that her subjects would 'resent being encased in the straightjacket of standardized questions'. They were far from acquiescent interviewees, seeking to direct the discussion and add 'almost endless qualifications' to what they said. Zuckerman's efforts to gather comparable information was soon abandoned, as free-flowing discussion revealed the complexity of processes through which work (and interactions with co-workers) occurred. This example points to a need to adapt methods to the questions and participants in a study. Hence, in identifying input supply and output destination for manufacturing plants, more standardized questionnaires work well. If the investigation is about decision-making autonomy in corporate hierarchies, strategies to increase market share or negotiations with trade unions over work-practice flexibility, then more nuanced understanding is probably required (Healey and Rawlinson 1993).

This perhaps provides one reason why oral history in the USA, which characteristically involves long interviews with respondents, has been dominated by projects on political and social elites (see Gittins 1979; Caunce 1994). This is despite the efforts of populist writers like Studs Turkel (e.g. 1967/1993, 1970/1986, 1992), or in Canada Barry Broadfoot (1973, 1976, 1977). In the UK this tendency has been less evident (albeit see Allen 1975, 1983), with oral history projects exposing one reason for in-depth interviews:

> *Oral history often has a rather subversive image in this country, which probably reflects the fact that documentary history already serves elite groups adequately. Oral histories about them will only add to what we know rather than radically alter it ... (Caunce 1994: 10)*

By contrast, documentary sources on the lives of 'ordinary people' are less bountiful (Chapter 4). Often written records were not kept, as with many farm servant contracts up to the 1930s, which tended to be oral. Yet the absence of 'other records' has raised criticism about intensive interviews (see Gittins 1979), for cross-checking sources is often not possible. This is not a sufficient reason to limit the use of this method. To accept that it should is to deny the right to investigate certain questions. Significantly, these are prone to be questions about the least powerful and articulate. In addition, we should recognize the potential for cross-validation using different informants – albeit communal memory issues are at stake here (e.g. Cappelletto 1998). When other data sources are not available, researchers should feel extra responsibility for ensuring they capture the essence of interview accounts. Yet the realities of intensive interviews are that great care is needed in all contexts.

Getting the most from an interview

Many texts present comprehensive (sometimes overly prescriptive) lists of interview procedures (Fink and Kosecoff 1985; Foddy 1993; Ritchie 1995). We will not rehearse their contents. Rather our intention is to highlight key issues that affect interview materials. Not unexpectedly, these start from awareness of what is sought and what is achievable. Of particular importance is recognition that in-depth interviewing is not an easy option. Newby's interviews (1977: 117) with farmworkers and their employers offer a vivid indication, even for a 'structured' procedure:

> *At the beginning each interview left me completely drained ... So much in interviewing depends upon handling the particular nuances of the situation, so that it can be authoritatively defined in a manner conducive to one's intentions. With the farmers, this meant convincing them that I was a serious researcher, with the requisite stage props to provide it: briefcase, printed questionnaire, formal demeanour. Farmers were more difficult to handle than their employees. Some resolutely refused to be mere respondents but insisted on questioning my questions.*

What Newby draws out is the need to be on guard constantly, to weigh words carefully, to control gestures (also Batterbury 1994). He goes on to stress the need to be well prepared. If interviewees sense an interviewer is not knowledgeable about a subject, the chances of gaining in-depth insight are much reduced.[2] Being well prepared *before* an interview is stressed repeatedly in the literature (see e.g. Nichols 1991; Healey and Rawlinson 1993). Going further, interviewers need to be conscious of the sort of information an interview is intended to gather. This message lies behind Plummer's observation (1983: 97) that: 'Designing an interview schedule for an unstructured interview is very largely a matter of designing ideas about the right probe at the right time' (see also McCracken 1988).

If the gains from intensive interviews are from 'opening up' insight on meaning and reasons for action, then the process of conducting the interview is fundamental. The experiences of established investigators make this point:

> *Qualitative methods are most useful and powerful when they are used to discover how the respondent sees the world. This objective of the method makes it essential that testimony be elicited in as unobtrusive, nondirective manner as possible. At crucial moments in the interview, the entire success of the enterprise depends upon drawing out the respondent in precisely the right manner. An error here can prevent the capture of the categories and the logic used by the respondent. It can mean that the project ends up 'capturing' nothing more that the investigator's own logic and categories, so that the remainder of the project takes on a dangerous tautological quality. (McCracken 1988: 21)*

There might appear to be an implication in this that we are advocating positivistic, 'value-free' research. This is not the case. Rather, the message is that mutual disturbance is going to occur. This should be of a form that encourages respondents, as opposed to closing down accounts: 'You are after what they have to say, and any interference by the interviewer prevents this' (Plummer 1983: 96). The message that silence is golden perhaps exaggerates the situation. The message that interviewer interventions should be kept to a workable minimum and be informed does not.

A key message from this relates to the nature of the enterprise conducted. Investigations using intensive methods of enquiry tend to engage in two kinds of analysis. This point needs emphasizing, for the literature too often implies that such approaches are driven by one research aim (namely, constructing meaning). But this approach is also used when investigators want to understand complex or little-

2 For Bouma (1993, 143): 'It is unethical to rush into the field to collect data before doing a literature review, to learn about what has been done before, what is known and what questions remain in the field of study you propose.' Newby (1977) records how he read trade magazines so he was aware of current farming issues. He still found he learnt more through the interview process itself, but the trade magazines did enable him to recognize key issues. Putting more meat on the bones, Johnson and Josyln's discussion (1995) of interviewing (political) elites points to the need to study documents on relevant political events, as well as biographical details on the person being interviewed.

known issues. Here in-depth interviewing might be for hypothesis generation, but it can investigate the interplay of complex decision criteria. The interview is a vehicle for elucidating interacting influences on people's lives – to get at detail. This kind of analysis need not be concerned with layers of meaning. There have been many studies of this sort, with Western's exploration (1992) of Barbadian immigrants in London as a good example (see also McDowell and Court 1994; Parry 1998; Doyle 1999). For the second type of study, it is the subtext of interviewee responses that is of interest. How interviews are conducted and how material is recorded need not vary between these kinds of study. It is in the analysis of material that the difference lies – as with documents (Chapter 4). Yet both benefit from researcher understanding of an interviewee's 'mental set', as this impacts on perceptions and interpretations of events. Put simply, there is value in interviewers being able to empathize with what informants tell them (Dean and Whyte 1970). But the process of establishing empathy has epistemological implications, for it denies the utility of positivist approaches to information gathering (Schutz 1960; Cooper 1967). Moreover, informants offer insight into their lives of uneven depth.[3] As Derman (1990) points out, the process of opening up to an interviewer depends on the confidence the interviewee has in the interviewer and the interview situation. Such confidence is affected by the manner in which interviewees are approached, how happy they are with the aims of the interview, the uses to which information will be put, how at ease they feel with the interviewer and so on. Since most intensive interviews involve people who either were not acquainted before the interview or are only slightly familiar, despite the claims to 'depth of insight', analysts should be aware that they only touch the surface of an interviewee's views.

What will you get?

Following initial extravagant claims about the capacity of quantification to transform geography into a 'science' (Billinge *et al.* 1984), some quantitative researchers questioned the merits of certain statistical investigations. The question phrased was DIDO or GIGO? Are we putting data in and getting data out or are we putting garbage in and getting garbage out (Berry 1971)? Although we have to rephrase the statement somewhat, the sentiment is as valid for qualitative as for quantitative research. Wilson (1992: 180) offers a pointed reminder of his experiences in this regard:

3 Scharfstein (1989: 33) notes that the French anthropologist Marcel Griaule studied the Dogon for 30 years before being initiated into the tribe's deepest secrets. At the other end of the spectrum, those who have completed in-depth interviews with people they do not know will recognize the sense of relief some interviewees feel when a meeting ends and they do not feel they have had to answer 'difficult' questions (or refuse to answer questions).

> *I would argue that as much damage has been done by poor academic quality as by poor ethics in relation to informants. Many researchers have contributed to negative stereotypes, incorrect but fashionable ideas, exacting but unfounded theories and over-simplified notions that have supported (directly or indirectly) ideologies, policies and programmes with negative results.*[4]

Our point is not to stress negative consequences, which (as with positive outcomes) often cannot be determined before work starts (see Madge 1994). The message is to emphasize the need to be self-critical in recognizing the need for adequate preparation. What you get out of a project will not just be determined by what interviewees tell you, but also comes from what you are capable of and what effort you put in.

Implicit within the last few pages is the message that intensive interviews have an artificial character. While this is the case, in the sense that they are deliberately structured, the interactions that result are not sterile or free from multi-scaler power relations. As with many discussions in everyday social interaction, such as purchasing goods in a shop or asking directions to a place, interviews have structure and purpose. The communicative interchange that occurs means interviews are intimately (and unavoidably) connected to norms of social interaction. Intensive interviews cannot be conducted under 'laboratory conditions'. As a result, attention needs to be given to the dynamics that shape an interview, in particular the researcher–researched dynamic.

The intensive interview is a process whereby interviewer and informant jointly 'create knowledge' (Bryman 1988). This occurs through the interaction of linguistic expression (forming, asking and answering questions), through understanding or misunderstanding (the interpretation of meaning and intent) and by way of societal positioning (the placing of research participant as 'subject' and the perception of the 'researcher' by participants). It is critical to be ever aware that intensive interviews do not have direct access to others' experience. There is always a gap between lived experience and communication. Giddens (1987) has termed this the 'double hermeneutic' of social science, where research is visualized as an interpretation of an interpretation of lived reality (see Fig. 1.2). It follows that intensive interviewing will always be selective. Some information will be unseen, some forgotten and some omitted (Ball 1984: 78). As McCracken (1988) outlines, for many elements of our lives we find it difficult to give a full account of what we believe and why we act in a particular way; long ago beliefs became assumptions and actions habits. Compared with standardized questionnaires, intensive interviews at least

4 As Derman (1990: 110) signifies: 'People in the developing world increasingly perceive that those who are studying them and providing projects are doing so for their own reasons, motivations and goals'. Commonly these are not local priorities. More broadly, the less powerful can be cautious over interviews. Gareth Shaw (1982) offers one example, in reporting that compilers of nineteenth-century trade directories found the 'uneducated' suspicious of providing information and short in their replies.

provide the interviewee with time to explore assumptions and habits, even if some aspects remain hidden. Accepting this position does not preclude the usefulness of the approach. It recognizes the boundaries of claims. As Healey and Rawlinson (1993) charge, there is little point asking if informants tell the truth. It is more fruitful to ask what statements reveal about feelings and perceptions, and what inferences can be drawn from the information provided (see also Dean and Whyte 1970).[5] If the research aim is to build an argument that convinces ourselves and others, rigorous analysis is required, with cross-validation within and across interviews, plus the exploration of other data sources to add light on conclusions (Baxter and Eyles 1997).

It also requires that we pose questions about our suitability for a particular research project. There is no point suggesting everybody is equally good as an interviewer, focus group moderator, participant observer or whatever. This is not the case (as with using documents, constructing and implementing questionnaire surveys or utilizing published data). Ask anyone who has experience of conducting intensive research and we will be flabbergasted if they have not experienced uneven interchanges with informants. Some are difficult to empathize with or draw out, such that they give up little information, while with others an immediate bond is struck with fruitful insight (apparently) gleaned from the outset. We find the literature peculiarly quiet on such issues, with a rather deafening silence on uneven researcher skill and temperament towards intensive (or other) research.

There are many personal aspects to such considerations, so we will not dwell on them. But it is worth noting one context – namely, that of language capacities. Consider the comments by Alice Keeler Harris:

> *I, for example, learned how to read and speak Yiddish when I started doing this [oral history], out of the feeling that I was really going to have to speak Yiddish to people whose memories were in Yiddish and listening to them speak in their own language is a very different thing from listening to them speak in English. My experience with Italian women, for example, speaking in English and my experiences with Jewish women speaking in Yiddish have been almost wholly diametrical. I've come to the conclusion that I can't interview Italian immigrant women, older women, in English any more because the language that they use is not one that represents their tradition. (Greele 1991: 78; see also Lewin and Leap 1996)*

As Bernard (1994: 145) reports, even well-established researchers are duped or unable to identify meanings because they do not know 'the language' of the group studied. Some are honest about their experiences, as with Barley's recognition (1983: 57) that his 'rather wobbly control of the language [of the Dowayo in Cameroon] was also a grave

5 Note Foddy's report (1993: 3) on La Pierre's 1930s work. La Pierre toured the USA for six months with a Chinese couple, staying at 66 motels or hotels and eating at 184 restaurants. Six months later he wrote asking if these establishments would serve Chinese people. Of the 50% who responded, 90% said no.

danger' to social relations in the village. Of course, it is feasible to have an interpreter (see e.g. Howard 1994) or research assistants (see e.g. Hanson and Pratt 1995). Both remove researcher from informant in some measure. But for in-depth investigations this weakens the strength of the approach. Rather than building your own interpretation of the interpretation of themselves that interviewees give you access to, you are left to build an interpretation from a gatekeeper's translation of what interviewees reveal of themselves (and, as Barley (1983) notes, the social position of that gatekeeper impacts on the information gathered). Some might hold that the ability to tape meetings gets around this. It does not. By the time you read the tape, 'damage' has been done. If there is an interaction effect from inadequate language capabilities (such that an interpreter is needed), then these affect what interviewees say, not necessarily how accurately it is recorded. If mutual disturbance occurs in interviews, having a third person makes for restraint in the articulation of an interviewee's position. We noted this earlier for work in Third World settings (see e.g. Derman 1990), but it applies as much for interviews with two informants (Valentine 1999). Yet, as Nath (1991) reported when interviewing Asian women in Newcastle-upon-Tyne, it might be that interviews cannot be undertaken without a third person involved. In this case the women required permission from their husbands to cooperate. With uncertainty over whether an interview is be undertaken singly or with another person, researchers have to be prepared for various options – do not assume you can control the environment of the interview.

So, beneath the skinlike veneer of clarity that appears to surround intensive interviews, there lies a melting pot of conditions that influence investigative conclusions. These are not easy to reconcile at times, and many imponderables raise significant ethical questions. If we take the points in the last few paragraphs for reference, there is the difficult question of what is adequate preparation or training for a project. There is the problem of how to be self-critical in evaluating your capacities to investigate an issue. These are not resolved easily. They are issues about which greater understanding often emerges if researchers experience different research methods. But the 'quality' of the research design for a specific project also depends on its epistemological foundations, and upon the researcher's awareness of (and response to) the same. These cannot be measured on any numerical scale. They require self-awareness and honesty in project design.

Summary: intensive interviews

- Attention to the context of interviews is essential to come closer to understanding research participants and their responses.
- The relationship between researcher and researched is a vital structuring factor in a productive interview relationship.
- Interviews are not neutral procedures. They are social interactions replete with power relations.

Focus groups

There are commonalties between the process of conducting focus groups and intensive interviews. Specifically both require direct interaction between research participants and researchers, albeit in different ways, depending on the form adopted (see Fig. 6.1). There are though important differences between focus groups and interviews. Fundamentally, these arise from the incorporation of social interaction in focus group approaches. The pertinence of this consideration is well articulated by Krueger (1994: 11), who noted that: 'Attitudes and perceptions relating to concepts, products, services, or programs are developed in part by interaction with other people. We are the product of our environment and are influenced by people around us.' In this section drawing out the particularities of focus group approaches allows us to address commonalties and differences with interviews. In doing so, we will consider the nature of research possibilities within the focus group genre, drawing out links between epistemological positions and research application.

Why use this approach?

The roots of the group approach are found in two different contexts: psychotherapy and market research. The difference between these is reflected in the nature of groups and the content of discussions. Within geography, group work that draws on psychotherapy has been conducted most comprehensively by Jacquie Burgess and colleagues (1988a, b), who drew on the pioneering work of Foulkes (1948/1983) to investigate the value of open spaces in urban areas. Foulkes was influenced by the Frankfurt School of Sociology, which saw the essence of self as social. He felt that a carefully constructed group discussion could replicate social relations and interactions. This is because communication within the group becomes multidimensional, intra-personal, interpersonal and transpersonal. This means that dialogic interaction can have meanings for an individual, between individuals, and for the group as a whole. As a result, group responses are more than the sum of individual responses; during conversation one set of ideas can set off other thoughts and exchanges. Through in-depth group meetings Burgess and associates promoted a dynamic of trust and support amongst participants, and between participants and researchers. This occurred over a number of meetings, so enabling expressed views and values to be revisited and qualified (on this approach in Third World settings, see Nichols 1991). This approach demands considerable commitment on the part of participants. It also requires a lot of skill to promote positive intra-group dynamics, and a supportive atmosphere, without creating participant dependency on a group that is likely to have a limited lifespan.

In contrast to research derived from psychotherapy, market research groups are not concerned with in-depth responses. Here the aims are speed, responsiveness and economy. In the geographical sphere this approach has been used to examine views of planning developments (Hedges 1985), to explore cultural values of landscapes (Little 1975) and to address the physical, intellectual, social and emotional benefits of participating in wildlife projects (Mostyn 1979). The drawback of this style of focus group is its reduced level of intensity, depth and interaction over time. However, even in a once-only format, groups provide a forum for people to share and test their views with others. This kind of focus group is most advantageous when time is short or policy recommendations are required quickly (J. Burgess 1996).

As with intensive interviews, the form, nature and structure of focus groups are flexible. This is reflected in the diversity of names applied to them (focus groups, group discussions, in-depth groups). While there are 'cookbook' publications on how to conduct focus groups (Krueger 1994, 1997a, b, c, 1998; Krueger and King 1997; Morgan 1997a, b, c; Greenbaum 1998), the burgeoning application of focus groups in geography challenges the prescriptive statements many provide (J. Burgess 1996; J. D. Goss 1996; A. R. Davies 1999). For example, it has typically been claimed that focus groups should be seen as a preliminary, exploratory tool, as in hypothesis formulation (J. D. Goss 1996). However, as Morgan (1988: 11) argues, focus groups have the capacity for broad application, as when 'orienting oneself to a new field; generating hypotheses based on informants' insights; and evaluating different research sites or study populations'. They can also be helpful in presenting research to obtain 'consumer' or 'community' validation. It follows that focus groups have become a popular method for investigating a range of issues, from political affiliation to product advertising. They are increasingly visible in academic and policy studies (Kitzinger 1994; Agar and MacDonald 1995).

In essence the focus group functions through gathering people together to interact about a given topic, the 'focus' of the discussion, in the presence of a moderator (generally the researcher). Group discussions have to be sensitive to the particular needs and situation of the participants, as well as to the aims of the research. As with intensive interviews, the focus group can be a research tool to gain insight into participants' vocabulary on a topic. The essential difference between the intensive interview and the focus group is interaction between participants. Interviews restrict interaction to direct communication between researcher and researched, whereas focus groups provide possibilities for multiple interactions. The researcher hears not only what people say, and how they say it, but how informants interact, whether views are challenged and how people respond to challenges. As a result, focus groups are particularly effective in capturing tacit or experiential knowledge, seeing understandings and feelings as socially situated rather than independent. As Jacquie Burgess and colleagues (1988a: 310) suggest: 'groups enable researchers, and group members, to explore together the embeddedness of environmental experience and values within different cultural contexts.'

While interviews reflect individual views, values and opinions, focus group discussions offer conversation, argument and debate through interaction. Morgan (1997a) rightly states that for every focus group there is uncertainty about what participants would have said on their own. As Stewart and Shamdasani (1990: 36) put it: 'it is not always clear to what extent individuals' behaviour is influenced by others, but … evidence from research indicates that people do, in fact, behave differently in groups than when alone.' Interaction among group members can draw new insight into informants' beliefs and values. This is because it can indicate points of dispute, as well as agreement. A real benefit is if participants gain confidence from being within a group. Group dynamics can instil confidence by revealing shared understandings, as well as encouraging participants to add to messages other participants provide. This can lead to the divulging of more insightful and sensitive information. Groups enable researchers to explore how views are supported in the face of disagreement, with interaction from participants potentially leading to insight that would not have been secured from one-to-one interviews (see e.g. Zeigler *et al.* 1996). However, for this process to work effectively, participants need confidence in the process. This is dependent on a positive group dynamic. Hence, many commentators emphasize the benefits of bringing together individuals with a shared interest in the topic and/or a similar social background (J. Burgess *et al.* 1988; Macnaghten *et al.* 1995).

Despite the potential to offer a different perspective on beliefs and values, commentators warn that the centrality of the 'moderator' in the focus group method is problematical. This is because the moderator is often solely responsible for making decisions about screening participants, recording or taping discussion, making supplementary notes, prompting questions and presenting results. Yet this places the researcher (if the moderator) in a position that is no different from that of an interviewer-cum-researcher. Despite this, commentators regularly note that moderators can direct discussion unduly, as they have a pivotal role in shaping interactions, influencing group dynamics and developing a positive moderator–respondent relationship. For this reason some academics employ professional moderators (see Macnaghten *et al.* 1995; Holbrook and Jackson 1996).

But, compared with interviewing, little research has been conducted on focus group methods. This makes concerns difficult to gauge. For Goss (1996), common sense and the preferred practices of a few researchers have too often been reified into rules about acceptable practice. The real impacts of uneven contributions from participants, the setting of discussions, participant value diversity, interest in an issue and moderator skills are difficult to assess. The small corpus of knowledge that has developed, alongside few researchers having more than passing experience with focus groups, should caution against *a priori* assuming the method is for everyone or for every topic. Group dynamics create particular demands, which remind us of the merit of a pilot investigation to assess suitability for research projects – just as Caunce (1994) highlights for oral history, or Nichols (1991) and Foddy (1993) suggest for questionnaire surveys.

Caution over assuming that focus groups inevitably lead to 'real' insight is also justified by potential problems of securing group participants. Here the literature acknowledges that getting people to take part and deciding on group composition are not straightforward. Strategies for choosing participants are very different from selecting people for questionnaires (as they are for intensive interviews). For focus groups purposive sampling predominates (Valentine 1997), so theoretical (as opposed to statistical) sampling is preferred (Glaser and Strauss 1967). This is when participants are selected according to the research issue, with selection linked to expected capacities to explore a theoretical (or policy) issue. Although not demanding consensual interaction, group discussions require a degree of compatibility amongst members (Morgan 1997a). Who participants are inevitably influences group dynamics. People have to be able to talk to each other about the research topic. This might sound straightforward. A practical difficulty is getting enough participants, or those who can add insight (J. Burgess 1996; Holbrook and Jackson 1996; Longhurst 1996). As Gow (1990: 150) reminds us, power relationships and resource inequities between informants and researcher are not avoided by focus group approaches. Both participation and expressed views can be affected by anticipations of personal benefit, as well as encouraging participants to express views they think moderators want to hear.

Getting the most from a focus group

'Successful' focus group research depends on many factors, most notably the aims and objectives of the project. Beyond this, 'getting the most from a group' requires that attention is devoted to participant recruitment, moderating and facilitating positive group dynamics.

The problem of recruitment is more significant for focus groups than for interviews, as negotiation is required with many participants. The time commitment, if more than once-only focus groups are held, is notable. What strategies maximize efficiency and effectiveness? There are no guaranteed mechanisms. Researchers have to be flexible in accommodating the needs of participants, as well as realistic in setting timetables for conducting meetings (Morgan 1997c). People are different and require different strategies of persuasion. Some might be intimidated by a group discussion. Others might find the group context reassuring. In terms of persuading people to take part, leaders of clubs and organizations or respected members of the community can be useful for making contacts, as with interviews. The note of caution to be heeded is that you might find that gatekeepers screen participants according to their own agendas, or because of what they see as your agenda.[6] One example would be where a head teacher selects the 'best' pupils to take part in a group discussion when you are interested in a range of students. Gatekeepers might also not pass on relevant information to participants, who

6 It is difficult to establish what 'troubles' an organization or group before a study starts. For Gerwirtz and Ozga (1994), access is eased if a researcher appears 'harmless'; Howard (1994), notes that government officials in Nicaragua thought she created less of a political threat as a woman. There is no doubt access is a critical research problem. The example of William Foote Whyte (1943, 1994) gaining access to youth gangs only after befriending 'Doc' now holds a legendary place in social science, but the reverse also happens. Jeffrey Johnson (1990) lists 'gatekeepers' who turned out to be socially marginal, which limits data collection potential. Other researchers express relief that initial contacts did not work out, as they later discovered the association would have damaged access to others (Palriwala 1991; Lakhani 2000), although sometimes realization came too late (C. Bell 1977).

then feel they have participated under false pretences. This has ethical implications for the question of informed consent.

If gatekeepers are not immediately forthcoming, it is common to find participants through lists of people and organizations found in libraries, telephone directories, community centres and the like. These can be a useful springboard, but lists can quickly become out of date, can be vague and are often undifferentiated, so beware. Sometimes when people are difficult to contact through lists or gatekeepers you have to be imaginative and pro-active. In community development terms this is called 'out-reach' work. It essentially means going to where the people you want to talk to spend time, perhaps community centres, sports societies, football clubs, church halls and the like.

On contact it is important to make the right 'first impression'. This requires that you know something about the people you hope will to take part in the research, so dress and act appropriately. As with interviews, you should be prepared, be confident and be aware. You will need to explain what your project is about, why you want to talk to this group of people and what the focus group process entails. Once participation has been agreed, setting a time and place for discussion needs to be negotiated. As people have busy lives, it is important to ensure participants are given reminders in advance of sessions.

The importance of moderating skills for effective focus groups has been touched upon earlier, particularly in creating a positive group dynamic. Krueger (1997b) devotes an entire book to moderating. Here it is sufficient to address some basic issues. Many of the skills required are equally applicable for any intensive researcher, such as being knowledgeable but not intimidating, encouraging articulation but not interrogating, and actively listening and probing. A key difference is the need to facilitate interaction between group members to draw out those who are not participating and to manage those who might otherwise dominate discussions to the detriment of inclusive debate. There are stock phrases for this (see Krueger 1997a, b), but whatever the words the approach has to be diplomatic and polite. The dynamic within an interview or group discussion is shaped by interventions by the researcher. The more structure that is imposed, the more data will focus around a researcher's ideas rather than those of participants. Such an approach requires certainty that the right questions are being asked, as there is less space for spontaneous generation of ideas and new pathways of thought. With focus group discussions, greater structuring by the researcher constrains intra-group communication. By contrast, less structured interaction will be more reflective of participants' interests. This can mean a loss of direction or focus on the topic that interests the researcher. Of course, at many points in group discussions, it is difficult to know if the conversation is leading to an interesting insight or wandering off the point. In reality the degree of structure can alter through a meeting, with times of more focused discussion mixed with free-flowing conversation. To seek some form of balance, Krueger (1994) prescribes a 'funnel design', which implements a moderate degree of structure. Here discussion is moved from broader to narrower topics. It begins with one or two broad, open-ended questions to let participants express their thoughts on a topic. This is followed by three of four central topics that pursue a set of predetermined, broadly defined topics. The session is concluded with sharply focused discussion on narrowly defined issues. These are both pre-designed and emerge in discussion.

Of course, group dynamics are not just about discussion, but are impacted on by group size (and composition). Deciding on how many are enough people to generate discussion, and not having so many that they cannot participate, is a fine balancing act. Advice about group size and composition in existing guides is often didactic. This can hamper imaginative and appropriate application of focus group methods. Experience suggests that smaller groups give greater opportunities for each person to contribute, so these are useful when the topic is complex, sensitive or controversial. When recruitment is difficult, small groups can be forced on researchers, although if a 'group' is only two people, as described by Longhurst (1996), it is questionable whether the process is a 'focus group'.

What will you get?

By their very nature, focus group data will be extensive and detailed. Transcription and analysis will be more time-consuming and labour intensive than for interviews, with more voices to identify and over-talking to unscramble. The epistemological challenges of interpreting focus group data are addressed in more depth later in this chapter, but it is apposite here to highlight practical issues of dealing with 'what you will get'.

The amount of textual material produced from group discussion varies according to the accuracy of transcription. This is not just a matter of deciding whether to transcribe small sections or the entire discussion. It also refers to the extent of annotation given to the written word. That is whether the 'umms and errs', the intonation and emphasis, and non-verbal expressions of everyday conversation are included. A text without any of these 'cues' could be misinterpreted (see Briggs 1986), particularly if it is analysed by another researcher or some time after the discussion. A decision has to be made about the detail that is necessary to supplement the text. In some cases non-verbal cues give useful information about people's reactions to sensitive or emotive issues. In other situations the focus might be on the words used and the way people articulate concerns. A simple coding system to indicate time and meaning can easily be applied to texts during the transcription process with symbols for emphasis (perhaps italics), capitalization for anger and underlining for sarcasm or irony. Even a basic system such as this helps limit misinterpretation.

Owing to the magnitude of data produced, analysis and representation of group products can pose difficult questions. It is a frequent criticism of focus group research that reports on findings rely on a few selections from a large body of information. Representing the richness of long discussions in a few paragraphs can be frustrating for the researcher and the reader. The researcher will want to indicate manifold meanings and interpretations within discussions, but is often constrained by word limits (especially for academic journals). For the reader more material can also make the author's claims more convincing. However, focus group research does run foul of academic traditions for reporting research. Previous reliance on quantitative data or the researcher's 'expert' voice made shorter papers in academic journals more feasible. The view that participant voices should be directly revealed is constrained by word limits that were imposed for these earlier styles of reporting. It is possible that new information technologies could supplement curtailed reports in mainstream dissemination channels. Thus, the Internet could be a means for storing and retrieving large tracts of data, although this raises questions about intellectual property rights, as well as potential confidentiality questions.

Summary: focus groups

- The uniqueness of focus groups lies in the possibility for group interaction amongst participants.
- Group dynamics are crucial in shaping interactions both between participants and between participants and the moderator.
- Conversational interactions produce rich and complex narratives. The detail at which these narratives are recorded affects the level of analysis possible.

Themes and concepts

The centrality of context

While the subjectivity inherent in the production and interpretation of intensive data is criticized by some, a common response is that the (intensive interview and) focus group is so useful 'precisely because of its subjectivity – its rootedness in time, place, and personal experience' (Personal Narratives Group 1989: 263–4). The importance of retaining contextual information in intensive studies is predicated on the view that the site, situation and circumstance of research influence the research process and thus the data generated. Put simply, the abstraction of research from its site of production can lead to a loss of vital linkages, so reducing nuanced understanding of people's experience.

The centrality of context underpins the following sections in this chapter. It provides the source of information for establishing positionality, reflexivity, power, politics, interpretative strategies and, finally, research ethics and responsibility. In this way, context is understood on a number of interrelated levels, from the pre-existing geographical, historical, political, social and economic background in which research takes place, to the flow of conversation within the research act, be that a group discussion or an interview. The retention of context in this multi-scaler sense demands that the intensive researcher revisits the site of data construction throughout the research process, leading to the potential for greater research humility and more sensitive interpretations. Context is significant not only in the construction of a research programme and its implementation, but also in strategies of analysis.

As the communication process is central to focus groups and intensive interviews, these methods require attention to complex, often frustratingly dynamic, rules of conversation and interaction, which are played out in specific contexts in relation to particular historical moments. Interpretative skills are particularly significant in unravelling irony, the use of humour, anecdotes, role-playing and paradox. As Waterton and Wynne (1999: 128) note, researchers are 'forced to try and see what people meant when they said what they said'. Here retention of context adds to validation by presenting a clear picture of research structure, agenda and purpose. Such open presentation facilitates trust with those being researched and those reading results (Baxter and Eyles 1997).

In the remainder of this chapter key themes and concepts will be illustrated using actual research situations. The centrality of context is visible in examples that span different research fields and different stages in the research process. The example used at this juncture usefully illustrates the difference context can make in understanding perceptions of risk and safety in the nuclear industry (Box 6.1). What is interesting about this research is its comparison of two research techniques, one extensive and apparently decontextualized (namely, opinion polls) and one intensive (focus groups). Both surveys were conducted for Cumbria County Council to establish opinions about a planning application by Nirex UK for a nuclear waste repository. The opinion poll results indicated more support for the industry in the area around Sellafield than in the country as a whole, with people indicating that they felt safe living near Sellafield. In addition they stated that the nuclear industry was safe. Less than half said they were apprehensive about safety assurances or expressed concern about health risks posed by Sellafield. Without further embellishment these results appear to give solid affirmation for the nuclear industry. The County Council was concerned these results did not tell the whole story. In response researchers employed focus groups to provide a more reflexive and critical environment in which people could explain their reasoning on risk.

The focus groups revealed a different story on a number of levels. In focus groups people's faith in the nuclear industry's safety was expressed as 'having to believe them'. In other words, people felt dependent on scientific expertise rather than possessing a comfortable sense of assurance (Waterton and Wynne 1999). A desire to trust the industry was combined with anxiety about being forced to trust it. The focus groups revealed other factors influencing responses to opinion polls, such as the sense of community and identity. A feeling existed of wanting to defend associations with Sellafield even if the association had a negative 'stigma'. Fear of unemployment or economic decline through lack of investment in the plant also affected responses. This was not apparent from opinion poll responses, which narrowly focused on perceptions of safety risk (Waterton and Wynne 1999).

The researchers concluded that the contextualized focus groups uncovered a richer sense of community views in contrast to the 'misleadingly simple and impoverished' findings of opinion polls (Waterton and Wynne 1999: 127). While this conclusion seems correct, it is naive. What this example reminds us is that different instruments have uneven capacities to extract particular types of information. Waterton and Wynne

Box 6.1 Research context in the study of risk

Opinion poll results

Question asked	Agree or strongly agree (%)	Disagree or disagree strongly (%)
I think it is safe living near Sellafield *(particularly Seascale residents and those in the industry)*	79	17
I believe the nuclear industry is a safe industry *(particularly Seascale residents and those in the industry)*	71	23
I am apprehensive about the safety assurances given by the nuclear industry *(significantly lower in Seascale)*	41	55
I am concerned about the health risks posed by the activities of the nuclear industry at Sellafield *(significantly lower in Seascale and those in the industry)*	40	58

Extract from Seascale Mothers Focus Group II

CAROLINE. I don't know. The Japanese are talking about reprocessing their own waste … so are the Germans … the French are already doing it … the French are already dumping into the Channel. Nobody goes on about them though do they? You know, I mean Guernsey, they're sounding off about it because they're right beside it.

ANGELA. They're talking about bringing it here though aren't they? Nobody knows though … at least we don't.

CAROLINE. It's just in the wind at the moment.

MODERATOR. Yeah. There are two things you're saying here – on the one hand this means jobs, yeah, therefore good; and on the other hand it might mean that … we might have more stations like Sellafield in other parts of Britain as well?

Source: Waterton and Wynne (1999).

correctly point out how the space of focus groups facilitated deeper understanding of how people respond to polls. Thus, while pollsters might have a specific question in mind when they construct an opinion poll questionnaire (say about risk and nuclear power), the context in which responses are framed can be different from pollsters' intentions. In the Sellafield case the opinion poll focused only on narrow questions about risk or safety. It could have asked why beliefs were held, which would have covered issues that arose in the focus groups. This could have been achieved using open-ended questions. Identifying the significance of points made in group discussions, and then incorporating them into a questionnaire, would have required in-depth elaboration of people's views prior to undertaking the opinion polls. This is the role some commentators favour for focus groups (J. D. Goss 1996). Their critical reservation on focus groups compared with social surveys is the potential for capturing a limited range of views within a community. However, extensiveness is accompanied by less depth of understanding. Extensive and intensive methods add particular insights, which are more useful in specific research contexts. The key is not to laud the merits of one over the other but to be sure methods are applied appropriately.

That said, it is pertinent to recognize that who respondents are, where they live and work, basically their 'context', shape reactions to questions about the nuclear industry, even in the decontextualized environment of an opinion poll. But the apparent finality of reducing statements to statistical frequencies does not reflect the on-going negotiation that occurs in people's social construction of risk. As shown by the extract in Box 6.1, focus groups enable the uncertainty and ambiguity of opinions to be expressed in the flow of conversation. They bring out how negotiation can differ by age, gender or location (Waterton and Wynne 1999: 138). In this way focus groups not only permit a broad range of possible positions to be expressed but can reflect their 'active formation' (Waterton and Wynne 1999: 141). They illustrate a 'telling snapshot of attitudes-in-the-making' (Waterton and Wynne 1999: 142). The focus groups suggested that respondents near Sellafield talk about risk in relation to a social context that frames their perception of risk (they potentially depend on the plant for economic security).

Context shapes perceptions and conceptualizations of nuclear risk holistically in relation to a host of other factors. This is what Waterton and Wynne (1999: 127) call the 'relational construction of beliefs'. They believe people predominantly express themselves in a relational way, as with responding to questionnaire surveys (Briggs 1986). The questionnaire format presumes beliefs can be abstracted from their context yet retain their meaning. Research using intensive methods suggests that complex social issues are constantly negotiated and renegotiated. Focus groups are adept at reflecting this process, given 'the explicit use of the group interaction to produce data and insights that would be less accessible without the interaction found in a group' (D. Morgan 1988: 12).

What emerges from this case is the autonomy of the respondent in 'setting the scene' in which research is interpreted. Researchers cannot assume that the objectives of respondents coincide with their own. Just because issues are not viewed as relevant or are unknown to researchers does not mean participants do not build them into responses. In this setting, intensive methods can facilitate a deeper clarification of aims and interpretations. In this case, focus groups exposed the social context of statements about risk, shedding light on power relationships between the nuclear industry and local communities. However, the importance of recognizing context does not end with methods of data collection. It is central to interpretation and analysis. Merely acknowledging the importance of context does not ensure research is rigorous and transparent, for contextual issues not only influence the researched, but also shape the perspective of researchers, as well as relationships between researchers and research participants.

Summary: the centrality of context

- Site, situation and circumstance affect the research process and the generation of data.
- Context is not only geographical, but historical, political, social and cultural.

Research visibility: recognizing reaction, interaction and positionality

A frequent criticism of intensive techniques is researcher bias. This critique is commonly voiced as researchers shaping interviewee responses or focus group dynamics, such that the data obtained 'are as likely to embody the preconceived ideas of the interviewer as the attitudes of the subject interviewed' (Rice 1931: 561). While the shaping role of the researcher is prevalent in all data collection modes, and at every stage of research, its potential is particularly visible for intensive methods. This is because of the prolonged interaction of researcher and research participants. If an epistemological position is adopted that celebrates the 'humanness' of the research process in its inevitable subjectivity, as with post-structuralism, then this visibility is not in itself problematic, as long as it is openly acknowledged and is not detrimental to the research process. Nevertheless, the researcher's 'face' in the creation of research stories is of prime importance. All researchers are gendered and culturally situated so they approach research with a world view (ontology) that stimulates questions (epistemology) that require examination in specific situations, using particular mechanisms (methods) (Denzin and Lincoln 1998: 23–4). As Mishler (1986: 83) proposed: 'How the interviewer's role is to be taken into account is of course a difficult problem, but it is not solved by making the researcher invisible and inaudible, by painting him or her out of the picture.'

To focus on the researcher alone underplays the significance of the interaction between researcher and researched. The researcher's influence will be partly constructed in relation to others in the research process. To complicate further the reality of research, relationships are not fixed from one interview or group to the next. They shift and mutate depending on context (for example, the timing and location of research and external circumstances beyond the control of those involved), as well as the personalities and positionalities of those involved. As a result the intensity of relations between researcher and researched demands a reflective consideration of individual or group politics. There needs to be a 'recognition of the positionality of the researcher and her/his subjects and the relations of power between them' (McDowell 1992b: 399). 'Whether we like it or not, researchers remain human beings complete with all the usual assembly of feelings, failings and moods … there is no method or technique of doing research other than through the medium of the researcher' (Stanley and Wise 1993: 157). The task of the researcher is to examine critically how these may influence the research process (Fine 1998), and question rigorously interpretations to ensure evidence is not used 'conveniently' to reinforce existing values (Schoenberger 1992).

The process of learning, critical evaluation and understanding – namely, 'reflexivity' – is 'self-critical sympathetic introspection and the self conscious analytical scrutiny of the self as researcher' (England 1994: 82). But it is important to go beyond self-centred examination, which can descend into narcissistic navel gazing, and consider how researchers might be perceived by those they are researching. This is not a simple process. Some characteristics are visible, others can be obscured. Indicating experiential commonality between researcher and researched is complex, as research empathy can be subsumed beneath layers of institutional association that create a perceived distance in the eyes of participants, even if this is not felt or desired by the researcher. Thus, in her work on post-war migrants from the eastern Caribbean, Byron (1993) recounts her experience of being both an insider (as a migrant from Nevis) and an outsider (in terms of age, education and institutional association) from those she worked with. Common locational associations appeared to Byron to be strong enough to overcome possible informant reticence. More broadly, researchers engaged in projects on 'their own' cultures report varied responses. For instance, in Panini's compilation (1991) of studies on women studying their own culture, researchers generally reported greater restrictions than 'foreigners' experienced, as they were expected to conform while being reminded that they were different (albeit, these researchers often experienced freedoms denied to local women). Similarly, in Dyck and associates' cross-cultural study (1995) of women and health care, they felt that being caught between two worlds was overtly problematic (being women and academics studying women; see also Doyle 1999). The complexity of individuals makes it difficult to predict whether having ties of association, in this case being female, will be strong enough to overcome differences, perhaps in age, class or ethnic origin, between researcher and researched. This point is brought out with real clarity in the essays collated by Lewin and Leap (1996). This explores the fieldwork experiences of gay and lesbian researchers. Even a superficial reading of the chapters in this book – which, as the editors make clear, provides an important contribution to the literature, given silence in most fieldwork chronicles about sex and sexuality – reveals that the response of potential informants is unpredictable. In some cases, gay/lesbian relationships and/or culture were frowned on (Williams comparing the USA with Yucatan, Java and Polynesia), in others access was eased (Walters in Yemen), while in some it was less important than other contextual circumstances (e.g. Goodman in North Yorkshire).

How can research progress in the face of such uncertainty? Recognizing the complexity of interpersonal relations need not lead to indecision or paralysis. Through a process of transparent reflexivity, research practice can be enriched and research findings applied to achieve greater insight into people–place relations. This holds as much for research analysis as for research design and implementation. Needless to say, as England (1994: 86) rightly states: 'Reflexivity can make one more aware of asymmetrical or exploitative relationships, but it cannot remove them.' Of course, there might be little commonality of experience between researcher and researched. But even when the researcher feels there is commonality, as with Michael Bell's investigation (1994: 77) of middle-class residents in an English village, this does not mean informants are won over:

> *Possibly villagers suspected me of being a left-liberal university type (which is true), and tailored their remarks accordingly. There may be something to this – although I tried to contribute, when asked, only those aspects of my political views that I thought fit those of the company I was in. In other words, I tried to be polite, without being untruthful. And I imagine the villagers usually tried to be the same with me.*

Going one step further, there may be little empathy with a respondent whose views are essential for a project.[7] How is the 'positioned' researcher to react in such situations? This ethical and political decision involves questions of researcher responsibility addressed below.

While the intensity of in-depth interviews provides food for thought in relation to positionality and reflexivity, focus groups add another layer of complexity through the multiplicity of interactions. Although focus group participants are selected using specific criteria (age, gender, location, experiences and so on), such selection processes cannot predict group dynamics or the influence of a moderator. Focus groups multiply issues of power, positionality and reflexivity, for relationships are not just between researcher and researched, but amongst the group. Even for interviews, positionality is a significant consideration in projects involving particular sections of society, such as women (McDowell 1992b; Herod 1993), ethnic communities (hooks 1991; Byron 1993) or marginalized groups (England 1994; Gleeson, 2000). More recently empirical and theoretical work in geography has engaged with issues of disability and ableism (Chouinard 1997; Kitchin and Wilton, 2000). In the illustrative research summarized in Box 6.2, England recounts her experience studying sexuality in Toronto. England's work is particularly useful as it conceptualizes the research process as dynamic and evolving, in which 'the only inevitability seems to be unreliability and unpredictability' (England 1994: 81). She calls for the research process to be given as much attention as research products through 'true reflexivity', which is an on-going learning experience.

The range and diversity of distinctions between people preclude an unproblematized application of intensive methods. While researchers may wish for a non-hierarchical research relationship, power processes are complex and not necessarily unidirectional. The researcher has to take responsibility for making clear the partiality of knowledge

7 As Lee (1993) points out, this raises problems for researchers who believe the research act should involve 'reciprocity', with multiple meetings to build non-exploitative relations, increase empathy and potentially empower (see e.g. Oakley 1981). As the problem of drop-outs is real with repeat events, those who disengage from a study might have had less empathy with a researcher. How far this occurs or biases results is unknown.

Box 6.2 Visible researchers: reflexivity and positionality

England (1994) explored the construction of lesbian identities in and through space. This was informed by wider studies on marginalized groupings that suggested their apparent ability to exist 'autonomously, or even anonymously more easily in cities than elsewhere' (England 1994: 84). The concepts of identity, territoriality and contested space were central to the construction of the methodological framework. Recognizing public places as 'heterosexed' spaces, England explored spaces constructed by gay communities as 'safe havens'.

England had a number of concerns about research ethics and the politics of research relationships. The first was related to identifying places to be studied, and the potentially negative impact identification could have, through either physical or verbal attacks, on the communities involved. The second related to the 'rightness' of doing this research as a 'straight, white … feminist academic' (England 1994: 4). England employed a research assistant, as she says, 'mainly because of her intellectual abilities, but also because she is a lesbian and, as such, provided me with another means by which I could gain entry into the lesbian world' (England 1994: 4). England did not want to reinforce power relations when studying marginalized groups. She was particularly worried about fetishizing 'the other' (Probyn 1993), of conducting 'voyeuristic research'. After careful thought and struggling with her concerns about inappropriate exploitation, she closed the research project.

In effect the desire to construct a research process that is dialogic, meaning it is the product of both researcher and researched, requires an acceptance on both sides of the implications of the research act. The researcher is 'a visible and integral part of the research setting' (England 1994: 84). Most importantly, England found that it is one thing to be sensitive to power relations in studying marginalized groups, but doing so does not remove them, nor make them easier to justify.

claims based on research. This is particularly so in relation to in-built structural inequalities and personal differences. These inevitably raise issues regarding the power to speak for others through the medium of academic research, alongside problems of essentialism when researching 'others'. The following section engages directly with the politics of placing that can be encountered using intensive methods. Many of the preceding issues of context, positionality and reflexivity are interwoven in these debates.

Summary: research visibilty

- Recognizing the impact of the researcher's 'position' and how that position is perceived by research participants is crucial, but not straightforward.
- Careful reflection about the positioning of researcher and participant is vital for positive research experiences.
- Recognizing positionality through reflexive awareness does not dissolve problems of power relations.

The politics of place and placing

It is clear that issues of context, positionality and reflexivity imply that there is more to intensive methods than talking to people. If you are an intensive framework, then you need to revise the ideas of neutrality and consistency that are embodied in the positioning of positivism. As hooks (1991) eloquently suggests, researchers involved in intensive studies need to think about how they speak 'of' and 'for' others. Research is not a passive exercise, but a social encounter extending beyond the research project. But before we look beyond the termination of a research project, we should recognize rules and norms that are consciously and subconsciously employed by those involved in the process. Here also there are questions of uneven power. But while it is one thing to recognize this, it is another to identify cause and effect. So what should a sensitive researcher do? It is not difficult to agree that researcher and research participants should not be threatened by a research situation, even if discussion may at times be difficult, emotional or challenging. But can communication be heightened to maximize insight without damaging intrusion?

On one level there are practical steps to put an interviewee at ease, such as starting with straightforward questions the respondent is likely to feel comfortable with, asking questions in an interesting and interested manner, listening attentively and reacting to responses appropriately, so support is given without introducing avoidable bias. In other words, the researcher seeks to develop rapport with the interviewee so that mutual trust is encouraged. However, these truisms require deeper consideration. Rapport is a theoretical and ethical issue that is not without contestation. Its adoption raises questions of positionality and reflexivity. Some theoretical positions see rapport problematically, as a loss of neutrality, holding that 'the unchanging researcher [should] make a unilinear journey through a static setting' (Hunt 1984: 285). Those adopting post-structuralist perspectives reject this, with many seeking an emancipatory or empowering approach (Oakley 1981; Herod 1993). But, regardless of epistemological leaning, researchers come to intensive methods with immutable characteristics. As Schoenberger (1992: 218) commented:

> *questions of gender, class, race, nationality, politics, history and experience shape our research and our interpretations of the world, however much we are supposed to deny it. The task then is not to do away with these things, but to know them and to learn from them.*

This returns us to England's consideration (1994) of studying people because they possess different characteristics or experiences from the researcher (Bondi 1990; hooks 1991). Such a distinction – that is, defining people as 'same' or 'other' – is not an easy one to make, as everyone is unique when the minutiae of life's rich tapestry are considered. Equally it is difficult to draw strict boundaries around single dimensions, such as race or gender, given stark differentiation of experiences within such categories. Yet the literature contains sharp messages about the appropriateness of people with particular attributes studying others. Derman (1990) offers one note of caution, which applies for questionnaire surveys and intensive methods, when observing that interviewers in West Africa were differentially engaged and produced uneven results depending on the ethnic group they interviewed. Suttles (1968: 10) offers a somewhat different angle in noting the difficulties of surmounting ethnic boundaries in his multi-ethnic Chicago study. Also signifying uneven relationships with informants, Robson (1994) reports that it was easier to interview Hausa women because she was a woman. Offering a broader frame of reference, Burgess and Wood (1988) claim that to decode advertisements you need to share the cultural values and beliefs of referent systems, while hooks (1991) asserts that those engaged in cultural studies too often do not interrogate their own positions within power structures, so maintaining distance between themselves and those they study. This is a partial list, for many others espouse the benefits of 'commonality' – that is, the sharing of characteristics with research participants, claiming its potential to create empathy and mutual respect (Dyck *et al.* 1995; Valentine 1997). Others document how the same background or similar identity does not guarantee 'easy access' (Byron 1993; Reay 1995). People can hold what some commentators have called 'split affinities' (V. Smith 1991), sitting within and across potentially alienating borders of multiple communities of association. Affinities are not static, but evolve. As Fine (1998: 152) suggests: 'if post-structuralism has taught us anything it is to beware the frozen identities.' In this context it is instructive to note researchers who honestly proclaim the difficulties studying what others see as 'their own group', for the complexity of identities alters the symbols that are critical for acceptability amongst informants. As one example, in Mohammad's investigation (2001) of young, working-class, British Asian Muslim women, a critical element in securing cooperation from informants was not that she was a British Asian Muslim. Rather, on meeting potential informants, she was treated with suspicion as her style of dress marked her as 'Westernized'. Only when she demonstrated fluency in Urdu did suspicion evaporate.

In this context, McCracken (1988: 11–12) recognizes the benefits of commonality but adds a stifling caveat:

> *It is precisely because the qualitative researchers are working in their own culture that they can make the long interview do such powerful work. It is by drawing on their understanding of how they themselves see and experience the world that they can supplement and interpret the data they generate in a long interview. Just as plainly, however, this intimate acquaintance with one's own culture can create as much blindness as insight.*

If, as we know they are, identities are fluid and fragmented, then each individual has not only many potential commonalties with others but also, despite sharing one dimension, many discommonalties. As Doyle (1999) reports, just because she is a woman does not mean she understands the situation of homeless women any better than a man. She found she had little in common with the homeless women she studied except gender. This situation is not uncommon, for in the vast bulk of cases academic researchers do not have strong ties with either elite groups or the most powerless. This does not mean that important insights are not gleaned, or that such groups should only be studied by their own members, although this is the implication of commentaries like that of hooks (1991: 123–33). But if we hold to a rigid demarcation, which, as hooks implies, makes only those with 'the same' attributes able to offer 'legitimate' insight on human action, then we invalidate powerful contributions to geography (or social science). For example, should we reject the insight of Fainstein (1994) and Massey (1995) on corporate decision making because they are women when most corporate executives are men or because they were public-sector not private-sector workers? Should we reject Stuart Hall's insights (1995) on British culture because he is not white? Rejecting such a naive, essentializing viewpoint does not mean hooks (1991) is not correct in drawing attention to real sensitivities (and difficulties in obtaining data) that investigations of 'others' provoke. But on this we should refresh our minds about the end product of intensive research. Hammersley (1995: 53) does this nicely, while highlighting how the insight of 'others' is valuable:

> *There is no doubt that those in different social locations will be able to draw on different experience and different cultural assumptions, and that this diversity can be extremely fruitful for inquiry; both in producing novel theoretical ideas and in generating criticisms of established ideas. However, we must beware of claims that one group or category of people necessarily has more valid insights than another. Since all experience is a construction, it always carries the capacity for error as well as truth.*

In concurrence with Fine (1998), we believe it is essentialist to presume that commonalties are essential for 'valid' research. This leaves the message that there is 'objective knowledge', in a positivist sense. The 'one truth', this view implies, can be tapped by a particular cadre of investigators, while others are necessarily blind to this reality. This is a peculiar stance to find espoused by researchers linked to post-structuralism. As identities are dynamic, the extension of this position is that we are incapable of researching anyone but ourselves. Moreover, if we deny a particular interpretation of social behaviour because the viewer is not black, male, old, British, a long-time city-resident, working class, married, childless, and so on, this suggests that incompetent researchers with these attributes have a greater capacity to inform us than competent, empathizing and experienced researchers who do not have them.[8] This point is similar to one made by Hubbard (1999) in considering the merits of qualitative and quantitative methods, which is that a well-crafted extensive (quantitative) study might be markedly superior to a poorly crafted intensive (qualitative) one (see Box 3.5). Taking research stands for reasons other than research quality, ethics and the appropriateness of the research design can lead to significant research failings. For this reason we emphasize the importance of adequate training for research projects, of critical self-examination of a researcher's suitability for a project, and of the responsibility researchers have towards future investigators. This calls for researchers to ensure they are suited to a project and are clear about the epistemological position that underscores it. Provided we recognize personal attributes are a factor in this equation, this does not deny potential gains from 'different' perspectives on an issue.

These are complex issues that are not easily resolved. But in all but a few circumstances the conventions of academia provide layers of distancing and rituals that separate researcher from researched. This does not mean the researcher can ignore the dangers of imperial translation to which hooks (1991) refers, nor should researchers romanticize marginalized voices in an attempt to rectify past prejudices. Following Buraway (1992) and Fine and Vanderslice (1992), there is a need to braid the critical and contextual struggles surrounding processes of placing. This arises because 'politics suffuses all social science research, from the micropolitics of personal relations in a research project, to issues involving research units, universities and university departments, and ultimately government and its agencies' (Punch 1986: 236).

Summary: the politics of place and placing

- The research act is political and suffused with power relations even when rapport with research participants has been established
- Researchers and research participants have multiple identities, so complete commonalty is virtually impossible. Commonalty on one dimension, perhaps age, gender or life experience, cannot guarantee easy access to desired populations, nor a power-free relationship.

Power relationships in intensive research

The process of placing and positioning is intimately interwoven with power dynamics between researchers and researched. The range of topics to which intensive interviewing is applied in human geography means that structures of power should not be taken as

8 The reader should not read too much into this, as the idea of an 'empathizing' researcher is problematical. Drawing on his own experience, Barley (1983: 56) is pointed on this: 'Much nonsense has been written, by people who should know better, about the anthropologist [or geographer] "being accepted" ... The best one can probably hope for is to be viewed as a harmless idiot who brings certain advantages to his [or her] village. He [or she] is a source of money and creates employment.'

given. For example, it is generally assumed that power lies in the hands of the researcher, but this is not always the case (see Ball 1994). For inexperienced researchers, for example, the process of interviewing can shift the balance of power. Moreover, interviews with the less powerful can be difficult, if informants are less familiar with the interview process (Gerwirtz and Ozga 1994). In such circumstances, an interviewee might not accept an interviewer's definition of the interview situation, which limits the interviewer's 'power' (Lee 1993):

> *Despite their good intentions, the villagers' perceptions and mine regarding data collection did not always coincide. The villagers, especially the women, would get bored and tired of what appeared to them as repetitious questioning. They saw no need for me to interview so many people and the same people repeatedly ... All through the period I was collecting data for my research I was also being thoroughly researched. I was questioned about my life, my hopes and future prospects and my reactions to their life style ... (Palriwala 1991: 32; see also Batterbury 1994)*

In any study of elites, such as professionals or politicians, the interviewees often have a great deal of interview experience (and if they are politicians or corporate executives, they likely to be trained to deflect 'difficult' questions). Just as language, particularly academic jargon, can set up barriers between the researcher and researched, professionalized discourses can obscure meaning for a researcher. Power relationships of this sort can be difficult to deal with, particularly when the interviewee is a gatekeeper for a research programme.[9]

Perhaps more often than in interviews, focus groups transform the researcher–researched balance in favour of participants, through weight of numbers. Indeed, in community development and environmental projects, focus groups have been used to foster empowerment and public participation in decision making (Davies and Gathorne-Hardy 1996). In an academic project that seeks to empower participants, power sharing can be developed by involvement in the definition of research questions, in collaboratively writing up projects or as moderators. Focus groups can be educational and informative, as well as research tools. This approach forms a 'new politics of knowledge' which can disrupt researchers' assumptions and encourage participants to explore issues, identify common problems and suggest potential solutions through

9 This points to an interesting issue that Healey and Rawlinson (1993) raise, for which there is no clear answer. This concerns the order you interview in hierarchical organizations. If senior officials are interviewed first, this can cast a 'halo' over the project, which can smooth information gathering at lower levels of the hierarchy. However, an advantage of starting 'lower' down is that it introduces the researcher to the ethos and professional discourses of an organization. To start at the top without knowing these, or about issues that trouble the organization at the time of interview, could limit insight from senior executives. Perhaps with a great deal of homework before beginning data collection, starting from the top is preferable. But it is a close call, as all research has the potential to throw up the unexpected.

Box 6.3 The politics of placing

'Access – you'll have to get access, that'll be the really difficult part' (Parry 1998: 2147). Parry's study of the elite networks that control the bioprospecting industry began not with the problem of 'access' as first imagined, but with the problem of 'location'. The new international trade in the acquisition and commercialization of genetic material did not adhere to the institutionally focused norms of corporate elites found in other academic texts (Schoenberger 1991; Healey and Rawlinson 1993; McDowell and Court 1994). As Parry (1998: 2148) suggests, it is difficult to pin down 'these new elites which typically exist not as formally constituted institutionally based entities, but rather as increasingly informal, hybridized, spatially fragmented and hence largely "invisible", networks of elite actors'. She found control over trade in genetic material did not follow conventional structures. Agents did not come from one section of the economy, 'nor were they located within any particular set of like, or even geographically proximate, institutions or organizations' (Parry 1998: 2149).

As the elite studied did not exist in a formal 'institutional' sense, an inventive strategy was needed to locate interviewees. After an intensive literature review, patterns of association were discovered in terms of acknowledgements, discursive footnotes and funding. Requesting promotional material and proposing a meeting (rather than an interview), then asking for direction to others who might provide useful information, this elite group effectively defined itself. As Parry (1998: 2155) states, 'by deliberately playing the "new-girl card", I was inviting them to help me penetrate their network by using that network itself'.

Parry provides a detailed discussion of her 'positioning' in accessing the elite network, citing her location at a prestigious academic institution combined with her ethnicity, which gave kudos in the eyes of interviewees: 'the interviewees recognized in my dress, my social background, and my education a fellow "professional", but what I felt was of greater significance to them was the fact that I, like them, had arrived within a particular elite on the basis of merit rather than via any advantage inherited through class' (Parry 1998: 2155). Parry highlights difficulty explaining why interviewees were willing to divulge potentially damaging information. She acknowledges the possibility of co-option, with interviewees attempting to draw her into the network; of the 'confessional' with respondents perhaps eager to unburden themselves of misgivings; and of supplication (McDowell 1992a, b), whereby the female academic is seen as unthreatening to established power structures (see also Howard 1994).

A major preoccupation in the research was to 'engage in a degree of dissemblence about where I was situated politically in relation to the research' (Parry 1998: 2157). Given tension over the commodification of genetic material, interviewees were sensitive to ethical and economic dangers of being 'interviewed'. As Parry (1998: 2157) notes, the 'imbalance in the power dynamic between researcher and researched are perhaps even more acute when interviewing elites, precisely because the researcher is always conscious that the future of the entire research project may well be jeopardized if the relationship between you collapses'.

As interviews are social interactions, there is a view that establishing 'interpersonal relationships is often as dependent on luck, chance and intuition as it is on any determinants which may be formally proscribed' (Parry 1998: 2157). To facilitate information flow, relations can expand beyond dispassionate encounters. In one instance the researcher received confidential documents from a research institute with a letter that read: 'So enjoyed meeting you over the summer and having the opportunity to discuss our mutual interests.' This placed her in a difficult position: 'However differently we may have conceived of them, these "mutual interests" formed the basis of a relationship which eluded categorization, constituted as it was precisely at the intersection of the objective and the subjective, the professional and the personal' (Parry 1998: 2158).

sharing and comparing experiences (Barbour and Kitzinger 1999). Group discussion can provide a perspective that transcends individuals' contexts and transforms personal troubles into public issues. This can foster collective identity and initiate community contact. Such groups can be consciousness raising. They have been used to this end in women's liberation and black power movements. In feminist research this process of educating by participation has been called 'demystification' (Schoenberger 1992), whereby people are exposed to ideas and made aware of structures they can utilize to make themselves heard. Yet power differentials are not dissolved and empowerment cannot be assumed. There is a difference between working 'on' people and working 'with' people.

Feminist researchers like McDowell (1992a) and Schoenberger (1991, 1992) have long been involved in debates about interviewing elites in the workplace, where hierarchies of power are prevalent. More recently the rise of new information technologies and scientific advances has created another layer of complexity. Box 6.3 illustrates the difficulties of dealing with new elite networks in what Parry (1998: 2147) refers to as 'the bioprospecting industry'. Box 6.3 illustrates the politics of negotiating a relationship with interviewees to obtain information. What is interesting in this case is that, while there is a general presumption of care and appropriate representation when researching disenfranchized groups, there is less written about care, confidentiality and representation when interviewees are members of what Parry (1998: 2159) called 'one of the most privileged and influential groups operating in the global community'.

Summary: power relationships in intensive research

- Although researchers generally have power of redirection in the research process, they can be powerless in the face of elites, strongly bonded groups or participants who do not accept their definition of a research situation.
- Access to powerful groups for student researchers can be particularly difficult, as they are often prohibited not by age, experience or the quality of the research, but by assumptions about 'students'.

The placing of data collection

The process of participant recruitment is a key site of negotiation, influenced by positionality, and generated inevitably (and unashamedly) by research agendas. Given an aim to provide depth, detail and nuanced understanding, the selection of participants is not guided by statistical sampling to provide 'representativeness'. Recruitment should be informed by theoretical concerns, but it will inevitably depend on gaining agreement from participants. The proliferation of direct marketing techniques, shifting patterns of community association, diversified work structures and the increasing privatization of people's daily lives, means that making contact can be fraught with problems. Securing agreement to participate is influenced by participants seeing the research as interesting and worthwhile. Textbooks often list processes through which access can be sought (Valentine 1997; Kitchin and Tate, 2000). These processes help identify potential contributors, but none is failsafe. Owing to difficulties of engagement, many researchers find it attractive to study issues close to home or allied to personal interests. Knowledge of groups, networks and issues undoubtedly provides an initial head start and familiarity with interviewees can facilitate access. But there is the danger that participants know the researcher's views and values, perhaps tailoring responses accordingly, so potentially robbing the research of dissenting voices.

While place has been mentioned in this chapter in relation to situation and positionality, there are more concrete issues of place to consider – namely, the where and when of interaction. There may be no choice in terms of locating interviews or discussions if you are interviewing people in their capacity as company executives or in studying the workplace, but with interviews amongst the general population there are more options. Whatever place is selected, it should be safe (for the researcher and the interviewee) and preferably quiet. There are no hard-and-fast rules about 'ideal' locations, but researchers should be aware of potential influences on respondents' contributions. For example, formal places may lead to formal responses. It is widely held that people are more relaxed in familiar territory, such as their 'home ground' (Holbrook and Jackson 1996; A. R. Davies 1999). In terms of timing, the researcher will have a schedule, but so will interviewees. Both have restrictions on time availability. For example, interviewing working/single parents might mean interviewing at 7.30 in the morning or after chores are done late at night. Qualitative interviewing is not a laboratory experiment; it demands adaptation, flexibility and accommodation.

Summary: the placing of data collection

- Selection of research participants is central in shaping data collection, analysis and interpretation.
- Securing participation is dependent on flexible recruitment strategies and accommodating interview practices to meet the needs of those participating.

The art of asking and listening

Discussion so far has indicated that context, positionality and power have a significant effect on researcher–researched relationships during intensive projects. Intimately related to these issues is the construction of questions, researcher reaction to answers and the impact these have on responses. This interaction is conceptualized here as the art of asking and listening. At one level the impact of questioning is well studied in the literature through the issue of question construction (see e.g. Belson 1981; Foddy 1993).

Box 6.4 The art of asking questions

Researchers involved in a study of attitudes to leisure in the countryside decided to test the epistemological limitations of attitude surveys. They began from the position that the countryside and leisure exist as 'highly contested social categories which reflect on-going contemporary public disputes' (Macnaghten 1995: 137). The researcher 'tested' ambivalence in public responses to the survey method by examining the hypothesis that people's expressed opinions to contemporary leisure depend on how the researcher frames the issue. Three different 'voices' were used to frame questions about seven central cultural disputes over the future of the countryside for leisure. Here one issue is drawn out for closer examination, which is whether local authorities should encourage new leisure developments in the countryside.

The three voices selected represented different cultural outlooks on countryside leisure. The first concentrated on the economic role of leisure and tourism for the rural economy (pro-development). The second used the lens of the countryside as a place of peace and beauty that needs to be preserved (pro-quietness). The final voice emphasized the countryside as a place for freedom and escape (pro-freedom). It was envisaged that this last voice would engender attitudes that were more empathetic towards unpaid and extensive access.

The three voices phrased the 'same' question as:

Voice 1. 'Allowing tourists to use their cars in the countryside will help the rural economy and therefore should not be restricted.'

Voice 2. 'The use of cars in the countryside should be restricted to keep it peaceful, safe and unpolluted.' (Scores were inverted to allow comparison with the other questions.)

Voice 3. 'People should be allowed to use their cars to explore the countryside with as few restrictions as possible.'

The survey was carried out by a national poll organization with a representative sample over three consecutive weeks with 1000 respondents to each survey. The results showed a considerable range of responses. On a scale of +2 and –2 of agreement, the pro-development voice received a mean average score of +0.04, the pro-quietness voice –0.65 and the pro-freedom voice –0.12. The conclusion was that the survey shows that people are both in favour of restricting cars in the countryside, but unclear as to whether cars in the countryside should be unrestricted. For Macnaghten, these results confirmed the hypothesis.

For the most part, however, this issue has been approached through the lens of positivism. Once the parameters of neutrality are breached and research relating to 'being' or 'perception' embarked upon, questioning and reacting become a minefield of complexity. The findings of research quoted in this section illustrate this in relation to establishing attitudes towards leisure use in the UK countryside. Here researchers sought to investigate the impact questions have on data collection by framing questions from three differing perspectives to see whether contrasting voices engender different responses (Box 6.4). Although this research raises contentious issues (for example, about the comparability of the questions posed), what is important is that it illustrates that 'the use of particular voices ... is not only selective, but also powerful through its selectivity' (Macnaghten 1995: 148). Effectively, a specific stance in questions can legitimate particular views and exclude other visions. The influence of tone, focus and wording means that researchers need to be sensitive to the context of their research. This relates not simply to participants but also to the temporal context in which questions are asked.

Limited knowledge of research participants prior to interview or group discussion often means spontaneous decisions about how to conduct an interview or focus group have to be made. Spontaneity means there is more freedom to explore avenues of discussion. Ambiguous answers can be clarified as and when they occur. The added flexibility of a 'loose guide approach' offers more scope for addressing issues. Hence, in interviews and focus groups, researchers commonly start with only a list of topics to cover (McCracken 1988). But tension arises as a more 'formal' structure lessens the likelihood of omission. For Rubin and Rubin (1995) the solution is like planning a holiday. You might know where you want to go, but you are not locked into a fixed itinerary. You are happy to explore what you come across along the way. You have lots of guidebooks and maps, but you are not sure which bits will be most useful and are willing to change plans as new things arise. Rather than being locked into one set of questions for all interviews, you adjust questioning so individuals are asked about what is most relevant for them and your project.

A key feature of the intensive approach is the opportunity to probe, cross-check or clarify questions. To probe effectively it is necessary to 'listen beyond', where the researcher retains a critical awareness of what is being said and is ready to explore issues in greater depth. Given the privileging of the participant under the intensive structure, there may be times when the researcher is unfamiliar with words or phrases the interviewee uses. It is important to clarify meanings. Moreover, as participants can interpret questions or prompts in varied ways, the intensive approach demands that the researcher ensures that understanding and meaning are as intended. At the same time, skill is required to draw out the meanings respondents give to phenomena or processes. There are many verbal ways to encourage people to say more about an issue. One important technique, which is often harder to do than it sounds, is not to fill all silences, so people can gather their thoughts before speaking. As indicated by the title of this

Box 6.5 Encouraging participation

Who else has some thoughts about this – maybe something a little different?
What else have people experienced in this area?
You've been discussing several different ideas; what haven't we heard yet?
Remember, we want to hear all your opinions; who has something else?

Steering the conversation

Let me jump in here ... (pause) remember how I said at the beginning that I might have to move the discussion along ...
I think I need to come in and do that thing I mentioned in the beginning, to bring us back to the main question ...

Follow-ups

What are the needs within our community? [then] Which of these is the most important?
What is the major problem in the community? [then] What are the causes?
What's the greatest challenge facing youth today? [then] What should we do about it?
Is violence a growing concern in our schools? [then] What should be done about it?

Source: Kreuger (1998).

section, the process of interviewing or moderating has as much to do with listening as asking. Box 6.5 indicates some common approaches to encourage participation in focus groups.

The centrality of asking and listening and its impact on responses is crucial. This is not only during the construction of data, for it is a shaping factor in analysis and reporting. Yet it is often during the process of interpretation that the visibility of the researcher is obscured and the generation of data abstracted from its context.

Summary: the art of asking and listening

- Flexibility and probing skills are essential to capitalize on the benefits of intensive research methods.
- Silences during an interview can be as significant as questions to prompt respondents to discuss difficult or sensitive matters. Participants need the space to participate.

The challenge of interpretation

Analysis was for a time too often hidden in the backstage processes of intensive work, rarely being dealt with in research reports (Silverman 1993; Bryman 1994; Rubin and Rubin 1995; Baxter and Eyles 1997). This has been criticized by those who favour quantitative methods. As these critics have it, in their investigations analytic frameworks are exposed and acclaimed, as evidence of procedural and technical rigour. While there is some credence to these claims, it should not be forgotten that such reports were often taken to be 'factual'. Yet the data they relied on were often built on undeclared assumptions and biased world views. Statistical accounts might seem easier to justify in terms of precision in 'technical' procedures, but if these procedures embody dubious assumptions and measurement practices, the onus on quantitative researchers to interrogate their interpretative strategies is no less pressing than for qualitative researchers (Chapter 3). A numerical 'solution' does not represent reality any more than a discursive representation. Exposing an analytical framework should add insight rather than obscure data. Here it should be grasped that a key role for research is to look at the world in innovative ways to offer alternative or improved understanding.

One of the key benefits of qualitative approaches is the rich, nuanced data sets they are capable of producing – a 'thick description' of events and experiences (Geertz 1973/1975). Yet the translation of these rich data sets into written records is not without pain and trauma. As Millett (1971: 31) notes, if we strip away the guff and confront the reality of our own capacities, in-depth understanding can promote a deep sense of dissatisfaction:

> *What I have tried to capture here is the character of the English I heard spoken by four women and then recorded on tape. I was struck by the eloquence of what was said, and yet, when I transcribed the words onto paper, the result was at first disappointing. Some of the wit of M's black and southern delivery had disappeared, gone with the tang in her voice ... J's difficulty in speaking of things so painful that she had repressed them for years required that I speak often on her tapes, hoping to give her support, then later, edit myself out. (cited in Riessman 1993: 12)*

As Riessman (1993: 14) herself goes on to state, about the end product of in-depth enquiries: 'In the end, the analyst creates a meta-story about what happened by telling what the interview narratives signify, editing and reshaping what was told and turning it into a hybrid story, a "false document".' There is no 'telling it like it is'; only interpretation, selection and an attempted directing of the reader (see also Geertz 1973/1975). As a result, data interpretation and analysis can be daunting (Feldman 1995).

In the bulk of methods textbooks, the analytic process is broken down into two stages: mechanical and interpretative. We do not like the word 'stage' here, for it leaves a lingering sense that research projects are neatly packaged into blocks of time in which one task is performed, and completed, before moving onto the next. As intensive methods select informants according to their theoretical relevance (Glaser and Strauss 1967), the investigator can expect to adjust informant selection as the project progresses; as complexity is explored and meanings clarified, so theoretical vision develops. This unfolding is a fundamental principle of intensive methods (see also Chapter 7), in which data have to be to analysed as the project goes on. Data collection and analysis should not be disentangled, as analysis prompts new questions, possibly new theoretical concepts, and commonly challenges interim interpretations and explanations (Miles and Huberman 1984; Strauss and Corbin 1990). So, when the literature refers to 'stage', readers should think of 'task' or 'process'. Viewed in this light, the mechanical task relates to organizing and subdividing data into meaningful sections. This can be done manually, through a simple word-processing package, or using a specially designed program for analysis such as Ethnograph or NUD*IST (see e.g. Gahan and Hannibal 1997; Kitchin and Tate, 2000; see also below, Chapter 7). The interpretative process relates to coding information and finding patterns within codes (see Chapter 4). This task groups information into analytically distinct segments and identifies illustrative quotations and special features. The challenge of intensive research is to retain the context, detail and complexity of data while allowing people to gain a deeper understanding of issues (Feldman 1995). For this to be successful, the information conveyed has to contain sufficient material on data generation.

The recording of data can be done through a variety of means. Audio-taping provides an accurate record of the verbal component of an interview. It is easy to do and allows you to concentrate on the discussion taking place. It provides a rich data set, but some people are uncomfortable with it (Western 1992; Byron 1993). One consideration is that recording can encourage informants to be less candid.[10] Offering the situation in reverse, when issues are sensitive informants might want to be sure their words are accurately portrayed, as when one of the authors conducted interviews on rapid housing association growth, which some interviewees taped. The then leader of the Greater London Council (GLC), Ken Livingstone, used to do the same when he gave public speeches during the heyday of the Thatcher governments. This was at a time when the pro-Conservative press was spreading wild stories about the GLC and its leader (Curran and Seaton 1988). In this context, videos provide added visual clues, as they offer insight on body language, facial expression and so on. However, people are likely to be even more self-conscious about video recordings than audio taping, and a video makes confidentiality difficult (Byron 1993). Again we are not in a position to offer firm advice. To accept the extreme view that taping (or videoing) is essential leaves the researcher up the creek without a paddle if a potential informant refuses to be taped. Clearly, if precise words are central for research, taping must be a high priority. However, as methods textbooks suggest, note taking can be appropriate if the researcher feels there is sensitivity amongst informants (owing either to the topic discussed or to lack of familiarity with interviewing). As researchers we have used both devices, depending

10 Geoffrey Walford (1994a) offers good advice here. It may make you look stupid checking the tape recorder before the interview, and it might enhance the interviewee's awareness of being taped, but you look even more stupid if the tape has not worked at the end of the interview. If the informant feels at ease with the interviewer, a tape commonly does not 'intrude'. As Plummer (1983: 93) states: 'On balance, provided one has mastered the technicalities of tape recording … a tape-recording is probably the most satisfactory since it is relatively unobtrusive, and allows the material to be directly transcribed into a manageable form.'

upon the context in which interviews are conducted. Our own preference is for tape recording, but we have preferred note taking on many occasions. If this option is selected, our experience is that notes should be taken for the purpose of recall. Attempting a verbatim record through note taking distracts from the interview process (McCracken 1988). Taking notes that help you recall themes and key 'facts' can usually be achieved without causing distraction. However, once the interview is completed, researchers should seek a quiet place as soon as possible to write up the interview in as much detail as possible. Even some days after the event, our experience is that points are recalled that were not initially written down. How much is forgotten overall is perhaps slight, in terms of main themes. But we have found it difficult to justify using 'factual' information that was not recorded in the interview or immediately after. As a large number of studies have shown, the human memory is very inferior to a tape recorder in storing data accurately (e.g. P. G. Gray 1955; Cherry and Rodgers 1979). That said, as Greele (1991) points out, the 'problem' of memory depends on the questions asked, and at times there is no alternative.

Of course, if tapes are used there is the slippery question of transcribing. Tapes, of whatever form, are invariably more difficult to transcribe for group discussions, because of conversational interaction (or 'turn taking') and over-talking in group dialogue. This can be made more difficult by unevenness in the clarity with which people speak, by varying accents and by soft tones. The level of detail in transcription also affects the time transcription takes. If all pauses and intonations are marked, it is generally assumed that the process will take up to 12 hours for one hour of conversation (Humphries 1984). However, full transcription might not be necessary if tapes are retained for checking and cross-checking (accompanying extracts with a location on the tape counter). Indeed, much of what is transcribed from a tape is not useful. In this context, Riessman (1993) is among those who hold that not all the material needs to be transcribed, provided what is transcribed is selected on a theoretical basis. Strauss and Corbin (1990) make the same point, although noting that more should be transcribed at the start of an investigation, when the categories respondents use to describe social phenomenon are less well known. As time progresses, improved understanding of theory–data links should reduce the need to transcribe all. What should drive decisions about transcribing is consideration of how to maintain the richness of the data and, for focus groups, insights from participant interactions, while synthesizing information. This is especially the case given challenges to the view that attitudes can be isolated from the conversational context in which they are formed (Kitzinger 1994; Agar and MacDonald 1995). Whatever the precise purpose of analysis, if partial transcription is adopted, original tapes should be kept for validification procedures and checking conclusions.

Conversation analysis

The importance of examining a complete discussion is especially evident in conversation analysis. This approach explores the structural organization of 'talk', including processes of negotiation and turn taking that inform attitude formation (Myers and Macnaghten 1999; Waterton and Wynne 1999). Advocates of this approach maintain that it is important to start by analysing group interaction rather than individual statements, striking a balance between the picture provided by the group and individual 'voices' within it. This vision draws on the assumption that 'what is said is not the way it is accidentally ... forms of words are not rough and ready make-dos, but are designed in their detail to be sensitive to their sequential context and to their role in interaction' (Potter 1996: 58; see also Sacks 1984: 1992):

> *Language has often been seen as a carrier of meanings or ideas such that, on receipt of an utterance, the messy stuff of particular phrasing, intonation and so on in which the meaning was packaged can be stripped off to leave the elegant good within. Survey research is often predicated on this kind of notion of communication where participants' untidy conversational 'responses' to questions are filtered and coded into a set of clear cut categories and positions ... Conversation analysis ... undermines any such distinction between the meaning and the utterance. Rather than the 'detail' of delivery, intonation and so on being a sort of fuzzy aura that can be stripped off, conversation analysts have tried to show that these specifics are there precisely because they serve the action that is being performed.*
>
> *Conversation analysts have argued that talk-in-interaction (as they prefer to call language use) is very far from being messy. In fact it is incredibly orderly; and the principal ambition of conversation analysis is to reveal and account for this orderliness. (Potter 1996: 58)*

In analysing whole conversations, points of disagreement within group consensus, or the construction of consensus, are drawn out. This involves drawing together discussions on similar themes and examining how they relate to variation between individuals, as well as between groups.

As proponents of conversational analysis, Myers and Macnaghten (1999) set out a number of premisses that underpin their position. The first is the conceptualization that social and individual identities are negotiated in discourse. This follows from Foulkes's idea (1948/1983) that the self is social, so focus group analysis needs to look at how the group establishes and plays out roles. Conversation is seen as a social contract set in a cultural context of accepted norms and values. Second, as conversations between participants are organized from moment to moment, sequential construction should be maintained when information is analysed. This position is intimately related to the premiss that, as talk is sequenced, a single sentence needs to be considered in relation to what comes before and after it (Myers and Macnaghten 1999; Waterton and Wynne 1999). The focus group process means that 'people say what they say', not in an abstract sphere of opinion but at a particular moment. From this perspective, as Asbury (1995: 418) notes: 'focus groups are not oral surveys: that is, participants' comments should not be tallied, counted, or otherwise taken out of context in which the comments originated.'

Links can be drawn between this approach and broader notions of communication in the work of Habermas (1984), Giddens (1991) and Patsy Healey (1997). Here the idea is that thought is argumentative and people's views are based on a process of opposition and support through evaluation of different sides of an argument. As a result, expressed attitudes are not static entities but dynamic flows captured in the moment of research. As Potter (1996: 66–7) asserts:

> *One of the great virtues of conversation analysis is that it has tried to convert theoretical or philosophical issues of fact and description into questions that can be addressed analytically through studies of records of interaction. It leads us to look at the conversational sequences in which descriptions are used, the kinds of activities that descriptions are part of, and how they are modified or contested in the course of interaction.*

From this perspective, the strength of focus groups is their ability to allow flows of conversation and knowledge construction within research processes. If this benefit is neglected, the potential of the method may be impoverished. Myers and Macnaghten (1999) illustrate this through work on conceptions of nuclear risk in the north-west of England. Box 6.6 contains two isolated one-sentence responses taken from separate focus groups. With no further information there appears to be concurrence between views on scientific assessments of environmental risk. Through contextual analysis this position seems bankrupt. One reason is that there is no background information about speakers or the conversational context in which extracts are embedded. In fact the first response in Box 6.6 emanates from a group of unemployed middle-aged men in a city (Preston) and the second from retired people in a small Lancashire town (Thornton). It would be tempting to discern from this that people from different backgrounds and locations hold the same attitude, sharing scepticism towards scientific authority. Yet, as Myers and Macnaghten (1999) show, getting to these statements reflects different attitudes and positions.

An examination of the contextualized extract reveals a particular pattern of interaction. Myers and Macnaghten (1999: 176–7) recount the process, with the moderator starting by trying to clarify a previous point made by respondent Ben and asking him to make a choice. Ben does not give a strict answer but interrupts anticipating the moderator's line of questioning. The moderator responds by repeating Ben's response as a question back to the group. This is a common moderator technique. After returning to the original question, the moderator provides a continuer, encouraging further explanation. It is only at this stage that Ben states his opinion, which is the line used on its own. What this shows is not just that Ben holds a particular position, but that he had to be pressed to state it. The idea that everyone has to make a choice about whom they believe emerges only when he is challenged. It emerges through a series of turn-taking conversational links. This could also happen in an in-depth interview. It should

Box 6.6 The challenge of interpretation

Extract I – B: So you've got to take a choice then of what you think

Extract II – J: So it's really … you've got to take your choice who you believe

Extract I

MODERATOR. so do you, do you think when, when one professor will say one thing, another professor will say the other thing=
BEN. =well, who do you believe?
MODERATOR. well – who would *you* believe
BEN. well you got two totally opposed different things
MODERATOR. Yeah
BEN. so you've got to take a choice then of what *you* think
MODERATOR. yeah, yeah

Extract II

MODERATOR. ok so so / people in ( )
JULIE. / yeah but they don't do they [several at once]
JOHN. cuz you get scientists who will have different views
TOM. / yeah
JOHN. /on the same *sub*ject
TOM. /exactly
JOHN. so it's really you've got to take
TOM. /yeah
JOHN. /your choice who you believe
MODERATOR. well yeah but I mean I mean, given that these things might be a very real problem – who should be *tel*ling us.

Transcription conventions

you stressed word
. pause
/ onset of overlap
= = turns following without any gap
( ) inaudible section
[ ] comments added to transcription

Source: Myers and Macnaghten (1999).

be noted that not all respondents are as 'reticent' as Ben, but if they are a 'non-response' might be recorded in a standardized questionnaire survey.

In the second example, the utterance emerges as part of a lively interchange after one man states that church officials should give the unbiased truth on environmental risks: 'you get somebody in a powerful position makes an opinion.' There is some competition to hold the floor in this discussion, but the final speaker continually gets support to take the floor from other participants (yeah, exactly). This is not necessarily a sign of agreement with what is being said. Rather it might indicate there is agreement that the person should continue talking. Here the statement emerges from a shared position, called a 'commonplace' (Myers and Macnaghten 1999: 178). It is primarily a collaborative production between two participants. The point is that the scepticism they hold is not a set position, as might be implied through survey style quotations, which portray it as explicit and accepted. The general theme of scepticism in these discussions emerges in different ways in relation to other people and different attributions of authority.

Segmentation of focus group transcripts – that is, using one-sentence responses, as shown at the beginning of Box 6.6 – loses the flow and formation of ideas and opinions. While extracts inevitably have to be extricated from the text, the contexts from which they are drawn need to be analysed and interpreted. Otherwise the impact of the moderator and the interaction of members of the group, in guiding and steering conversation, can be invisible. Decontextualizing information compromises conversation analysis. Content analysis tells us how frequently people used words (see Chapter 4). Conversational analysis tells us how people talked about an issue. Attitudes in this way are not 'objects' (static things) located 'out there' in 'subjects' (participants), but are utterances in specific situations. A conversational strategy requires that large segments of conversation be used in presented data so the context of remarks is retained. As Riessman (1993: 4) explained: 'Precisely because they are essential meaning-making structures, narratives must be preserved, not fractured by investigators, who must respect respondents' ways of constructing meaning and analyse how it is accomplished.' Examining sections can highlight how conversation has ebbed and flowed, as well as how consensus is formed or broken down through conversation. However, a systematic counting of themes and issues at an initial stage can identify common threads in large volumes of text and prevent impressionistic assumptions. Having a broad awareness of proportions can usefully set the scene for more sensitive and detailed analysis.

One emphasis in conversation analysis is on reflexivity. Researchers are encouraged to consider conversations in relation to the action or event under discussion, as well as in relation to what conversationalists are doing at the time. The latter especially applies when the analyst is exploring conversations outside a formal research setting, as happens if this mode of analysis is used to examine ethnographic data (see Chapter 7). In these broad terms, critical questions to ask include:

- What is the context?
- How is the description occasioned?
- What actions are the conversations part of?

Offering more detailed guidance on preparing transcripts and undertaking conversation analysis, see Potter (1996: 57–66), Heritage (1997) or Silverman (1998).

Summary: the challenge of interpretation

- The richness and depth of intensive data sets are major benefits. This richness needs to be maintained in analysis and interpretation.

Research ethics and responsibility

As with other research methods, questions of ethics and responsibility are important considerations for intensive research. While every project has associated 'ethical issues', it is particularly important with intensive methods where the legitimacy of the work stands on the transparency of the 'stories' told. Ethical issues are relevant to research design, implementation and presentation. This involves ethical and moral responsibility to (a) the participants, (b) the general public, and (c) your own beliefs. As with so many aspects of intensive research, there are generally no right and wrong answers to ethical dilemmas. Explicit, reasoned justification for decisions and actions is paramount.[11]

Many methods books list ethical dilemmas that can occur during the research process (Eyles and Smith 1988; Rubin and Rubin 1995; Kitchin and Tate, 2000). Some provide models for good ethical practice (see Denzin and Lincoln 1998). Issues of consent, deception, confidentiality and trust frequently appear. Initially there is the question of informed consent (Byron 1993). This is particularly relevant in 'natural' focus groups, where groups exist prior to the research (Holbrook and Jackson 1996) and when group pressure can override individual consent, or dominant gatekeepers work behind the scenes to coerce members to participate. Moderators have to ensure there is individual agreement to participate and that participants understand what is involved in participation. Unlike one-on-one interviews, focus group participants cannot be given an absolute guarantee that confidences shared in the group will be respected by others. This problematizes confidentiality. The content of interaction can also lead to the voicing of opinions that upset others. Another problematic interaction is the potential for participants to exchange 'mis-information', which the moderator knows to be false (Barbour and Kitzinger 1999). The moderator has to consider whether to step in and correct the statement, and if so when intervention is most appropriate. This is particularly important when talking about issues that are often considered to be matters for

11 As Lee (1993) points out, some are scornful of ethical debates. They see deception as part of everyday life, with research no different from other social action. For others, whether benefits outweigh potential damage is an important consideration (but assessing this before is problematic and damage could be long term). We are not comfortable with these standpoints. But, as Madge (1994) asks, who defines what is ethical? A key element of research is to 'scratch below the surface'. If researchers analyse, and so do not simply report what people say, then they report 'beyond' informants. Even if confidentiality is maintained, ethical questions can arise. Hence, irrespective of the 'accuracy' of the insights, Vidich and Bensman's *Small town in mass society* (1968) led to an outcry in the community investigated. Their 'sin' was to reveal that the community was not 'classless', as many residents claimed, but deeply penetrated by social distinctions. Are such revelations not part of good research?

'experts'. There is the view that, if the moderator does not step in to correct information known to be false, this implicitly legitimizes incorrect, potentially damaging, statements. One example would be if a group discussion on HIV suggested that the virus could be passed by casual contact. In such cases it is usually seen to be the responsibility of the moderator to provide accurate information. Another difficult example is when group discussions are on an issue that is inherently contested or uncertain, perhaps something like global warming, where there is no necessary agreement on the 'facts'.

These issues are intimately related to power relationships and the context in which discussions occur. Ethical problems, particularly in fieldwork, are not clear-cut. As Arlene Daniels (1983) has suggested, they may not be readily or even finally resolvable. However, while there may not be one 'right' answer, there are many questions that demand attention. In actual research situations ethical stances often merge with one another; they are not 'separate' as some academic typologies imply. Tightly argued models of ethical positions rarely map onto research experience when the researcher accepts conditions of situated knowledge, processes of societal mystification and the power of personality. While the 'ideal' of research conduct may be mutual respect, non-coercion, non-manipulation and support of democratic values and institutions, researchers are part of a society in which decisions about means and ends are constantly renegotiated. As the literature makes clear, these issues are particularly pointed when (relatively) wealthy Westerners undertake investigations in low-income countries (e.g. Derman 1990; Sidaway 1992; Howard 1994).

There has been much work into the transformative social process of constructing data using interviews. Yet there is a lack of systematic analysis of the ethics and motivations that infuse focus group methods. Some ethical issues are common to other forms of qualitative research (Winchester 1996), but others differ. Thus, dangers in researching the 'other' as a single group, as for the process of interviewing women (Cotterill 1992) or the powerful (Schoenberger 1991; Parry 1998), are complicated by focus groups involving many 'others', as with target groups including the elderly, young people, female and male groups, professionals and the unemployed (Macnaghten 1995; A. R. Davies 1999). As a result 'expanding the range of purposes for focus groups will also widen the range of ethical issues that are confronted' (D. Morgan 1988: 238).

Ethics and responsibility do not only come into play during data collection. With the increasing use of audio-visual aids by geographers to record and *analyse* intensive research, more reflection needs to be given to privacy. The work of Byron (1993) on Caribbean migration, which is summarized in Box 6.7, discusses some of these ethical issues. What this shows is that audio-visual material complicates the researcher–researched relationship, both directly in terms of access and indirectly in terms of how respondents are perceived by wider audiences.

If inequality in power generally favours the researcher (Winchester 1996), moderators will commonly 'retain the power of redirection' (Reay 1995: 212) within the layering of power relations, which often flows from the researcher (Dyck *et al.* 1995). This can be

Box 6.7 Ethics and responsibility

During the fieldwork construction of oral histories, Byron (1993) initially employed a camcorder and tape recorder to improve the portrayal of her work on post-war Caribbean migration. However, she later reflected carefully on the implications the recording tactics had for her respondents and for the audiences by whom the recordings were viewed.

Byron found that using a tape recorder initially weakened the relationship between her and the respondents, because people either were intimidated by its presence or tended to put on a 'performance', 'clearly embroidering both their language and the content of their accounts' (Byron 1993: 382). In addition, there was concern amongst respondents about their lack of control over material once taped. A camcorder provoked even more intense reactions. Only two out of 40 households were filmed in this way. In these cases the video session was discussed in detail prior to filming. The use of film was a key part of this discussion. It was established that the video would be used for the researcher's thesis, to illustrate migration from the Caribbean to student audiences and to show to Caribbean migrants in Leicester. As Byron (1993: 383) states: 'once information is obtained, the researcher is faced with the perennial problem of protecting the interests of the subjects.' This is particularly important when the reaction of audiences to which tapes are shown is unknown.

In this case the researcher used the videos in courses on oral histories and social geography. While students were not formally prepared for the videos, the courses provided contextual information that supplemented their positionalities as viewers. A number of students commented on the 'typicality' of the participant employment, as bus drivers and manual workers, and how that accorded with their preconceptions about that section of UK society. Yet the same students failed to acknowledge the different career paths of the interviewees' children, who were all in professional occupations. Byron (1993: 384) describes her feelings following the student viewings, given their responses: 'In a way, I felt that by screening the film I had contributed to reinforcing their negative stereotypes of the black population, in the sense that the film was used selectively to strengthen their vague and inflexible preconceptions of this group.' The researcher was dissatisfied with the outcome of showing the film. Byron (1993: 385) concluded that 'the reaction of external audiences will always be an unpredictable factor … Protection of the subjects may involve making compromises which limit the use of techniques and the dissemination of research results.'

manifest in setting boundaries relating to the research topic, in the selection of participants, in analysis and interpretation, in the initiation of the research act and in the presentation of findings. However, there are ways in which the flow of power can be reversed. There are times when the researcher can be distinctly powerless. As already noted, people cannot be forced to participate in a study; and it is difficult to rationalize the sense of rejection and vulnerability that is felt when access to vital groups is denied. There are times when it may be necessary to employ invasion tactics and a commodification of the research act through the payment of expenses, but this raises ethical questions about pursuing reluctant participants, exploiting power hierarchies and negotiating access by befriending gatekeepers (Punch 1986). Pile (1991) suggests this could be resolved by the construction of a research alliance where, in spite of power relations, an understanding of what is taking place is formed. This has been criticized as overly optimistic (McDowell 1992a), even if it might be possible in some circumstances to achieve this partially through open negotiation of research aims.

By way of a summary, it would be an unusual research experience if the researcher were not 'up to the waist in a morass of personal ties, intimate experiences, and lofty and base sentiments as our sense of decency, vanity or outrage is tried' (A. K. Daniels 1983: 213). This is particularly true for research involving people with different world views and political priorities who make assumptions about shared politics and beliefs (Smart 1984). This can put the researcher in an uncomfortable position, as when remarks made on this assumed consensual basis are sexist, racist or otherwise offensive (see e.g. Doyle 1999). As a researcher you can, in some instances, offer contradicting views, but this does not help when the respondents' behaviour and comments are personally directed. Intensive research is a site of ethical negotiation that needs to be continually revisited.

Summary: research ethics and responsibility

- Intensive researchers have responsibility to research participants, the general public and other researchers.
- This responsibility goes beyond the research act and is important during analysis, interpretation and dissemination of research findings.
- Issues of confidentiality are complicated in focus groups, as there are multiple participants.

Concluding remarks

Intensive research does not offer a unified set of principles developed by a homogenous group of researchers (Denzin and Lincoln 1994). Any application of intensive methods will be infused with tensions and contradictions. To reiterate a theme developed by Ely and associates (1996: 66), it seems that conducting rigorous and sensitive intensive methods:

> is like walking a tightrope without a net while juggling sharp swords. It is far from easy and there is a tremendous amount of pressure and responsibility ... There should be a balance between designed questions, ad-libbing, and not leading a respondent down the 'expected' paths to knowledge. All this while maintaining a calm, relaxed and confident exterior and keeping the respondent from feeling threatened or coerced.

Yet intensive methods have many advantages. We stress they do not work for everybody or for every topic. For example, in focus groups some people are too shy to interact and some topics are too sensitive to address carefully or without offence unless there is a therapeutic purpose, such as gathering people together to share traumatic experiences (which raises the question of whether you are the right person to conduct such meetings). Undertaking intensive interviews is not straightforward. It is full of trials, tribulations and heartaches. Perhaps a fool or someone happy to produce low-quality work might claim it is easy, but those who have engaged with the process seriously are far more reticent. Self-critical reflection should be a constant companion before, during and after data collection. As a device, intensive analysis enables detailed exploration of individual standpoints, with participants having more control in taking discussion into realms that are important to them. This option is less secure in focus groups, as participants need to persuade others to tag along. The strengths and weaknesses of these methods are intertwined.

This does not mean anything goes; in fact it means researchers have to think harder about why they employ one method rather than the other (or even another). They must be prepared to defend and justify their decision. Owing to concerns about 'representativeness', especially for policy-related questions, intensive methods are commonly justified more carefully than standardized questionnaires. This is a positive situation, as it requires researchers to engage critically with the political, theoretical and practical issues associated with their research. In essence, intensive research is a human construction that is conducted in a social context. The recognition of researcher positionality and constantly renegotiated researcher–researched dynamics means intensive research products are framed by discourses and ideologies. These epistemological skeletons are not necessarily explicit or even internally consistent. They are influenced by site, situation and context. In this way there is an inevitably political subtext to intensive research practice. What this means in terms of how method is embedded in epistemology is that the ties are not one to one. In-depth analyses can be used in research grounded in a positivist framework, as often occurred when in-depth interviews were used to generate hypotheses. They can be grounded in a realist framework, in which generating abstractions about the structures that frame action are critical. In the 1990s, in-depth studies were often closely associated with post-structuralism. This is a real association, in the sense that many employed such methods from within this framework. But it is not a fixed association. If we take Geertz's vision (1973/1975) of researchers producing representations rather than 'reality', then such representations can be explored through a variety of research approaches, not just face-to-face, in-depth ones. Each epistemological stance might have dissimilar expectations, in terms of what an in-depth research method is likely to produce, in the nature of the 'evidence' it yields, but they all have a place for this analytical form. The key for researchers is to draw out explicitly the linkages they are exploiting (or hypothesizing).

This point has parallels with the conspiracy of silence that has obscured backstage processes of fieldwork in the past (W. F. Whyte 1984). Fortunately these are being broken down through the work of theoretically engaged researchers who maintain a self-critical edge in their research. Methods, philosophy, theory and data are intimately interlinked and form part of the whole research process. As Sharon Macdonald (1997: 173) has articulated: 'By writing, we inevitably write ourselves into a particular and politicized context. The challenge is to disrupt easy positioning and to highlight the semantics and politics of representational practices themselves.' These ideas are just as pertinent for the materials examined in Chapter 7, where we explore research styles in which the researcher takes a more involved role in the everyday operations of the population under investigation.

7
Part of life: research as lived experience

> *An important element in making the method of ethnography less mysterious and more accessible to both practitioners and outsiders (e.g. funding agencies) is a more open and explicit discussion of how we, as ethnographers, come to know what we know.*
>
> (J. C. Johnson 1990: 15)

As part of the growing awareness that research processes do not merely describe social life but construct a particular 'reality', there has been considerable interest in the relationship between researchers and their research. In this chapter we discuss two methods that focus on this relationship – namely, participant observation and, one of its relations, Action Research. Participant observation differs from the detached observer studying a situation, because the observer is a player in the scene investigated. So far little participant observation research has been undertaken in geography, but this method has become of increasing interest. Taking a more active part in the research process, and producing effective research, requires real skill, indeed a number of important skills. In this chapter we discuss the merits of participant observation and explore the limitations of the approach.

Participant observation can be *overt*, as occurs when the observer makes known what he or she is doing, or *covert*, when the observer does not tell subjects they are being studied (see e.g. D. Rose 1987). The type of participant observation that is focused on in this chapter is *ethnography*. In this method the researcher tries to understand the world through the eyes of the participants in a social situation (organization, community and so on). To undertake this type of project, different approaches can be used, including *autoethnography*, which is an ethnography of ourselves, and *cyberethnography*, which is participant observation on the Internet, in chat rooms and so on. These approaches differ from Action Research, for which a key feature is that the researcher (and sometimes participants) seeks to alter a situation, by taking an active role. This method is action, rather than user, oriented.

The research methods discussed in this book so far have involved the researcher keeping a certain distance from the everyday lives of informants. In participant observation the researcher has personal involvement in the everyday. But the reduction in distance between researcher and researched creates challenges, alongside potentialities:

> *A methodology that offers little in the way of prescription to its practitioners and has no formula for judging the accuracy of its results is vulnerable to criticism from methodologies such as surveys, experiments and questionnaires that come equipped with a full armoury of evaluative techniques. In the face of these critiques the popularity of qualitative methodologies, including ethnography, is based on their strong appeal as ways of addressing the richness and complexity of social life. The emphasis on holism in ethnography gives it a persuasive attraction in dealing with complex and multi-faceted concepts like culture, as compared with the more reductive quantitative techniques. Ethnography is appealing for its depth of description and its lack of reliance on a priori hypotheses. It offers the promise of getting closer to understanding the ways in which people interpret the world and organize their lives. By contrast quantitative studies are deemed thin representations of isolated concepts imposed on the study by the researcher. (Hine 2000: 42).*

The reader might note slippage in the wordage used in the chapter so far. This slippage shifts from references to participant observation on to ethnography. The seeming collapsing of these terms should not be taken at face value, even though most commentators see the core element in ethnography as participant observation. As an outcome of this close association, reference to autoethnography or cyberethnography does not jar against images of the containers into which different research methods are placed. Referring to auto-participant-observation, by contrast, brings forth the kind of chill down the back that sharp nails inspire as they claw their way down a blackboard. Yet we need to be a little careful in equating the two. This is seen if we look at the roots of the word 'ethnography'. This is taken from the Greek word *ethno*, meaning people, and from *graphein*, meaning depict. As such, ethnography is in many ways as old as human curiosity. It reflects an approach to understanding behaviour that goes beyond observing while being a participant.[1] As Hammersley and Atkinson (1995: 1) remind us, ethnography is:

> *a particular method or set of methods which in its most characteristic form involves the ethnographer participating, overtly or covertly, in people's daily lives for an extended period of time, watching what happens, listening to what is said, asking questions – in fact, collecting whatever data are available to throw light on the issues that are the focus of the research.*

The emphasis on multiple methods creates a distinction between participant observation and ethnography. This is brought out in Hammersley and Atkinson's *Ethnography* (1995), in which participant observation is one of a multitude of methods reviewed under the

1 But note Jeffrey Johnson's observation (1990) that ethnography is an ambiguous term, as it is not clear whether it refers to an end product (detailed studies) or a process (method). We use it in the latter sense.

heading ethnography (see also Atkinson *et al.* 2001). In a sense, all the methods in this book can be involved in ethnography, with the caveat that participant observation should be a core feature. Critical for this chapter is grasping how the character and contribution of participant observation can deepen understanding of human societies. In seeking to provide that understanding, readers might find we are somewhat loose in the manner we glide between using 'ethnography' and 'participant observation'. The primary reason is that quality research using participant observation inevitably draws on other research methods, not simply in preparing for a period of participation but also in providing triangulation to strengthen validity claims (Chapter 2).

The tradition of participant observation research

The roots of contemporary research using participant observation are found in three veins: anthropology,[2] the Chicago School of Sociology and humanist geography. For a long period of time, participant observation was the main research method in anthropology. Undertaking studies of predominantly non-Western societies, anthropologists held that researchers had to immerse themselves in the culture or society they were studying (e.g. Barley 1983). The model was of the anthropologist going 'native'; living, eating and sleeping with the investigated population. Famous early monographs that used this method were Malinowski's *Argonauts of the Western Pacific* (1922) and *Coral gardens and their magic* (1935/1978) and Evans-Pritchard's studies of the Azande (1937) and the Nuer (1940). These studies were based on the researcher spending long periods of time living with subjects. According to Kuper (1983), the 'Malinowskian revolution'[3] united the fieldworker and the theorist, for the fieldworker travelled, looked and reported as well as analysed and interpreted. Travel often still plays an important part in distinguishing ethnographic approaches, even if many more investigations are of 'home' locations (e.g. D. Rose 1987; Rapport 1993).[4] Yet, as van Maanen (1988: 3) noted: 'Whether or not the field worker ever really does "get away" in a conceptual sense is becoming increasingly problematic.'

Participant observation was long sacrosanct in anthropology. Yet the power relations that underscored one portion of humanity selecting, valuing and collecting the 'pure' products and artefacts of others were not deconstructed, criticized nor transformed until relatively recently (D. Rose 1990: 38). Traditional approaches to anthropology were characterized by critical distance and emotional detachment from research outputs, the

---

2 Franz Boas's American 'salvage anthropology', with its emphasis on culture-history and the collection of texts and text-analogues, was a great influence on Carl Sauer and the Berkeley School of Landscape Morphology (cultural geography). This later impacted on the 'new' cultural geography (Skeels 1993). Boas's analysis of historical texts, folklore and oral history, alongside interviews and observations (sometimes with film), can be seen in the ethnographic practices of many cultural geographers today.

3 Rob Burgess (1984) notes that Malinowski first advocated direct observation using participant observation. Van Maanen (1988: 10) holds that *Argonauts of the Western Pacific* is the Genesis of the fieldworker's Bible.

4 Although early ethnographies did involve some studies of 'home' communities in the UK (e.g. A. D. Rees 1950; W. M. Williams 1956; Littlejohn 1963) and the USA (e.g. Park 1925; Lynd and Lynd 1929/1956; W. F. Whyte 1943).

research process, and those researched. Barley (1988: 9) offers a somewhat exaggerated account, but does bring out that investigators commonly failed to appreciate bias in their own 'authoritative' accounts:

> *Traditionally, anthropologists have written about other peoples in the form of academic monographs. The authors of these somewhat sere and austere volumes are omniscient and Olympian in their vision. Not only do they have a faculty of shrewd cultural insight superior to that of the 'natives' themselves, but they never make mistakes and they are never deceived by themselves or others. On the maps of alien culture they offer there are no dead ends. They have no emotional existence. They are never excited or depressed. Above all, they never like or dislike the people they are studying.*

Early accounts commonly gave the impression of 'telling it like it is'. Overtones of positivism are allied to a view of the anthropologist (or more broadly the ethnographer) as a skilled, trained observer who produces objective reports. This form of ethnographic enquiry and writing has been much criticized.[5]

The other two roots of ethnography, the Chicago School and humanist geography, were premised on an interactionist approach. These drew on a humanist theory of knowledge, experience and reality developed in the late nineteenth- and early twentieth-century writings of Pierce, Dewey and James. The pragmatic philosophy that underpins this perspective is concerned with the social construction of meaning; that is how meaning is constructed in the practicalities of everyday life (see S. Smith 1984). Pragmatists view knowledge construction as fallible. The challenge for pragmatists is to correct, as much as possible, this fallibility. In these terms, the research process is a form of knowledge construction. As such it is a fallible process. From this perspective knowledge is a trial-and-error process in which individuals make choices based on their attitudes and beliefs as they search for 'truth' (see Jackson and Smith 1984: 71–9; S. J. Smith 1984). Both in the urban sociology of the 1920s and 1930s Chicago School and in behavioural and humanist geography in the 1960s and 1970s, pragmatism played a significant part. For pragmatists, understanding is rooted in experience; as such it has to be inferred from situation and behaviour. Research attention thereby focuses on the interaction of individuals (society) rather than individuals as such.[6] By exploring the lives of people, analysts discover beliefs and attitudes about society (and everyday lives).

---

5 As Bell and Newby (1971) note, there has long been recognition of investigator influence on research conclusions, as commentaries comparing work by Redfield (1930/1973) and Lewis (1951) on Tepoztlan, Mexico, portray. But the absence of a strong epistemological and/or overtly theoretical base in early ethnography was a major flaw. Awareness of such shortcomings changed the nature of ethnographic studies. As van Maanen (1988: 5) notes: 'Exotic-mongering ethnographies of a remote but romantic wind-rustling-through-the-palm-trees kind are … out of favor these disenchanted days, replaced, by and large, with more focused, technical, cold and puzzle-solving varieties.'

6 This perspective is different from later humanist approaches, where the subject/individual is the focus.

For Susan Smith (1981) the Chicago School marked the inception of a genuinely 'humanistic methodology'. The Chicago School used ethnographic methods to undertake first-hand studies of city life. One of its founders, Park (1925: 15), appreciated the merits of the patient methods of field observation undertaken by anthropologists. Park and the Chicago School undertook numerous ethnographic studies of social relations and urban structure from Chicago's Gold Coast to the city's slums. Park studied the middle classes, ghetto-dwellers and hobos. Park's work reveals the pragmatist's view of history, where change is seen to be inevitable. Thus he viewed race relations as progressive and irreversible (Jackson 1984: 171): 'Customs regulations, immigration restrictions and racial barriers may slacken the tempo of the movement; may perhaps halt it altogether for a time; but cannot change its direction; cannot at any rate, reverse it' (Park 1950: 150). Park's perspective proved highly influential, with Chicago attracting high-calibre graduate students who undertook what for the time was highly original empirical research. As an illustration, Nels Anderson, one of Park's students, undertook a landmark participant observation study. In order to write *The hobo* (1923), Anderson lived with hobos for 12 months and collected over 60 life histories. This work revealed how understanding the language hobos used was an important and strategic part of interpreting their lives. Insights such as this came in abundance from the Chicago School, which introduced a series of approaches and beliefs that still influences contemporary ethnographic research (Box 7.1).

Box 7.1 Chicago School legacies for ethnographic research

Geographical research in the early twenty-first century utilizes ethnographic approaches in which the following legacies from the Chicago School can be found:

- the situational approach;
- the use of the personal document (letters, diaries, oral histories);
- the case study method;
- content analysis;
- the belief in the socially constructed nature of phenomena;
- the belief that people can make decisions about their own lives (they are not all predetermined);
- the idea that the observer (researcher) only gains partial insight, the whole truth is never revealed to a single observer;
- the importance of language.

Second-generation Chicago sociologists, such as Suttles, have also been influential. His *The social order of the slum* (1968) revealed the complex interplay of social order, ethnicity and territoriality. This was expanded in his *The social construction of communities* (1972), in which he sought to reveal how people build collective representations with communicative value. The reader just has to look at Kay Anderson and Fay Gale's text (1999) on the 'new' cultural geography to see how collective representations have a communicative value. This understanding has been employed in classic geographical studies that employed participant observation and other methods to construct local ethnographies. Thus Ley showed how Philadelphia street gangs used graffiti to claim territory, with graffiti more prevalent and aggressive (communicative) at boundaries between gang turfs (Ley 1974; Ley and Cybriwsky 1974). As Jackson (1984: 179) rightly states: 'Discovering the social and moral order in such communities requires unusual patience, sensitivity and skill. For undertaking this task, the ethnographic tradition of the Chicago School provides a rich heritage and a fertile source of inspiration.' This emphasis on patience is illustrated by third-generation Chicago sociologists, such as Elijah Anderson (1978), who spent three years as a participant observer studying the interactions and exchanges that produced social order at Jelly's bar and liquor store in Chicago's South Side. The depth of human insight that can be derived from such investigations offers a key advantage for ethnographic approaches.

Such work was admired 'from a distance' by human geographers, but they were slow to adopt the approach. Even when humanistic geography sought to draw closer lines of understanding with ethnographic approaches, this was a cause of disappointment, Thus, in criticizing Ley and Samuels (1978), Prince (1980) argued that the work tended to chart spaces 'populated by identical automata and sociological marionettes', whereas Prince wanted the new humanism to study 'passionate, thoughtful, active men and women' (Prince 1980: 295); he sought deeper exploration of the agency of ordinary men and women. Such criticism was not restricted to humanistic geography. By the early 1970s so-called behavioural geography was also placing greater weight on human agency, but influential research under this umbrella carried a sustained ethos of positivism (see e.g. Cox and Golledge 1969). The universalist model of science that underpinned much behavioural work was rejected by those who engaged with humanist geography. Perhaps, as Prince hinted, this work was not well developed, but early humanist geographical studies did react against limited interest in the complexities of social and cultural processes. Leading from this, humanist geographers incorporated ethnographic methods into their research repertoire (see Ley 1988).

Box 7.2 Four types of 'participation' and 'observation'

Junker (1960) outlines four combinations of 'participant' and 'observer' for empirical investigations. These all involve participation and observation, but only participant as observer and complete participant are generally accepted to come under the umbrella of the participant observation method. The combinations Junker identified are:

- *The complete participant.* This is often associated with covert research. Some commentators see in this strategy the dual risks of 'going native' and of data collection being restricted owing to expectations associated with the role the researcher plays. In addition, for a full participant, writing up conclusions can create pressures not to reveal events or practices that cast the group or organization in a light others might see negatively. After all, as complete participant, the writer often continues to be involved in a group or organization after a write-up is in the public realm. Even publication after retirement (or death) can cause anguish for ongoing participants, as seen for noteworthy early publications in this genre – for example, Richard Crossman's Cabinet diaries (1975). Studies by geographers who were ongoing participants in processes written about include Blowers (1980) and Gilg and Kelly (1997).

- *The complete observer.* Here observation is prioritized, with social interaction avoided (for example, recording data from behind a one-way mirror). This method does not equate with participant observation, quite the reverse. Its strengths (and weaknesses) and its varied forms are outlined in Webb and associates (1966/2000), who provide a sweeping overview of non-reactive measures in social research. As one illustration, to assess the popularity of radio stations while avoiding the status biases of social surveys, record the stations that car radios are set to when they are brought in for servicing. Such information can aid participant observation by checking potential biases from uneven engagement with social groups. In its own right this strategy raises difficulties in 'interpreting' the significance of data, which can lead to problems of ethnocentrism (albeit even complete participation cannot remove this potentiality, as the actions of others still have to be interpreted).
- *The participant as observer.* This emphasizes participation and social interaction over observing. This is what commentators most commonly mean by 'participant observation'. Priority is given to participating in the everyday life of a group, event or organization, with data recording commonly undertaken 'under cover' (that is, not in front of other participants). In some cases the recording is deliberately in secret, for fear discovery will threaten the ability to continue participating in group activities or even result in violence towards the researcher (see e.g. Friedland and Nelkin 1971; Wilkinson 1981). In other cases, while informants might be aware of the researcher's purpose, recording observations when away from informants is seen to mitigate the intrusive influence of the researcher on 'normal' social activities (C. Bell 1977; M. Evans 1988; W. F. Whyte 1994).
- *The observer as participant.* Here the balance favours observation over participation. This can be associated with the utilization of a variety of data collection mechanisms (for example, in-depth interviews or focus groups). However, it is not unusual to find references in the literature to this kind of work as 'participant observation', when it is really observation with some participation. This sense is conveyed in Bernard's observation (1994: 140) that: 'At the extreme low end it is possible to do useful participant observation in just a few days.' With information collection commonly taking substantial allocations of time if the researcher adopts a participant-as-observer stance, many researchers are driven by preference or lack of time to sample events they observe (or engage with via interview and so on) or to focus on interpreting activities going on around them rather than in which they are active agents.

However, even today, there is criticism that attention to human agency has been inadequate in human geography. As one illustration, Jackson and Thrift (1995: 210) assert that the 'residual influence of Marxian political economy' in the study of shopping malls has led to a rather one-sided view of the mall as 'an essentially threatening presence, able to bend consumers to its will'. They argue research should focus on the active role of consumers in negotiating messages, meanings and lines of action in the mall, instead of seeing them more passively. After using ethnographic approaches to investigate shoppers, they found consumers were not passive agents, duped by advertising and commodities. As Jackson (1999: 402) put it: 'Landscapes of consumption are revealed as peopled by skilled and purposeful human beings whose behaviour is governed by a variety of moral concerns rather than solely by the cold economic logic of the marketplace.' That this point has broader resonance is exemplified by Thrift's answer (1997a) to the question 'what is place?' As Thrift notes, too commonly cultural geographers have looked at place as something 'to be animated by culture', as if place is made before it is lived in. By contrast, Thrift conceives of place as 'a part of us', as 'something that we constantly produce, with others as we go along' (Thrift 1997a: 196–7). These arguments embody a deep criticism of that band of cultural geography that complacently restricts itself to representation and signification. As Goss (1999: 48) argues concerning geographical studies of shopping malls, 'the ethnographic approach risks banality by reproducing the obvious finding that consumers make their own meanings, without engaging in positive or negative critique of the politics of meaning'. Lees (2001) takes up Thrift's call by driving this point further, noting that geographers should concern ourselves with the inhabitation of space as much as its signification. The implication is that the situated and everyday practices through which built environments are used needs to be better understood. Participant observation provides one way to explore how built environments produce *and* are produced by social practices within them, yet human geographers have been slow to recognize this. Despite regular appearances of the word 'ethnography' in the geographical literature, much under this heading is rather superficial. Especially when analysts limit investigation to issues of representation and signification, the sense is that ethnography is not seen as 'participant as observer' but as 'observer who happens to be participant' (see Box 7.2 on this distinction). Offering a stronger sense of omission, Thrift (2000a) calls for researchers to be *observant participants* rather than *participant observers*. In effect he calls for recognizing that representation and interpretation are insufficient, that greater emphasis on the materiality of human action is needed to move us towards theories of practice. This call has clear links with ideas on Action Research, yet it is grounded in an appreciation of the limitations of earlier participant observation research. In this Thrift and others are pointing toward a new critical ethnography.

Summary: the tradition of participant observation

- Participant observation is not the same as ethnography but is a core element of it. Ethnography includes multiple data sources.
- In anthropology the 'Malinowskian Revolution' united the fieldworker and theorist in one body, with a critical distance often maintained between observed and observer, and much work in 'foreign' lands.
- Work within 'home' environments was associated with the Chicago School of Sociology and humanist geography. Such work was slow to take off in geography, with contemporary criticism still levelled at undue emphasis on observing rather than participating.

Towards a critical ethnography

The roots of participant observation lay in white ethnographers entering 'foreign' cultures. From post-colonial and anti-racist viewpoints, this approach was problematical. It came to be challenged in the fall-out from decolonization (Katz 1992: 497) and growing recognition of the one-sided representations of former colonial societies (see e.g. Chakravarty 1989; Banthiya 1994). But in a global political economy with vast social inequalities it is not surprising that commentators are still concerned about bias in contemporary reports. Thus hooks (1991) has noted a continuing link between traditional ethnography and the 'new' ethnography when discussing the image on the front cover of Clifford and Marcus's book (1986). This shows a white male doing fieldwork in India, sitting at a distance from darker-skinned people 'located behind him'. For hooks (1991: 126–7):

> *As a script, this cover does not present any radical challenge to past constructions. It blatantly calls attention to two ideas that are quite fresh in the racist imagination: the notion of the white male as writer/authority, presented in the photograph actively producing, and the idea of the passive brown/black man who is doing nothing, merely looking on.*[7]

What hooks (1991: 124) draws attention to is the 'cultural context of white supremacy' that she sees as embedded in writings by white scholars about non-white people. She urges white scholars to interrogate critically their own positions, to realize the difficulties

7 We have mixed views on this commentary. It embodies a powerful critique of ethnographic bias, but to extend this critique because of a book cover is less convincing. No evidence is provided to link the cover with the editors' views. Authors might be given little or no choice on book covers by publishers. If Clifford and Marcus provided the picture, hooks's criticism is valid. If they did not, the hooks critique might fit 'popular' (or publisher) images of anthropology but not the editors'. How important, then, is one book cover? Or one vision? (Examine S. Daniels 1992, and Kinsman 1995.) The front of Sanjek (1990) illustrates nothing except a male and a female (white) anthropologist writing at a desk 'in the field'. Does this signify anthropologists never see the people they write about – contriving a magical fiction from an enclosed verandah?

of translating ethnographies of the 'other'.[8] She illustrates these difficulties by outlining her grandmother's 'talk story' or philosophy of being and living:

> *One of her favourite sayings was 'play with a puppy he'll lick you in the mouth' . . . These lectures were intended to emphasize the importance of distance, of not allowing folks to get close enough 'to get up in your face'. It was also about the danger of falsely assuming familiarity, about presuming to have knowledge of matters that had not been revealed. (hooks 1991: 23).*

An 'outsider' would probably have difficulty translating this. The point is that we are not always able to understand other 'languages', even if this involves the language we speak. In any translation, meaning is lost and invented. Despite the intensity of participant observation as a method, it cannot be assumed it reveals 'reality' as experienced by studied populations.

With such insights located at a time of significant challenge to dominant modes of thought in the social sciences, especially regarding claims that knowledge is independent of practices of knowing (see Chapter 1), the basis on which ethnography offers insight into human societies has been increasingly questioned:

> *Rather than being the records of objectively observed and pre-existing cultural objects, ethnographies have been reconceived as written and unavoidably constructed accounts of objects created through disciplinary practices and the ethnographer's embodied and reflexive engagement. (Hine 2000: 42; see also Dorst 1989)*

Denzin (1997: 3) has called this the 'triple crisis of representation, legitimation and praxis'. This triple crisis threatens ethnographers claims to authentic knowledge. As Marcus (1997: 399) noted:

> *Under the label first of 'postmodernism' and then of 'cultural studies', many scholars in the social sciences and humanities subjected themselves to a bracing critical self-examination of their habits of thought and work. This involved reconsiderations of the nature of representation, description, subjectivity, objectivity, even of the notions of 'society' and 'culture' themselves, as well as how scholars materialized objects of study and data about them to constitute the 'real' to which their work had been addressed.*

8 The implication is that white people should study white people, black people black people, and so on. Yet it is a long time since Dollard (1937/1957) found that trying to study the black population in a southern US town could not be separated from appreciating white actions and concepts of whiteness, both of which were critical to black people's self-imagery and behaviour. Multi-member research groups might help get around this problem, although this does not circumvent other weaknesses in the hooks argument, as discussed in Chapter 6 (and, for warnings and advice on team ethnography, see Erickson and Stull 1997).

The possibilities within the constraints/limitations of the ethnographic method are well demonstrated by van Maanen (1995), who offers a paper that students and researchers should benefit from reading. The message is that, although participant observation offers a prospect of deeper insight into the social fabric of a group or organization, the researcher should not forget he or she is at best a *marginal native* (Walsh 1998: 223).

hooks's critique should not be restricted to white researchers and non-white study populations, for the problematic applies as much when 'First World' investigators undertake research in the 'Third World' (Sidaway 1992), as it does when middle-class investigators study 'subordinate' cultures in their own country (Jackson 1989: 73). As explored in Chapter 6, the issue is not that 'others' are studied, for this is the heart of most research (if 'others' means people who are different from the researcher). The critical issue owes more to the embedded power relations that lie in representing the lives of others. This issue is raised by all research. Rapport (1993: 57) captures the mood, for an Englishman amongst the English:

> *as I unwittingly barged my way into the local social milieux and, in my eagerness to learn more names and news, exposed my ignorance of the manners of the people, I became more suspect. I was a 'mystery man', I had come from nowhere, without a family or a name, and yet I was English, or claimed to be, so why should I be so naive and odd? It was a challenge to their Englishness.*

As many postmodernists note, there is a bias in the background of researchers, who are commonly white, male and middle class, with disproportionately low ratios between researcher and researched for non-whites, the poor, women and so on. Research often involves those with more economic and cultural capital studying those with less (see Haraway 1988, 1989).

Recognition of this led critical researchers to engage more openly with their own positionality, while emphasizing the necessity for more reflexivity over the research process (see e.g. Box 7.3). This recognition has become embedded in more profound ways in researcher psyche than earlier calls, like that of Myrdal (1969), for the researcher to declare his or her values. Within a more reflexive research process, issues of researcher impact on study conclusions are now conceptualized and appreciated in more profound ways. Thus, while studies examining interviewer impacts on questionnaire responses have been conducted within a positivist framework for a long time (see e.g. Ferber and Wells 1952; or, more recently, Leal and Hess 1999), the postmodern challenge offers a deeper critique than technical deficiencies. In this regard, Clifford and Marcus's *Writing culture* (1986) was a significant marker in its attempt to redefine ethnography, to produce a more reflexive research style. *Writing culture* is a foundation text on the representational crisis in ethnography. Clifford and Marcus argued that ethnographies were not true or transparent representations of culture. They

Box 7.3 Positionality

As participant observers at a feminist geography workshop in New Zealand, Lees and Longhurst (1995) demonstrate the positions of Longhurst the insider (which is reworked as inside/embodied) and of Lees the outsider:

Insider (Inside/embodied)

'I want to discuss my inside-embodied-experiences rather than my "insider" experiences by way of my institutional positioning within the academy in Aotearoa/New Zealand ... The brief round of introductions that evening did little to reduce the "knots" in my stomach as I realized that the participants had come to the workshop excited and with high expectations. I could not quite shake the feeling that I was somehow solely responsible for the success or otherwise of the unfolding events ...'

Outsider

'a number of issues around the New Zealand feminist geography scene, and my own positionality, emerged as the conference opened with a greeting in Maori from Robin Peace ... Robin suggested that her welcome in Maori might itself be a topic for later discussion since few Maori people were present at the workshop and the implications of her greeting Maori/Pakeha academics in Maori seemed complex. For me identity politics reared its head and I was unsure with what/whom I was identifying ... So I decided to identify, and indeed position myself as the "other", the "outsider", non-Pakeha and non-Maori.'

critiqued and deconstructed the concept of culture, alongside the fact the anthropological project was/is built on implicit cultural comparison (see Marcus and Fisher 1986). Ethnographies to these authors were accounts, or what van Maanen (1995) calls stories, that rely on the experiences of researchers, as well as on how engaging or authoritative an author's writing is. Central to this view is the understanding that ethnographic stories are selective; or, as Paul Atkinson (1990) puts it, textual constructions of reality. In this regard, Geertz (1973/1975, 1983, 1988) questioned the *author*-ity of ethnographic accounts, whereby participant observer narratives laid claim to special insights that were inaccessible to others. Offering a thought-provoking pointer, Clifford (1986) charged that ethnography is not about representing cultures but inventing them. As Wolcott (1994: 15) expressed it:

> *Never forget that in your reporting, regardless of how faithful you attempt to be in describing what you observed, you are creating something that has never existed before. At best it can only be similar never exactly the same as what you observe.*

This point is now recognized to hold even when investigators focus on social groups who are 'similar' to themselves. In any translation from the observed to the written, meaning is both lost and invented (Ely *et al.* 1997), which, as Keith (1992: 554) implied, reminds us of the responsibilities of the reporter and the ethics of the reporting process:

> *in the architecture of the academy there is an internal relation between researcher and researched in the production of power-knowledge that makes all ethnographic writing, in part, an act of betrayal ... it is in the very nature of representation that misrepresentation occurs.*

This idea, alongside the term 'inventing' (or 'imagining'), is now common parlance in geography. Take, for example, Anderson and Gale's *Inventing places* (1991), Gregory's *Geographical imaginations* (1994) and Westwood and Williams's *Imagining cities* (1997).

The contentious nature of representing others, which such book titles imply, prompt the deeper question of who has the right to write about whom? Additionally, who are we writing for? These are fundamentally questions of *positionality* and *situatedness*. For geographers these questions are not new. They certainly extend beyond participant observation. Most evident in this regard, these issues were explored in radical geography in the early 1970s (Dickenson and Clarke 1972; D. W. Harvey 1973). One geographer who became aware of the need to reposition himself away from books and onto the street, to become more personally involved in the research process, was Bunge (1971; see also Merrifield 1995). But Bunge found it difficult to make the change, and questioned the ability of researchers to empathize with disadvantaged 'others'.[9] In particular he was concerned he could not locate himself outside his own past, his own upbringing. For Bunge, part of the research process was learning to cope with this.

Bunge's ideas did not capture the imagination of a broad spectrum of geographers, whose work continued to be grounded in then-dominant research philosophies. It required a Kuhnian (1970) type transformation for self-reflexive approaches to gain currency. This entailed acknowledging the limited capacity of researchers to empathize with 'others' and recognition that accounts were partial, not simply because informants reveal only some of themselves but because investigators' cultural baggage conditions interpretations and values allocated to them. As Katz (1994: 68) reminds us 'only in "The Wizard of Oz" do women descend on other lands without obvious cultural baggage' (see also Coffey 1999). This ethnographic recognition has drawn attention to the way Western, male, heterosexual, adult and able-bodied biases pervade studies of 'others'

9 More broadly Foddy (1993: 16) exposes why Geertz (1973/1975: 1988) is critical of the *verstehen* approach, which calls for investigators to 'empathize' with those they study: 'even if qualitative researchers honestly strive to "absorb" the culture (i.e. the organizing concepts and rules which govern perceptions and behaviour) of the member of a social group, in the final analysis they have either to infer or to guess the nature of the cultural elements.'

(on postmodern ethnography, see Denzin 1997).[10] This has led to a questioning of researchers' motives, as regards whether the benefits of investigations are primarily for themselves or for the people investigated. In this regard, most social scientists concur that research is not a non-exploitative process. Action Research and, more recently, the critical geography movement (see Blomley 1994; Castree 1999; Lees 1999), have sought to tackle these issues. In social science more generally, suggestions have grown on how to be more sensitive to power relations in research. Some suggestions include:

- framing research questions according to the desires of the oppressed group, doing work that others want and need (see Mascia-Lees *et al.* 1989). Overt efforts to move in this direction have come through participatory rural appraisal (Chambers 1994a, b, c);
- exposing 'colonizers', or more broadly those with power, who mistreat 'others' (J. Thomas 1993);
- studying our own cultures instead of other cultures (Bourdieu 1990);
- analysing our own practices, rather than those of others (Katz 1994);
- looking at global relations between peoples (Marcus and Fisher 1986);
- playing with the language (textuality) we use (Clifford and Marcus 1986).

As Katz (1992) argued, to counter oppression in the research process researchers must renegotiate and engage with the situatedness of their knowledge and positionality. This involves continual questioning of a researcher's social location (gender, class, ethnicity, age, abilities and so on) and of the physical location of the field area. For Katz (1992: 495):

> *As intellectuals operating in a postcolonial world, we must take seriously Spivak's admonition about representations as a staging of the world in a political context and begin to connect the 'micrological textures of power' with larger political-economic relations. In this expanded field, we can no longer valorize the concrete experience of oppressed peoples while remaining uncritical of our role as intellectuals. Neither can we presume to speak for or about peoples and nations as if they were outside of the contemporary world system, refusing to recognize that our ability to construct them as such is rooted in a larger system of domination.*

Going further, 'the ethnographic encounter is premised largely on the intent of the ethnographer to represent some constituted "other" to a particular audience' (Katz 1992: 496). Katz's solution is the excavation of 'a space of betweenness' (see also

10 Feminists point out that new ethnographers, such as Clifford, Marcus and Geertz, have been silent on women as 'other' (Strathern 1987). The postmodern turn in anthropology, it seems, has provided limited insight on gender or sexuality, as indicated by Clifford and Marcus (1986) including no feminist authors in their collection.

Anzaldua 1987). She explores this by trying to find connections between the lives of young people in rural Sudan and Harlem (her two field areas). From this work, she found 'striking parallels' between the two groups. Children in both experienced inadequate preparation for adulthood owing to socio-economic change. Through making this comparison, Katz offers an understanding of how capitalism affects social reproduction.

One solution that has been advanced to avoid the 'othering' of the subject of enquiry is autoethnography (see e.g. Anzaldua 1990). There are few examples of autoethnography in geography but Lakhani's Ph.D. research (2000) offers one example (for anthropology, see Okely and Callaway 1992). This study explored social constructions of identity amongst young British (South) Asians using participant observation. This involved engaging with young Asian people in a youth club, in employment contexts, on college courses, as well as in home, street and community locales. At a surface level the approach Lakhani used was not so different from those used in more traditional participant observation studies. A key difference that underscored both her research approach and the presentation of her conclusions was that Lakhani reflexively engaged with her own experiences as a young South Asian in Britain to inform and to underscore the interpretations presented.

A further mechanism through which researchers have sought to mitigate influences of uneven power relations arises in strategies for writing up research. Whatever the approach used, researchers inevitably employ textual strategies, and/or textual experimentation, to make their research appear plausible. When quantitative methods are employed, we see this in the utilization of statistical significance indicators, which give an air of authority to conclusions. Similarly, when Eagleton (1990: 384) discussed Foucault's use of a detached prose style, juxtaposed with the gross treatment of the body in *Discipline and punish*, he argued that this textual strategy (bordering on pornography) is used to convince readers of the validity of the arguments. Recognizing the links between authority claims and presentation styles, some self-reflexive researchers are experimenting with different textual strategies. The aim is to show that ethnographic texts are structured multi-vocal exchanges in politically charged situations. According to Katz (1992: 498), they have entered 'the hermeneutic or postmodern ethnographic project'. Some of these textual experiments are viewed as 'poetry'. As Tyler, (1986: 125) put it:

> *A post-modern ethnography is a cooperatively written text consisting of fragments of discourse intended to evoke in the minds of both reader and writer an emergent fantasy of a possible world of commonsense reality, and thus provoke an aesthetic integration that will have a therapeutic effect. It is, in a word, poetry.*

As Keith (1992) indicates, the nature of this reflexive poetry means the audience it is written for is of particular importance (see Fig. 1.2). Thus Keith has sought to write differently to avoid reinscribing racist representations in his work on race and racism. He has argued that

> *the texts through which cultural accounts present themselves are important and too often a neglected element of the research method ... An academic tradition of putative critical distance tied to methodological objectivity obscures, through mystification and naturalisation, the process of research itself. (Keith 1992: 556)*

We find this perspective worth exploring, but acknowledge there is no agreement on the advantages of such textual strategies. Thus, while some have praised the 'poetry' of textual experiments for their reflexivity, others critique them for being self-indulgent (Thrift 1996). But, whatever the perspective on how we *write* about our research, this should not stop us thinking about how we *do* that research.

The practical activity of *doing* research, and looking at how people *do* things, is the concern of geographers like Thrift (2000a, b) who are interested in moving researchers towards 'non-representational styles' of working (see Chapter 1) – that is, away from the obsession with representing representations. This movement in geography has entailed revisiting ideas on pragmatism, the idea that we only know through practice. Rooted in human subjectivity this movement is tied to the development of psychoanalytic geography, which focuses on the inaccessibility of the human unconscious (see Pile 1996). As a result of this 'non-representational turn', some ethnographers (although few as yet in geography) have adopted performances such as storytelling, dance, music and drama, by either themselves or those they are studying, to counter criticisms of ethnographic practices. Here the ethnographer concerns him or herself with practices that are not necessarily articulated linguistically. The ethnographer is not a participant observer but an observant participant.

Summary: toward a critical ethnography

- There has been critical self-evaluation of the products of participant observation studies. A 'triple crisis of representation, legitimation and praxis' has been linked to recognition that ethnography does not reveal 'reality' as experienced by studied populations.
- Strategies for the researcher to be situated in 'spaces of in-betweenness' have been encouraged, with autoethnography, textual strategies and 'non-representational styles' of working of note.

Approaching 'the field'

We are ready to enter the 'field'. Preparing to enter the field can be a big challenge for research that seeks to be sensitive to power relations and ethical issues. Yet fieldwork training in ethnography has been and still is largely absent in the social sciences. Punch (1986) offers a variety of indications of this. These include the renowned Chicago anthropologist Everett Hughes telling fieldworkers to go and 'fly on their own', and the widely quoted 'training' offered to Evans-Pritchard, which was little more than to 'play it by ear' and be cautious about taking quinine and sexual relations (see also Bernard 1994). Perhaps of more help as a guide, even if suggesting somewhat Herculean a task, Malinowski is reported to have argued that everything seen or heard should be written down, as it is impossible at the start of a project to know what may be significant (D. Rose 1990), while Park is said to have told Nels Anderson to write things down like a newspaper reporter (van Maanen 1988). Today more advice can be given, but it is difficult to imagine training that is not deeply penetrated by a multitude of cautions and calls for flexibility, rather than precise instructions. The reality of participant observation is that the nature of the experience is difficult to predict (Hammersley and Atkinson 1995: 23). We noted this in Chapter 6 with regard to the unpredictable consequences of gaining access to social groups through 'gatekeepers' (J. C. Johnson 1990) and how personal attributes have uneven consequences for the research process (see e.g. Lewin and Leap 1996). As Rob Burgess (1984: 31) summarizes the situation: 'Accounts by researchers have revealed that social research is not just a question of neat procedures but a social process whereby interaction between researcher and researched will directly influence the course which the research programme takes.' In this context one has sympathy with van Maanen's view (1988: 139) that: 'On advice to students of fieldwork, my feelings are traditional. There is, alas, no better training than going out and trying one's hand at realist tales.' But if being flexible and sensitive means most researchers learn most through trial and error, this does not mean guidance cannot ease the learning process.

In this context, Cook and Crang (1995: 13–14) emphasize the importance of groundwork before data collection begins:

> *As a first step in any ethnography, it is important to develop early contacts in the organisation/industry/community in which you are interested, to find out what research may be possible within the constraints of access, time, mobility and money available for 'fieldwork', and to undertake methodological, theoretical and linguistic preparations accordingly.*

These considerations have commonly led researchers to 'familiar territory'. For example, the decision of one of the authors to investigate the meaning of urban community in Brooklyn Heights, New York,[11] was eased because the area is adjacent to a neighbourhood in which research had been conducted before. With contacts already there, a base existed for developing a network of new contacts.

11 As Katz (1994: 70) rightly notes, the act of 'choosing', deciding and travelling in such contexts overflows with the arrogance of research.

But personal knowledge of research sites is not always available. In such cases we can learn from the implied messages in Cook and Crang (1995: 15), who note that: 'Ethnographic projects do not emerge in the form of pristine hypotheses to be tested later "in the field" but require a fusion of knowing what is interesting and what is accessible.' Implicit in this message is recognition that there is a politics to gaining access to informants. This is why Cook and Crang (1995) suggest 'a combination of reading *and* doing' in preparation for fieldwork.[12] As Keith (1992: 553) reminds us: 'the context in which the research takes place is structured by moral, political and ethical considerations.' Such considerations require judgement prior to entry into the field, as well as when there. This is signified in his discussion on the problematic of access in research on racism in the police:

> *Several attempts on my part failed and I eventually succeeded because the family of a partner at that time came from a small mining town in Wales ... and, it transpired that the partner was a cousin of a Deputy Assistant Commissioner of the Metropolitan Police. The officer concerned could not have been more helpful when I wrote to him. Yet the presentation of myself, the researcher, was mediated by a double link; quasi-family and an academic reference. My 'critical distance' and 'academic objectivity' were staked out from the outset, but the disclosure of 'self' was selective; several facets of my personal history were obscured and not even mentioned. Deceitful? Yes, certainly at one level. Defensible? I'm not sure, though neither am I sure what the disclosure of myself without dissembling would have looked like, short of an imperfect biography ... the trust that was extended to me was considerable and the fact that the authorisation for fieldwork came from on high [made] access that much easier. In the hierarchical world of policing, my senior officer endorsement on several occasions conferred on a doctoral research project the appearance of an official investigation, at times radically altering some of the most finely balanced research relations. Clearly there were both gains and losses from this, but then it is hard to pretend that such status was not in and of itself deceitful. (Keith 1992: 554)*

As Newby (1977) asked, how many projects have been 'won' or 'lost' by luck in being associated with a key informant? How many researchers, as Howard (1994) reports for

12 In preparatory work, practicalities should not be forgotten. For overseas research, a visa, immunization, foreign exchange and a local bank account are often essential (noting Barley's caution (1983) that local arrangements may not work). In addition there are personal concerns about health: 'I had water purifying tablets, remedies against two sorts of malaria, athlete's foot, suppurating ulcers and eyelids, amoebic dysentery, hay fever, sunburn, infestation by lice and ticks, seasickness and compulsive vomiting. Only, much, much later would I realize that I had forgotten the aspirins' (Barley 1988: 15). Do not let the amusing endnote distract here from the seriousness of health dangers. Even a glimpse at standard works on ethnography reveal regular references to Howell's survey (1990) of US fieldwork experiences, with frightening figures like 70% of fieldworkers in the South-East Asia region reporting liver disease and about 40% malaria (see e.g. Bernard 1994: 157). To this add that a quarter were accused of being a spy. Death after being attacked by someone from a study population is not unknown (Lee 1995: 3, 31).

Nicaragua, need to make extra efforts to gain information from certain population segments because they were seen to be closely associated with another segment? How many, as Bell (1977) reports for a multi-investigator study of social life in Banbury, fail in this quest? There are no easy answers to such questions (for valuable insights, see J. C. Johnson 1990). But access will be influenced by the nature of a research project. Researchers who seek access to physical sites like a prison, a youth club or a library are likely to experience different degrees of difficulty. The same can be said for access to specific activities, such as a football match, movie watching or prostitution. Often unevenness is found in the same project, for the social milieu of an investigation can be a multitude of sites, like home, the street, a youth club, workplaces and so on (see e.g. Lakhani 2000). Each of these sites poses different access issues, which are informed by the personal attributes (and personality) of the researcher and the researched. In any event, making contact with people is inevitable. It is at this point that fieldwork 'formally' begins (albeit much preparatory work should have taken place previously to prepare for initial contact).

Recognize the need to be reflexive about 'doing fieldwork'. As one consideration, recognize that fieldwork has attained a status of its own historically. As James Duncan (1993: 42) put it, empirical investigation came to be called 'field*work* … in order to professionalize it and thereby elevate its products above the representations of amateurs such as colonial administrators and travellers'. This point is taken further by Driver (2001: 12–13), who notes the significance of 'field' and 'fieldwork' in the development of geography as a discipline:

> *If we think of geographical knowledge as constituted through a range of embodied practices – travelling, seeing, collecting, recording, mapping, and narrating – the subject of fieldwork becomes difficult to escape. The field in this sense is not just 'there'; it is always in the process of being constructed, through both physical movement – passage through a country – and other sorts of cultural work in other places. It is produced locally by the spatial practices of fieldwork, and discursively through texts and images …*

Doing fieldwork encompasses a complex politics. This is discussed in an excellent 1994 special issue of the *Professional Geographer* on women in the field. In her opening remarks in this volume, Nast (1994: 54) argues that little attention has been given to how a 'field' is defined, nor to the politics of choosing and working in a particular 'field'. The authors in this volume are concerned that the 'field' is not naturalized as 'place' or 'people', but is located and defined in terms of political objectives, which cut across time and place. The feminist objectives of these authors embodied working towards critical, liberatory ends (not saving the exotic 'other') and overcoming shared experiences of oppression through racism, patriarchy and capitalism:

> *This is not to say that we can 'move unheralded into just any field situation and become an effective part of its struggle for change just because we believe in its political ends' ... There also needs to be a recognition that some historical and material realities are beyond our personal and social reach. Such reflexive analysis requires that we link larger-scale political objectives to smaller-scale methodological strategies which break down hierarchical objectivistic ways of knowing in the field. (Nast 1994: 58)*

The reflexivity such ethnographers seek involves 'processes of locating, reaching out to, and working within a particular field' (Nast 1994: 59). The field for these authors is a social terrain through which researchers can forge a bond between the academy (also a field) and the world at large. 'Thus fieldwork allows "fields" of everyday bodies and problems "out there" ... to be incorporated into and thereby to subvert what has traditionally been the preserve of the white, the masculine, and the abstract – the ivory tower' (Nast 1994: 57). As Katz (1994: 72) argued: 'I am always, everywhere, in the "field".' Put another way, the participant observer is always positioned simultaneously in a number of fields, in a constant state of 'betweenness' (see e.g. Lewin and Leap 1996). In this context, in 'the field' the researcher is neither an insider nor an outsider in an absolute sense, but an interlocutor. As such, 'the field' is a place 'in between'. It is politically situated, contextualized and defined. Its social, political and spatial boundaries shift with changing social circumstances and in different political contexts.

Observer positioning in the research process

A decision ethnographers have to make about their role in the research process is whether to be overt, by explaining their research role to subjects, or to undertake research in a covert manner, by concealing their purpose (see Fyfe 1992: 131). Both covert and overt research are fraught with political, moral and ethical dilemmas. But covert research raises the potential for danger and violence (in both the metaphorical and actual sense of the word), as well as personal heartache. While totally covert research is rare in ethnography, the words of those who have undertaken such projects are poignant. One of the most vivid expositions is provided by Dan Rose (1987). Living undercover for two years as a car mechanic, Rose provides heart-felt diary entries at the end of his book. He also indicates the emotional pain of the negative experiences of his fieldwork:

> *The fact that I did not say explicitly that I was conducting anthropological fieldwork led rapidly to a disintegration of any assumption about what information I could gather. I feared that what was occurring was a complete lack of match up between what I had read in graduate school and entries in my field notes. At the time I could think of no greater anxiety. (D. Rose 1987: 13)*

Less traumatically, discomfort and loss of rapport resulted when Bax (1976) revealed he had used a hidden recorder to tape conversations during his fieldwork. But, despite some classic covert studies (e.g. Griffin 1962; Friedland and Nelkin 1971; Wilkinson 1981), it is more usual for researchers to tell at least some informants about their research. Many researchers argue that ethnographers should always be open about their research role, explaining their project to the people who are to be researched. Yet other ethnographers accept that it is not always possible, desirable or safe to outline one's position to everyone involved in the project. For one thing, as Hammersley and Atkinson (1995) argue, relaying too much information before the ethnography begins can introduce unwanted potential bias into a social situation, by influencing the behaviour of the people towards you (potentially by misinterpreting intentions). In addition, the nature of the 'lived experience' of ethnography means that many people on the margins of a project will be encountered during its passage (marginal in the sense of not being primary informants). If we stick to the stipulation that everyone has to be informed, this will require that every time researchers meet new people in study sites they tell them they are researching a particular topic (presumably this ethical stance would demand this be done before the person has a chance to speak, so they do not talk before being informed). One can imagine what effect this would have on primary informants. If they do not think you have lost a few screws, they will certainly become bored or embarrassed by the format spiel, with the potential result that they will avoid you or seek not to introduce you to new people.

There are no easy answers to these issues, for the context of the research can be critical. Thus Lee (1995) notes that many researchers see deception as permissible, if not laudable, in highly stratified, repressive and unequal social contexts. But, as Jackson (1983: 43) observed, whether circumstances favour overt or covert research is a question 'every social scientist must face personally'. In this regard we should not automatically class overt research as good (or better) and covert as bad (or worse):

> *Both research methods involve complex personal positionings (when 'doing research' and 'writing research'), differing 'explanations' of the self, immersion in local geographies, and observation and recording of ethnographic materials ... Overt research is not assumed always to be free of covert practices, since much consensual participant observation involves the recording of social moments and processes, of which the research 'subjects' are often unaware. (Parr 1998: 29)*

In this regard Parr (1998) offers an interesting discussion of her covert research on mental-health problems. In this discussion she seeks to politicize the body as an ethnographic tool, including discussing how she acted and dressed:

At first, I would attend the drop-in in appropriate dress, and then be uncomfortably aware of how my hair smelt of shampoo. The people with whom I was sitting might not have washed their hair for months, and shampoo set me apart from the other people in the centre. In later visits, I learned not to wash before I arrived, perhaps not as a means of ensuring intimacy, but certainly in order to limit my own 'otherness' in that situation. The ambiguity of the body (how we adorn ourselves with soap, scent and deodorant to disguise real bodily odour) was highlighted in this research setting, and my body, in this sense, had to become much more physically present ... It was important that I walked without purpose and authority, also that I sat slumped, often staring into space for hours at a time, as this was the norm in the drop in ... (Parr 1998: 32)

Similar accounts can be found in a variety of social contexts. Notable examples include ethnographic studies of 'down and outs' and exploitative work environments (e.g. Brody 1971; Friedland and Nelkin 1971; Wilkinson 1981). Encapsulating what this can mean for a researcher, Punch (1986: 17) advised that:

Continued involvement in the field can be likened to being consistently on stage. The role has to be played without dropping your guard, and researchers frequently comment on the strain this causes, not only on themselves, but also on their families.

Capturing this sentiment, Batterbury (1994: 63) explains how:

During my fieldwork in Mossi communities of Burkina Faso, I was constantly involved in self-preservation and my conscious and unconscious actions – greetings and departures, body positioning, clothing and appearance, as well as general social conduct and behaviour, were closely observed.

Our cautions over all this are twofold. First, we emphasize the need to be prepared for the unexpected – be adaptable and anticipate needing to accommodate to situations you have not come across before. As Lewin and Leap (1996: 17) note, on gay/lesbian acceptance in new settings: 'The question of whether being lesbian or gay spans cultural boundaries must be answered anew in each situation.' Second, be prepared for hard work. Ethnographic accounts tend not to dwell on real research problems. This gives an un-real sense to some reports: 'One often has the feeling ... that reports on fieldwork experiences gloss over the problems, or treat them in a semi-comical manner' (Punch 1986: 14). The work of Barley (1983) comes to mind, although the forthright way he specifies research problems is commendable. Also of help, and an essential read for students who wish to prepare for the emotional and physical discomfort of fieldwork, are writings on the difficulties of undertaking fieldwork (e.g. Howell 1990; J. C. Johnson 1990; Lee 1995), compilations of actual fieldwork experiences (e.g. Bell and Newby 1977; Panini 1991; Robson and Willis 1994; Lewin and Leap 1996) and autobiographical accounts of fieldwork living and learning (e.g. Malinowski 1967; Mead 1977; W. F. Whyte 1994).

A final thought on entering the field is to think through the implications of your research strategy and topic for the ethics of project implementation and exposure to potential

Box 7.4 A typology of ethnographic dangers

- *Legal danger.* This might be when a researcher enters a graffiti underground or marijuana-growing subculture. A dilemma is that deep involvement is vital for detailed understanding, yet an observational perspective is necessary to analyse and critique – to do critical ethnography. The solution is inevitably a personal one (with potential legal consequences likely to vary across countries). There are also differences across issues. Thus, the researcher of urban graffiti might choose to participate as fully as possible, as he or she sees any risk to be minor – perhaps with a short jail stay, fine or probation as the legal punishment. Participating fully would be more risky for the marijuana researcher, who might face a long prison sentence, as well as loss of employment and personal property.
- *Stigma danger.* This occurs when the researcher is the brunt of prejudice and discrimination owing to his or her identity or group affiliation. Critical feminist research on the sex industry is illustrative. Using participant observation, a female researcher of female prostitution will inevitably suffer the indignities of the whore stigma and the male gaze. This could even cross over from the world of participant observation into academia – where colleagues might react inappropriately or with disgust at the nature of the research. To overcome this the researcher must draw on emotional power, sense of humour and adventure (on issues a male researcher faces when researching female sex work, see Hubbard 1999).
- *Ethical danger.* Hamm and Ferrell (1998) cite the words of Ned Polsky, a criminological ethnographer: 'If one is effectively to study adult criminals in their natural settings, he [*sic*] must make the moral decision that in some ways he will break the law himself. He need not be a "participant" observer and commit the criminal acts under study, yet he has to witness such acts or be taken into confidence about them and not blow the whistle. That is the investigator has to decide that when necessary he will "obstruct justice" or have "guilty knowledge" or be an "accessory" before or after the fact, in the full legal sense of those terms.' This raises pointed ethical issues, such as maintaining the confidentiality of informants. For one thing, if the police request your field notes and interview tapes, what is your primary duty? As Lee (1993) outlines, there are documented cases of a subpoena being issued for researchers to hand over field notes. Responses to such requests (or demands) will no doubt be impacted on by the researcher's initial decision on whether to tell subjects that he or she was undertaking the study, as well as by the nature of the incident that provokes the call for releasing records and by personal values. Such potentialities bring into the open the myth of value-free social science.

Box 7.4

- *Emotional danger.* Developing emotional attachment to one's subject during the research process can be dangerous, as it can lead to emotional pain. If the reader wishes to grasp the potential for emotional disruption resulting from the research process, then Dan Rose's account (1987) of his participant observation study in Philadelphia is a good place to start.
- *Physical danger.* Take the researcher who immersed himself in a motorcycle gang and paid a heavy price. He immersed himself, not just in mind-altering substances with the gang, but in their survival instincts (this tactic was also used in research by Brody 1971). We join him as he zooms down the highway at over 100 miles an hour, the passengers close enough to touch one another across the roaring windblast: 'the experiential equivalent of a researcher shooting heroin in order to describe yet another ineffable state'. His risk paid off and he was accepted into the gang. But this acceptance came at great sacrifice: 'And with blood filling his punctured lungs, slipping into the darkness with broken bones protruding from his leather jacket', he brings us fully into the 'sheer terror of the ethnographic moment'. Included amongst those who undertook research in the anticipation that violence against themselves might accompany their work, see Friedland and Nelkin (1971) and Wilkinson (1981).

Source: We have drawn the topics and some of the examples here from Hamm and Ferrell (1998: 256–66). The reader is also recommended to look at Lee (1995) for a broader commentary on danger in fieldwork.

dangers. The decisions ethnographers make about participation are important in this regard (for examples, see I. Cook 1997). Researchers who engage in higher levels of participation run more risk of placing themselves in ethically difficult, if not potentially dangerous, situations (Box 7.4). This raises the question of how far researchers should involve themselves in actions that violate the law or take risks that lead to injury. There are no easy answers to such questions, but some social scientists do believe such steps can be necessary in order to understand groups whose behaviour has important public policy implications. For example, some researchers who study crime and deviance argue that we cannot understand the taste for danger, pleasure and excitement that is a part of crime and deviance without experiencing this ourselves. As a complementary thought, it can be argued that to understand cults we need to participate fully in their activities. Joining an urban street gang, an all night graffiti crew or the Ku Klux Klan might similarly expose the researcher to 'dangerous' situations, just as potential health risks should be acknowledged if you live as a 'down and out' (e.g. Brody 1971; Wilkinson 1981). Such research steps move the researcher's personal experience from the periphery to the centre of the research process. Moreover, these experiences can give rise to 'intellectual insights that blossom from the application of particular human survival skills in times of personal and methodological crisis' (Hamm and Ferrell 1998: 256). That most academics reveal their dual-identity as participant and observer in such situations is commonly exposed when those who have 'gone native' 'save themselves' by 'going academic' (see G. Armstrong 1993, as a football hooligan).

Summary: approaching the field

- Approaching the field is structured by moral, political, ethical and safety considerations, which need serious self-reflexive evaluation.
- You are always 'in the field', in a constant state of betweenness.
- Preparation for participant observation research requires serious attention to the practicalities of entering and undertaking research.

Data collection

Participant observation is a peculiar research approach, in that the researcher is the primary tool for collecting data. It is an approach to learning about the social and cultural life of communities, institutions and other social settings. It is social scientific and investigative and can be inductive. It emphasizes and builds on the perspectives of people in a research setting (both the researched and the researcher). When undertaken well, it employs rigorous methods of data collection, which try to avoid personal bias by questioning researcher values as interpretations are formulated and assessed, and as far as possible seeks to ensure the accuracy of information used (namely, cross-checking and questioning validity claims).[13] In the past, participant observation was seen as a tool that permitted researchers to learn about new situations from the perspective of 'insiders'. Some still see this as a key feature, given that researchers become involved in social settings and acquire knowledge through hands-on experience (LeCompte and Schensul 1999: p. xv). More recent accounts conceptualize the approach in a more refined manner. As Madge (1993) has put it, fieldwork is an interactive text that is created from dialogues, in which researcher and researched not only learn from one another but potentially change one another as the process evolves. As such, while perhaps starting as an 'outsider', as fieldwork progresses there is a transcending of 'insider'–'outsider' status. Unless subject to extraordinary bad luck or ineptness, the interpenetration of positions associated with being a participant who observes (not an observer who participates) should set the researcher apart from both 'insiders' and 'outsiders' – albeit the balance requires sensitive handling (Madge 1994). For Duncan Fuller (1999: 221), 'mixing and manipulating the researcher's various identities, and the transparent and overt recognition and awareness of multiple positionality (as person, academic, as activist) can ... benefit geographical ethnographic research (and researchers) in three main ways'.

1. Transparency of thought and reflection, which improve the design, implementation and documentation of ethnographic research, should be promoted.

---

13 For Sanjek (1990: 395): 'Validity lies at the core of evaluating ethnography.' He lists the canons of validity as: (1) theoretical candour; (2) the ethnographer's path – a detailed portrait of what was done and how it was done; and (3) presenting field-note evidence.

2. Owing to this reflexive practice, the potential for a further layer of professional accountability should be added as the researcher reflects on his or her positioning and roles.
3. Critical engagement means researchers can play a greater role in social change, which links critical engagement to the potential for conducting Action Research.

Set in a context of critical self-reflection, learning and a 'fusing' of researcher and researched social environments, the investigator should continuously ask a series of principal questions. What is happening? When is it happening? Where is it happening? Who is (and is not) engaging in what kind of activities? How are people responding to what is happening? And so on. These questions are used as tools to obtain basic information about social events, social structure, cultural patterns and the meanings people give to these patterns. But to be effective participant observation has to go beyond being part of and observing social action. The dictates of conceptual validity point to the need for a triangulation of data sources, in order to challenge emergent interpretations and support claims that are made. As Susan Smith (1988: 22) reminds us, the intention 'is not, however, simply to discover the "rules" of the social game, but to become conversant with them to the extent that one feels safe angling for advantage and seeking privileged access to information'. For many this means drawing on extensive research methods in order that the researcher is confident when entering the more socially and culturally exacting environment of intensive research methods. In this combination of research techniques we find the conflation of 'participant observation' and 'ethnography'; good participant observation almost inevitably requires engaging with a bundle of other research methods. For one thing, it has long been established that, if researchers restrict themselves to participating in social activities without calling on other data forms, then the information that is gathered will be selective. As research for *Small town in mass society* revealed, despite efforts to be inclusive, participant observation gathers uneven information across social groups (Becker and Geer 1957; Vidich and Bensman 1968). Issues such as caste, ethnicity and sexuality, which were not part of the *Small town in mass society* investigation, can be expected to add to this potentiality (see e.g. Panini 1991; Mohammad 2001).

Prompts for participant observation

But, as study populations are often socially diverse, between groups, communities and organizations, let alone within them, and as efforts by researchers to be angelic pegs in local social scenes depends a great deal on the researcher him or herself, there can be no concrete step-by-step guide to participant observation or supporting data collection procedures. If the reader will indulge us to restate implied or actual messages in this chapter, participant observation is a method that necessitates flexibility, patience, a willingness to bite your tongue, a need to think before you speak, a potentially burdensome necessity to be charming (or at least to try to be) and a willingness to take an active part in activities you loath (a bit

Box 7.5 Bias, 'objectivity' and ethics in ethnography

The observant (or irritated) reader might have noted that we have referred to the need for unbiased data collection at several points in this book. We anticipate this might be provoking comments of the form, 'Hold on, what about the argument that "objectivity" is impossible, that we inevitably provide interpretations that draw on our own past experiences and by positionality?' This is a fair point, so we need to clarify what we mean. Our meaning is not that we believe that data can be collected in a value-free way, being capable of bad interpretation but incapable of inaccurate reading if 'approached' correctly. But this should not be taken to mean that researchers should wallow in their own value dispositions, seeking out and selectively reporting 'evidence' that 'conveniently' reinforces their existing prejudices. Look back at the introduction to Chapter 4, when we related the outcome of the David Irving libel trial, if you want a reminder of the slippery slope that this approach can lead you down; and remind yourself of your ethical duty to fellow researchers and the damage revelations about poor research can do for public confidence in the research community. What we mean by researchers seeking to be unbiased is embodied in investigators being conscious of their own value dispositions and engaging in a self-conscious process of seeking countervailing evidence to challenge the interpretations they are developing, in order to reassure themselves and their audience that what they report is not just pre-existing prejudices promoted under a halo of in-depth research (for example, calling on politically correct arguments, of whatever ilk, to gain reader support rather than convincing through offering argument and evidence). One aspect of this is to strive not to close down data collection avenues because researchers do not like the values or personalities of potential 'gatekeepers'. You might think the vicar is pompous, but, if your body language and manner of speaking to him make this clear, then you are likely to reduce his willingness to offer you insight into your research problem, as well as possibly ensuring that doors he could open for you will stay shut. We are not saying that this is easy, but no research is, if it is done properly. What you must never forget is that you are beholden to the people you collect information from. Most evidently, if you are living in a community, conducting interviews or in some other way imposing yourself on the time of others for data collection purposes, you are already beholden to them. For us, the ethics of this situation are both that you seek not to cause distress amongst those who are providing you with information and that you strive to the best of your ability to provide a high-quality research project, which does not simply trundle out the prejudices you came into the field with. In this regard, readers are well advised to read materials that draw out the ease with which ethnographic investigators reveal their existing prejudices and so close down potential insights. Accounts like those of Rapport (1993) and Michael Bell (1994), or many of the chapters in the Robson and Willis collection (1994), provide us with honest assessments of the problems pre-existing ideas and values have for fieldwork. You will not be able to eradicate these, but you can strive to minimize their effects. Quality research calls for approaching the field with a preparedness to change and challenge your existing ideas, not to act like a supercharged zealot determined to find class conflict, difference, distinction or discrimination in every social setting.

Our obvious caveat to all this will become clearer later in the chapter, for in Action Research deliberately calling the vicar pompous might be part of a coherent research strategy that is driven by a desire to bring about change as well as developing theorizations.

like eating over-cooked brussels sprouts with a smile on your face). This is not a research method everyone feels comfortable with (Box 7.5). We are not going to ease this situation by providing the equivalent of a Lonely Planet guide so readers can rush to do the same things. We are not going to try to cram the previous chapters of this book into the next few pages. We expect the reader to have grasped messages about the potential for participant observation to focus the researcher's attention on specific sub-populations. On this basis, we emphasize that participant observation studies can be enhanced by less intensive methods that cover a broader slice of a population. But rather than examining items like questionnaires, our intention below is to offer prompts on how to prepare for participant observation, alongside techniques we have not explored so far that might unlock insight. There are two essential elements in preparing to use these and other data collection methods. The first is timetabling (and the personnel who will execute them), for if more overt methods are employed early in a project this diminishes prospects of blending with a study population (albeit this is less likely if a researcher is not at 'home' with the study population). The second is how material will be written up. As a key thought here, we remind readers of Pollitt and associates' warning (1992: 60) of the need to convince readers of the validity of research outputs: 'it has been a weakness of some ethnographic projects that they have not been able to describe their own methods of observational enquiry in sufficient detail for internal validity to be assessed.'

Survey the community, social group or organization

It is useful to begin research by seeking basic information on the group or population to be studied. If the study is of a particular community, this could involve collecting information on age, gender, occupation, education composition and so on. These might be available in statistical publications, annual reports (of companies, local governments and so on), or official plans (such as land-use plans). Such information can be used later in your write-up as well as helping you get a feel for the community before starting your participant observing. Such written materials can be complemented with interviews with those not directly involved in the issues you study but who can offer insight on the study population. These might include planning officers, journalists, national NGOs and so on. But there is a caution to observe over such interviews. This is to treat the information collected critically. If a planning officer informs you that a locality has significant social divisions between one neighbourhood and another, this might be interesting to explore during data collection. Yet it also runs the risk of directing attention towards this issue and away from other, perhaps less overt but more fundamental social divisions. Such interviews provide prompts but you need to maintain a critical stance towards *all* the information you receive (this extends into the participant observation period). Making contact with people in key (knowledgeable) positions in an investigated community, group or organization offers an extension to these less immediate interviews. Here you are advised not to jump too soon. Preparatory work is critical before you make formal contact with someone who might prove to be a critical informant; indeed, it might be that a formal interview is inappropriate with such people, depending on the nature of the research problem you are investigating.

Collect relevant documentary material

Relevant documentary material might be in the form of newspaper articles, minutes of local government meetings, and so on, but could include letters, diaries, films, photographs or audio tapes (for example, as found in the UK's National Museum of Photography, Film and Television at Bradford; see http://www.nmpft.org.uk/). Familiarity with such sources can give insight on previous practices, issues or events in an organization, group or community, as well as highlighting sensitive topics, potential research problems or 'local' peculiarities. Gaining information of this sort often involves visiting local (or national) archives. As one example, the London Metropolitan Archive offers access to films and tapes of radio programmes on London, with many taken from private companies, so films of working practices can be obtained. As part of the European Visual Archive, which is funded by DG-XIII of the European Commission, the Metropolitan Archive also collects and digitizes photographs of London from 1900 onwards (with 20 000 images to be displayed eventually on the World Wide Web). As the Metropolitan *Archive News Release* (4 (Summer 1999)) states:

> *The dream would be for a researcher interested in a particular subject and time span to have ready access via the Internet to images held in many different archives across Europe. Thus someone researching Docklands pre-1914 could locate images in Antwerp and London and compare installations regardless of language or national barriers*

Photocopies or notes from preparatory work should be retained, as their pertinence can be difficult to assess until you are writing up. The importance of documentary evidence in ethnographic research should be recognized, for they provide key insights in many substantial contributions (e.g. Geertz 1963; Douglass 1984).

Use a field notebook or diary

When you start to engage with your study population, the best way to record your ethnography is in a *field notebook or diary*, recording what you have observed, heard, smelt or physically felt. In doing so your observations will need to be shaped by theory, which, even if undeveloped initially, directs you towards social events or situations you need to record. Your notes will be of greater value if they are maintained with meticulous attention to detail. This requirement is especially pronounced at the outset, when interpretations of events and processes are more tentative. At this time the significance of particular observations can be lost on you. If you do not record them accurately in detail, your chances of picking up sequences, periodicities or co-variations is diminished. Whether you employ an inductive research strategy or not, if you wish to make full use of the primary advantage of participant observation – detailed contextualization for interpreting human action – then consistently question the theoretical ideas that underpin your interpretations and guide the way you conceptualize what you observe. As Feldman (1995) observed, ethnographic interpretation tends to yield clusters of data that stick together, with uncertainty over how far these clusters are determined by researcher preconceptions or emanate from how investigated populations organize their culture. Researchers need to challenge their interpretations constantly, to test their interpretations by seeking events or processes that test the understanding they are developing (Glaser and Strauss 1967; Strauss and Corbin 1990). In doing this, observing settings and tracking events or sequences are two key forms of observation. The challenge for the ethnographer lies in the transformation of such observations into field notes. The more complete and accurate your field notes, the easier it is to catalogue and code your data. Field notes are best written on a regular basis and in a detailed fashion. You should begin to analyse your data early, in order to develop procedures for identifying patterns and connections over time and space, as well as across social contexts. Early analysis is also important to identify if there is bias in the social contexts you observe, which is a worrying potential distortion in participant observation studies (Vidich and Shapiro 1955; Becker and Geer 1957). Early analysis of your data should also be undertaken so you can engage in an iterative process of refining emergent interpretations against actual behaviour in appropriate 'test' situations.

On a more practical note, the suggestions of Schensul and associates (1999: 115) on making good field notes are appealing:

- Behaviour should be defined behaviourally, rather than in terms of what actions mean for the observer. For example, fidgeting with a pencil and keeping eyes downcast in a meeting may mean several things: boredom, disagreement, lack of understanding, anger, frustration, or preoccupation with another matter. Researchers should describe behaviour and avoid attributing meaning to it in field notes. Discovering what a particular mode of behaviour communicates to others in one setting might not be discernible until the same behaviour has been observed in a variety of other settings (and/or by many other people).
- Descriptions of a person should include details of appearance – clothing, shoes, carriage, items carried, and indicators of the status of material items. For example, rather than describing a person as 'poor and dishevelled', it would be more accurate to describe the person as 'dressed in blue jeans with shredded edges, an army jacket with dirty spots on the back and a torn collar, no belt, a white T-shirt with red smudges around the collar, and shoes with ripped edges, carrying a bulging backpack and a paper bag full of newspapers'. People at the farewell party for the local newspaper editor might be elegantly dressed, but unless the ethnographer described what they were wearing, we would have no idea of the meaning of elegance in that environment.

- The physical state of the environment should be described as if through the lens of a camera. A classroom might be described as bright and warm with a lot of visual stimulation, but readers would understand the description better if it is noted that:

> *... the walls are painted in warm shades of yellow and orange. Three of the four walls have collections of between 10 and 20 photographs, posters, children's drawings, and writing samples. Some of the writing samples are on coloured construction paper. Others are accompanied by colourful outlines, frames, or drawings. (Schensul et al. 1999: 115)*

As Schensul and associates (1999: 115–16) go on to suggest: 'Inferences and personal observations, reflections, hunches, and emotional reactions of the field researcher should be recorded separately from the stream of field notes that describes the event or situation'. In this context, the researcher might adjust to his or her particular needs a format for recording field notes that looks something like that in Box 7.6:

Box 7.6 A basic framework for recording field notes

Date/time:	
Location:	
Field researcher's observations:	Field researcher's comments:

A snippet from the research undertaken in the summer of 1998 by Lees on youth in public space in Portland, Maine, is provided in Box 7.7 (another illustration of field notes can be found in Cook and Crang 1995). Most ethnographers do not make notes when they are with informants. There might be opportunities to make some abbreviated notes (for example, during 'toilet breaks' or other short times on your own), but many researchers use a notebook or a small tape recorder to record material as soon as possible, when on their own (such as at their place of accommodation). It is important to transform abbreviated notes into lengthy descriptions as quickly as possible, as

Box 7.7 An entry from Lees's ethnographic notebook, Summer 1998

Date: August 6th 1998
Location: Tommy's Park
Notes:
There's at least three sets of kids hanging out in T's Pk today – one of these is a group of 3 homeless kids sitting on the wall opposite Starbucks under the tree. They look tired and withdrawn. I approached them to start a conversation but felt nervous – what on earth are they going to make of me. I plucked up the courage however and asked them if I could buy them coffee and donuts – and asked if they'd be prepared to talk to me. The middle guy (looked by age) said OK – he was wearing a black Metallica T-shirt and really baggy jeans with chains and a noticeable hoop nose ring. I went to

SB

Comments:
Why did I pick the obviously homeless kids to talk to first?
Ethics – is coffee and donuts enough recompense for their time – I'm annoyed there isn't much I can do for these guys.
Taping – they were not into being tape recorded – so I had to write down what they said afterward – have I got it all I wonder. I suppose

memory soon fades. If notes are word processed at this time, this should make them more amenable to coding and analysis.

What researchers should bear in mind when they write or type up notes is that field notes are the primary mechanism through which others can evaluate your research. Sanjek (1990: 400–1) is forthright in his criticism of past practice in this regard:

> *One who questions the validity of the historian's conclusions knows what places he or she decided to visit and which documents were used. The sceptic may then examine the historian's sources. And here the anthropology/history parallel ends (in most cases). Headnote evidence is manifested in the ethnography, but rarely are field notes open to anyone's inspection.*

For Sanjek, showing the relationship between field notes and the ethnography based on them is crucial for convincing accounts. In this regard, making field notes available to other researchers is critical for rigour in the research process.

Mapping socially significant patterns

As a cartographic exercise, mapping socially significant patterns is an underutilized technique, although it occurs regularly in Participatory Rural Appraisal (see e.g. Box 2.1). Another example of its use is Ley's mapping (1974) of graffiti and territoriality in a black inner city neighbourhood of Philadelphia, which illustrates an effective way of recording and (re)presenting your observations on a significant social marker. As Webb and associates' commentary (1966/2000) on innovative ways to conduct unobtrusive data collection signifies, a vast array of markers can be mapped (or recorded) to demonstrate socially significant events, whether these be where empty heroin syringes are found or where particular people congregate at specific times of the day. Calling on documentary sources, mapping also illuminates social patterns, as with the home locations of couples married in local churches or the home locations of children attending particular schools. The researcher needs to write notes associated with a map, in order to explain conventions, ways of handling uncertainties and so on. Depending on their length, such notes might be better placed on the map itself or in the researcher's fieldbook.

Mapping images

Mapping images can be employed to get a handle on how informants (differentially) conceive of their social world. It is rare to find this technique employed in ethnographic studies. Yet the technique can confirm interpretations a researcher develops from verbal reports. In this regard, this technique might not be employed until the end of a project and might need careful explanation (even if informants were aware the researcher was investigating their community, group or organization). Perhaps the most well-known publication in this genre is Lynch's assessment (1960) of key landscape features in images of US cities. Another example is Eyles's maps (1968) of how residents demarcated the boundaries of Highgate village, with resulting drawings revealing ellipse-like shapes that (from whatever direction) generally encompassed their own home and the village core but failed to extend the boundary to a similar distance on the opposite side of the core from their home. More recently, Johnston and Valentine (1995) asked lesbians to draw their homes in their research on the performance and surveillance of lesbian identities in domestic environments. Although this mapping procedure has not been used extensively in geographical research, its potential is well recognized by Cook and Crang (1995: 73–5), who provide a useful summary of how to construct information from visual data:

- concentrate on the content of the image(s);
- treat the image(s) as a form of language so they can be interpreted according to the codes of representation they embody; and,
- identify patterns and regularities that provide insight into social processes.

Photography

As a record of life, photography has a long history in the social sciences (see Emmison and Smith 2000: 22–45). As one example, photography was used to explore the inner city in Riis's work (1971) on the Lower East Side. Photographs are not just useful for recording the present, nor should their use be thought of as limited to explorations of the past, for they can be valuable indicators of how landscapes have changed (presuming comparable photographs at different time-points exist). In drawing on photographs to support participant observation work (or indeed for other research approaches), researchers can: (1) collect photographs from the locality that is being researched (on the analysis of existing images, see Emmison and Smith 2000: 46–50); (2) take photographs as part of their field notes; and/or (3) encourage those they are observing to take photographs – sometimes known as autophotography, which is often used in Action Research (see Schartz 1989). Issues of authorship and positionality will be different for these options. In terms of researcher assurance about the context and content of photographs, the most comforting option is (2), photographic ethnography. If this is used, the researcher needs to note the setting in which the photograph was taken, what interaction took place between those involved and so on. As Cook and Crang (1995: 68) point out:

> *Each photo is an act of (self)-presentation that involves the photographer, the photographed, and the expected audience. Both the people taking the photographs and those in them are already aware of the social contexts that will determine the meaning of their actions.*

This reminds us that the meaning of a photograph changes with viewing context. Thus, while photographs can be used in ethnography to find out more about a community, culture, situation and so on, they can also be used as part of a critical text. Baudrillard's *America* (1986) provides a good illustration. Baudrillard uses photographs, like Ronald Reagan's face and smile, juxtaposed with a chapter title to establish an ironic dialogue between text and image. What Baudrillard is seeking is the use of image to elaborate on the ideas in his book. But photographs are always embedded in a narrative context. For instance, if you take photos from your family album, you are likely to find that one photo can trigger discussion about family life at different times and places. Prosser and Schwartz (1998: 119–28) offer further discussion on photographs as part of data collection, as well as commentary on the analysis of photographs. More recently, Gillian Rose's *Visual methodologies* (2001) explains the methods available for reading visual culture, for interpreting visual objects.

Audio-visual techniques

Audio-visual techniques include filming, audio taping and video recording,[14] all of which provide an alternative to the written or drawn record. Audio-visual aids have the notable advantage of taking both visual and vocal acts seriously. Neither filming nor audio taping is new to ethnography, with traditional anthropologists like Franz Boas using them in work on native Canadians (his recordings can be seen in the Museum of Anthropology at the University of British Columbia). However, this tradition of 'film as science' has

14 For a practical handbook for researchers who wish to use text, image and sound in their investigations, see Bauer and Gaskell (2000).

given way to the idea of 'film as experience', which is valued for giving insight into the experience of being a participant in another culture. This audio-visual based ethnomethodology* has increased in sociology but has not gained the same popularity in geography. Given the focus of most ethnographers on 'interaction', the video recorder is particularly useful – in that both verbal and corporeal communication can be recorded. Audio-visual techniques are especially useful for the observation part of participant observation, and, given the prevalence of video recording, can often be employed without sending overt messages that the taping is for research. This does not mean the technique is not subject to 'distortion' from people knowing they are being filmed (would people who pull ridiculous faces at the camera at football games do the same if there was no camera?). Moreover, as the researcher (usually) operates the video camera, this can narrow his or her field of vision, with research events perhaps missed because they occurred out of the line of vision. The fact you record moving images and sounds does not mean your evidence has captured the 'reality' of social context, although it can offer a record of observed events that can be used to reveal your method of interpreting events (for example, for a Ph.D. thesis it could be seen by examiners). We should guard against a temptation to see film as an infallible notebook, so privileging the empirical and visual (Slater 1995). As Banks (1995) reminds us: 'while film, video and photography do stand in an indexical relationship to that which they represent, they are still representations of reality, not a direct encoding of it' (cited in Henley 1998: 42). This caution is especially noteworthy in the context of Crang's warning (1997: 368–9) that: 'Given the pressures to produce "scientific" material it is easy to see how the camera became invested with the weight of making the human, fallible process of observation "scientific".' This must be avoided, even if visual methods open up new windows for analysis.[15]

Audio-visual techniques introduce new analytical and ethical problems. We will not repeat the ethical issues raised in Chapter 6 here (see Box 6.7), but focus on the capacity of audio-visual taping to empower informants. This can be seen when researchers have given video cameras to individuals who do the filming. A good example occurred in Vancouver, where journalists working for the Canadian Broadcasting Corporation gave two street kids – Spice and Shorty – a video camera to make a diary of life on the streets (see Lees 1998). Titled *Home street home*, this video: (1) enabled these children to provide a personal representation of street life; (2) alleviated problems of voyeurism camera crews have often been criticized for; and (3) empowered (albeit temporarily) the

---

15 This point can be made generally for image-based research using CD-ROM technology and hypermedia software. These offer a practical way to explore relationships between text and images, for the researcher can quickly move between media and cover a large volume of information. However, care is needed over the tendency to see CD-ROMs as 'authorless aggregates of objective information which the user can wander over at will constructing his or her own narrative threads. In this sense, these CDs are more akin to an encyclopedia' than they are to an authored research text (Henley 1998: 55). Authorship and narrative need to be clearly established. Good use of multimedia in 'writing up' can be found in Walker and Lewis (1998) and on their website: http://www2.deakin.edu.au/hathaway/.

subjects under investigation. Readers who wish to consider the possibilities of using film might look to research papers based on this medium (e.g. G. Rose 1994; Lawrence 1998) or explore Cook and Crang (1995: 67–75) for a useful espousal and critique of filmic approaches. More broadly, Prosser (1998) offers a series of useful reflections on image-based research, Fiske and Hartley (1978) offer insights on 'reading' television and Gillian Rose (2001) discusses visual research methodologies.

Writing up

It might seem strange to include an entry on writing up in a list of ethnographic research methods. However, it is especially important that researchers self-critically examine their strategies for producing reports based on so intensive a research design. If we envisage writing up as a distinct stage in the research process, this implies a spatial separation between 'field' and 'home'. The first is the place where data were collected and the second where analysis is conducted and research written up. This framing of research seems to be associated with two types of writing (Gupta and Ferguson 1997: 12). In the 'field', there are field notes, with recordings as close to the research experience as possible, including raw documentation of surveys, questionnaires, interviews and observations. At 'home' the writing is polished, reflective, theoretically tied, intertextual, dare we say 'cooked' into journal articles, masters'/Ph.D. theses and research monographs. As Gupta and Ferguson (1997: 12; see also Clifford 1990) illustrate with reference to anthropology:

> *The former is done in isolation, sometimes on primitive equipment, in difficult conditions, with people talking or peering over one's shoulder; writing at home is done in the academy, in libraries or studies, surrounded by other texts, in the midst of theoretical conversation with others of one's kind. Moreover, the two forms of activity are not only distinct, but sequential: one commonly 'writes up' after coming back from 'the field'. Temporal succession therefore traces the natural sequence of sites that completes a spatial journey into Otherness.*

The spatial distanciation associated with such distinctions is linked to entry and exit narratives that have been seen to authenticate and authorize written material (M. L. Pratt 1986). This strategy was traditionally linked to an emphasis in ethnographic writing as objective, coming from a distanced observer. As Katz (1994: 67) explains, such artificial boundaries are commonly taken further:

> *between 'the research' and everyday life; between 'the fieldwork' and doing fieldwork; between 'the field' and not; between 'the scholar' and the subject … between recognition of the artificiality of the distinctions drawn between research and politics, the operations of research and the research … the researcher and the participant …*

Katz warns against constructing such synthetic separations, seeing a (field) site as a vessel holding certain attributes. 'Field' and 'home' are shifting locations rather than bounded spaces. The research process builds what Haraway (1988) calls 'web-like interconnections' between social and cultural locations (see also Gupta and Ferguson 1997: 39). Most evidently, if we undertake research in a self-reflective manner, in which our own positionalities, prejudices and comfortable associations are laid bare, then any distanciation from the rawness (realness) of 'the research process' introduces false distinctions. One reason for making this point is that, even in a post-'crisis of representation' time, when geographers are supposed to be more sensitive to the subjective nature of the research process, we find criticism of the clinical separation of 'field' and 'home' to be too silent. Researchers need to ask how they might engage with writing up 'in the field' (which is not always straightforward), how they can transpose the 'rawness' of what they encounter into the written word, and 'whether seeing fieldwork' as an integral part of everyday life changes their approach toward it, in ethically and theoretically positive ways.

Particularities of poorer countries

We do not wish to add to the list of considerations above, but it is pertinent to draw attention to a companion list associated with research in Third World settings. One reason for particular attention to these settings is because research by those from advanced economies heightens concerns about power inequities. Much of the commentary in this book focuses on settings in Europe and North America, with references to poorer nations providing a sense of the difficulties of generalizing about research methods, as well as drawing out similarities in dissimilar contexts. But only a short library search reveals that the research process is seen in a different light in countries with many poor people. One look at research methods books that focus on Third World settings confirms this impression (e.g. Bulmer and Warwick 1983; Dixon and Leach 1984; Casley and Lury 1987; Finsterbusch *et al.* 1990; Nichols 1991; Devereaux and Hoddinott 1992; Mikkelsen 1995). Much as early participant observation studies came to be associated with studies of poorer communities, so calls for research to empower local communities, to merge theory and practice, found early expression in less developed countries. Here, however, in a reversal of the previously common practice of engaging in lengthy, participant observation studies, analysts have sought quicker multi-faceted data collection methods (see Box 2.1). In doing so, researchers seek to be transparently honest about their intentions, involving the community as much as possible in data collection, while 'handing over the stick' to the community itself, so data can be employed for community benefit (Chambers 1994b). Initially instigated under the heading 'Rapid Rural Appraisal', a later change in title (and approach) to 'Participatory Rural Appraisal' reflects greater emphasis on placing people first.[16] Readers

16 Although now more common in academic research, it is poignant to note that Rapid Rural Appraisal came from the university world, whereas Participatory Rural Appraisal traces its roots to the NGO sector (Chambers 1994a).

might catch some similarities here with community profiling, which is commonly associated with poorer areas in advanced economies (Hawtin *et al.* 1994). Critical to such approaches is recognition of power relations in data collection, with weight placed on triangulation of research methods, to provide more valid information than formal surveys or participant observation alone (e.g., Buzzard 1990; Derman 1990). A key consideration is that the information gathered should be of value to the people in the research site; the researcher, as outsider, adopts more of a facilitator's than an investigator's role (Gow 1990; Honadle and Cooper 1990; Chambers 1994a). Through this researchers seek a more positive research contribution. Although based on different principles, precisely this idea underscores investigations adopting an Action Research framework.

Summary: data collection

- For an effective participant observation study, data collection needs to start before the researcher enters 'the field'.
- Multiple data sources have the particular advantages of tapping different perspectives on a study population, as well as encompassing different segments of that population.
- Data collection and analysis should always be undertaken with an eye on strategies for writing up.

Action Research

In Action Research it is the responsibility of the researcher to get involved in research that matters to the people affected by it. Action Research is often thought of as an effort to democratize research processes. The proclaimed aim is to see that those who are part of the research process benefit in some way from it, for Action Research is intended to be participatory and cooperative: 'Action research provides broad participation in the research process and supports action leading to a more just and satisfying situation for the stakeholders' (Greenwood and Levin 1998: 4). This kind of research practice is associated with a social change agenda. This can be tied to Marxist or radical geography (see *Antipode*), to feminist geography (see *Gender, Place and Culture*) and more recently to the critical geography movement, as allied to critical theory (see *ACME* – the new e-journal for the critical geography movement: http://www.acme-journal.org/). Viewed in this way, Action Research refers to the conjunction of three elements – research, participation and action (Fig. 7.1). As Action Research is proclaimed to take its lead position from the aim of democratization, the researcher must decide what is meant by democracy. If Action Research seeks to change society through the research process, this could be undertaken from a variety of political standpoints. One has only to think with horror of vile research practices perpetrated in Nazi Germany under the guise of betterment for society to realize the potential abuses of such a vision. In this context it is appropriate to be clear about the underlying vision of Action Research. Here the

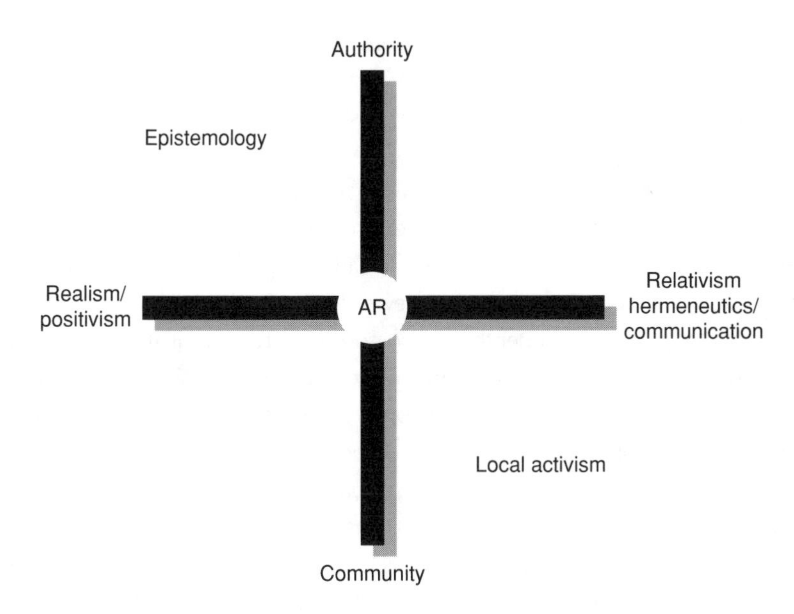

Figure 7.1 *A map of Action Research*
Source: Greenwood and Levin (1998: 90)

standpoint on democracy has been neatly captured by Iris Young (1990), who sees it as being associated with respect for diversity and the power of the disenfranchised to act on their own behalf.

The person generally thought to have coined the term Action Research is the social psychologist Kurt Lewin. Some of his slogans are worth repeating: 'Nothing is as practical as a good theory' and 'The best way to try to understand something is to try to change it'. In the 1930s and 1940s Lewin:

> *set the stage for knowledge production based on solving real life problems. From the outset, he created a new role for researchers and redefined criteria for judging the quality of an inquiry process. Lewin shifted the researcher's role from being a distant observer to involvement in concrete problem solving. The focus was/is on the world of experience, with its complexity, historicity, frictions and assumptions. The quality criteria he developed for judging a theory to be good focused on its ability to support practical problem solving in real-life situations. (Greenwood and Levin 1998: 19).*

Lewin's ideas were taken on board, post-war, by the Tavistock Institute of Human Relations in London. The Tavistock Institute (or Clinic, as it was then known) worked through psychoanalytic thinking and an action orientation. Amongst its outputs was the now famous study by Trist and Bamforth (1951) on the social and psychological consequences of longwall coal getting. This represented a break with conventional

Tayloristic approaches to work, where research focused on finding technically efficient ways to organize workers. It was associated with the industrial democracy movement as an alternative to the conventional hierarchical structure of organizations (see Pateman 1970).

Philosophically Action Research rejects the separation of thought and action, an assumption that has underlain social science research for some time. In doing so Action Research is not prescriptive on method. It can be quantitative or qualitative, using statistical analyses, social surveys, interviews, focus groups, ethnographies and life histories. It is as a form of in-depth participatory research that it fits the remit of this chapter. The aim of action researchers to overcome the separation of thought and action is mirrored by their aim to overcome the separation of science from action. Action researchers turn their backs on 'conventional' disengagement (whether intended or not) with the phenomena studied, equating this disengagement with naive assumptions about objectivity and impartiality. Instead, action researchers seek to narrow this distance as much as possible. They pursue constant interactions between thought (science) and action. This has led some to argue that this research style resembles practices in the physical sciences more than in 'conventional' social science (Greenwood and Levin 1998: 56).

The test of theory for action researchers is the capacity of the research process to solve problems, alter situations, aid social justice and transform power relations in the direction of greater democracy. Other research methods are perhaps less testing of theories that enframe their research design. The action researcher needs to have or learn certain skills – primarily they must know how to be a friendly outsider – for this external role is vital in opening up local groups to processes of change. They must be able to reflect findings about a group back to group members in a supportive, non-critical way. They must be expert (as with interviewers, focus group coordinators and ethnographers) at opening lines of discussion. At the same time the action researcher must enable people to access resources that could be utilized to bring about change. The action researcher must be self-confident in social situations without being arrogant, for people are good at picking up a person's sincerity. Playfulness and irony are said to be useful tools for the action researcher (more broadly the role of humour is explored by Rorty 1980). For Greenwood and Levin (1998: 107):

> *Someone who is unremittingly serious and dour and carries the burdens of the world on his or her shoulders energizes no one. Humor and playfulness have an important role in social change processes. This is because AR projects attempt to suspend business as usual and try to produce unlikely but positive outcomes. In these contexts the powers of irony, absurdity, and humor are considerable precisely because they cause ordinary thought to stop momentarily, creating juxtapositions that can provoke both amusement and openness to change.*

In Chapter 1 we pinpointed the key role of feminist social scientists in outlining the crisis of representation. Feminists have developed a strong commitment to articulate 'the view from below', to hearing the voice of the silenced (women) and to bring these henceforth to the table (see e.g. Townsend 1995). This is an intention that fits well with Action Research in the sense that is seeks to halt and turn around gendered silences (similarly, Marxist, neo/post-Marxist, radical and critical social scientists focus on class silences). The critiques of positivism, essentialism, oppression and the separation of theory and practice that are central to feminist research are central to Action Research as well. Viewing oppression as the norm, research practices are directed by a belief that the status quo must be overturned in favour of more liberating social relations/conditions. Feminist social scientist Patricia Maguire (1987) goes as far as to argue that feminist research cannot but move into the realm of Action Research (see also Reinharz 1992). We have sympathy with this view, although we add a rider. This is well articulated in the work of Catherine Riessman (1993: 80), who warns:

> *Qualitative researchers often seek to depict others' experiences but act as if representation is not a problem. Feminists, for example, emphasize 'giving voice' to previously silenced groups of women by describing the diversity of their experiences ... [but Riessman is more cautious] We cannot give voice, but we do hear voices that we record and interpret. Representational decisions cannot be avoided.*

The implication, that we should not forget, is that action researchers need to do more than give voice, for the remit of this approach is to bring about change; and there is a different politics to who reads academic literature, how it gets onto policy-making agendas, and whether or not it is listened to (Cantley 1992). Moreover, we should not forget that the voice that is 'given' by researchers can be penetrated by career considerations (on this problematic, see D. Rose 1990).

This second point is implicit in the conclusion of Kitchin and Hubbard (1999: 195) about research in human geography: 'it appears that many social and cultural geographers are happy to survey (and "map") the exclusionary landscape, but rarely do much to change that landscape apart from the occasional nod to "planning and policy recommendations".' For Kitchin and Hubbard the roots of this situation lie in the separation of the pristine 'ivory tower' and the messy world of the 'streets'. This distinction 'reinforces notions of modernist, rationalist science and seeks to maintain privileges attached to certain types of (academic) knowledge production over alternative ways of gaining understanding'. They further argue that Action Research may offer 'a route for geographers to combine a role of activist with that of putative academic' (Kitchin and Hubbard 1999: 196). However, this is not unproblematic, for the researcher/activist is often placed in a complicated 'third space' (see Routledge 1996). Perhaps linked to these problems Action Research has not made regular appearances in geographers' toolboxes. Indeed, some work that has adopted this style has not led to publications in academic venues – even if work by Bill Bunge and Bob Colenutt has informed other outputs they have produced (e.g. Bunge 1971; Ambrose and Colenutt 1975; Merrifield 1995). Yet Action Research projects are being undertaken as Ph.D. theses. A geographical example of this was Louise Ackers's investigation (1985) of racism in the Labour Party, which involved active steps to increase anti-racist awareness, alongside deliberate, pointed interventions in party meetings aimed at making policies more race aware. Potentially, research like this can make a powerful contribution to social change. For a window on Action Research in contemporary geography, read the papers in Kitchin and Hubbard's special issue of *Area* (1999). These papers provide examples of Action Research and draw out the pitfalls of the method. A wider perspective on the approach can be found in William Foote Whyte (1990).

Summary: Action Research

- Action Research is associated with learning about society through efforts to change it.
- Changing society involves more than 'giving voice', which has uncertain capacities to bring about change. Geographers have so far shown more concern with 'giving voice' than generating change, although there have been advocates of Action Research within the discipline for some decades.

New field sites for ethnographers

So far we have been dealing with participant observation as a social process in which 'face-to-face' interaction occurs between the researcher and the researched. We now want to turn to a relatively new field site for ethnographers, which is not grounded in this kind of social interaction. This new option arises from the emergence and greater availability of access to the Internet. From this the prospect of undertaking *cyberethnography* or *virtual ethnography* is feasible. While traditional ethnography is associated with travel, often to foreign places, the Internet provides opportunities for social engagement without leaving 'home'. On the Internet, interaction is commonly faceless (Chapter 5). As the ethnographer can be at his or her desk whilst exploring the Internet, the relocation is experiential rather than physical. Cyber or virtual ethnographers do not need to get their hands dirty 'in the field'. They can sit in a sterile office entering the field in a matter of seconds. But such investigators do 'enter' the field, as they have to negotiate access. That said, when we consider the importance of face-to-face interaction in ethnography, we are drawn into considering the crisis of representation in the social sciences (Chapter 1), as brought into focus by the role of 'travel' in the construction of ethnographic authority vis-à-vis fieldwork.

The idea of 'the field' in ethnography has been analysed and critiqued in both anthropology and geography – the two disciplines in which 'fieldwork' is most central to professional and intellectual identities. The field is a space in which (other) culture is waiting to be observed and written (Gupta and Ferguson 1997: 2). One of the issues with the traditional notion of fieldwork is how it fits a post-colonial, mobile, globalized world. As Appadurai (1996: 191, 196) points out:

> *As groups migrate, regroup in new locations, reconstruct their histories, and reconfigure their ethnic 'projects', the ethno in ethnography takes on a slippery, nonlocalized quality, to which the descriptive practices of anthropology will have to respond. The landscapes of group identity – the ethnoscapes – around the world are no longer familiar anthropological objects, insofar as groups are no longer tightly territorialized, spatially bounded, historically self conscious, or culturally homogenous ... The task of ethnography now becomes the unraveling of a conundrum: what is the nature of locality, as a lived experience, in a globalized, deterritorialized world?*

The Internet seems to represent this (over-travelled) new world. An ethnography of, in, and through the Internet can thus be seen as an adaptive and partial approach that 'draws on connection rather than location in defining its object' (Hine 2000: 10).

Given that communicative interaction is the root of ethnographic research, what challenges do new technologies of communication pose for ethnographic research? How does the Internet force us to rethink traditional ways of studying society and culture? For some social scientists the Internet is a logical extension of modern society's obsession with control and rationality. For others the Internet is distinctive by virtue of its uncertainty. It is seen as the embodiment of a postmodern mode of *dis*organization 'characterized by the fragmentation of concepts such as science, religion, culture, society and the self' (Hine 2000: 6). From this perspective it heralds a new information society with possibilities of a radical social change (see Castells 2000), with social restructuring across space and time (for an espousal and critique of this viewpoint, see Lees 1997). On a methodology note, the Internet and other new communications technologies are implicated in the postmodern project, for they cast doubt on authenticity, representation and reality, the unitary self and the separation of self from society. Thus, Poster (1990: 1995) has argued that the Internet enables the decentring and dispersal of the subject, the blurring of boundaries between human and machine, between real and virtual. 'In the Internet postmodernity seems to have found its object, in an "anything goes" world where people and machines, truth and fiction, self and other seem to merge in a glorious blurring of boundaries' (Hine 2000: 7). On the Internet:

> *Communication can be synchronous or asynchronous, it can consist of private messages between known individuals or discussion among large numbers in relatively public forums, and it can be textual or audio or visual. Talking about 'the Internet' encompasses electronic mail (email), the World Wide Web (www), Usenet newsgroups, bulletin boards, Internet Relay Chat (IRC), Multi-User Domains (MUDS) and many other applications. (Kollock and Smith 1999, cited in Hine 2000: 2)*

Geographers can now undertake ethnographic research online. This can be achieved through participant observation in, for example, a 'chat room' or in virtual communities.[17] Using the Internet in ethnographic research draws attention to specific research questions (Box 7.8). Yet the relative absence of ethnographic research on the Internet makes for difficulties in articulating experiences and good practice, even if some points have been rehearsed in arguments against fieldwork-based ethnographies in favour of written and visual representations (see e.g. Dorst 1989). We recommend that before embarking on Internet-based research, especially if this is seen as a step towards a virtual ethnography, be sure to read Hine (2000). Her informative ten principles of virtual ethnography, and especially her ethnographic enquiry into the Louise Woodward law case, are particularly noteworthy (Hine 2000: 63–5).

The Internet has already become a powerful research tool. Although researchers might be cautious until more evaluations of its limitations and strengths for ethnography have been undertaken, the potential for reaching new, especially geographically dispersed audiences, is appealing. However, to achieve high-quality research using the Internet, researchers will have to approach their task in a rigorous manner, critically evaluating whether this is the right, as opposed to an easy, medium through which to obtain data. As those who already receive a lot of e-mail questionnaires will be aware, there is a real possibility of informants tiring of being inundated with calls to participate in online research. Here again the responsibility of researchers towards those who follow needs stressing. As we have indicated often in this book, poorly thought-out projects have the potential to do great harm.

Box 7.8 Research questions raised by virtual ethnography

- How do the users of the Internet understand its capacities? What significance does its use have for them? How do they understand its capabilities as a medium of communication, and whom do they perceive their audience to be?
- How does the Internet affect the organization of social relationships in time and space? Is this different from the ways in which 'real life' is organized, and, if so, how do users reconcile the two?
- What are the implications of the Internet for authenticity and authority? How are identities performed and experienced, and how is authenticity judged?
- Is 'the virtual' experienced as radically different from and separate from 'the real'? Is there a boundary between online and offline?

Source: Hine (2000: 8).

17 Mann and Stewart (2000) discuss collecting data through participant observation in virtual communities (along with virtual focus groups, online interviewing, etc.).

Summary: new 'field' sites

- The Internet is providing access to geographically dispersed, specialist populations, that allow new insight into social behaviour.

Analysing data collected from participant observation

For the ethnographer primary data come in *written form* (text from a field notebook, field audio tapes or an audio-visual recording) or in *visual form* (photos, video images, film, drawings, paintings, maps). Some ethnographers may choose to include another category – namely, music (or more broadly sound). We have not dealt with the analysis of sound in this book, but for a good example refer to Susan Smith (1997) on the soundscape of brass band music (see also S. J. Smith 2000). To analyse written and visual data it has not been our experience that different approaches are necessarily needed. For many of the methods ethnographers use, it is possible to adapt principles and practices to analyse both visual and written data, provided the investigator remembers that different data types are being used.

Central to such analyses is the requirement that researchers do not fall into the 'theory' versus 'practice' binary that feminist geographers have long criticized (see Sayer 1989, 1991 on British geography). This is a binary that Bordo (1986) terms 'the Cartesian masculinization of thought' – that is, the idea that masculinity has been associated with mind, rationality and reason, and femininity with body, irrationality and unreason (see G. Rose 1993 on Anglo-American geography). Indicating the specificities of such ideas, Berg (1994) finds that, in New Zealand geography, this binary is reversed: theory is gendered feminine and empirical research masculine. But it is not the gendering of a binary that is important. What is critical is the fact that empirical investigations are already theoretical, 'just as our theories are infused with experience' (Berg 1994: 254). As Geertz (1988: 129–30) put it:

> *However far from the groves of academe anthropologists seek their subjects – a shelved beach in Polynesia, a charred plateau in Amazonia; Akabo, Meknes, Panther Burn – they write their accounts with the world of lecturns, libraries, blackboards, and seminars all about them. This is the world that produces anthropologists, that licenses them to do the kind of work they do, and within which the kind of work they do must find a place if it is to count as worth attention.*

The reader may wish to replace lecturn with lecture theatre and the 'place' of academia with a degree. Whatever, the point is that you cannot falsely construct a dividing line between fieldwork and brainwork. To summarize: we should explicitly acknowledge that the articulation between theory and empirical investigation involves flows in both directions – 'empirical investigations' are always already theoretical, just as 'theory' is always touched by our empirical experience. It is also necessary to point out, however, that this does not mean that theorizing and doing empirical research are identical activities. (Sayer 1991)

Box 7.9 Analysing ethnographic data using computer software

For field notes of say 50–100 pages, we suggest researchers code text manually, sorting it by hand. For larger studies, we suggest the researcher(s) enter the ethnographic data into a computer text management program. Most of these will read WordPerfect or MicroSoft Word. For the main part the use of computers does little more than increase the speed of retrieval of information from a large volume of transcripts. Software may, however, further assist by helping develop more refined coding schemes and counting the number of times something occurs (namely, content analysis).

It is certainly worth thinking about the rationales that geographers have put forward in urging us to analyse qualitative data using computer software packages. Crang and associates (1997: 772) point out that the academic boom in computer-aided qualitative data analysis software involves '*both* a quantitative expansion *and* a constriction in prior freedoms'. It returns us to those old 'skirmishes between more freeform "interpretation" and scientized "analysis" ' (Crang *et al.* 1997: 773). Using the computer to analyse qualitative data can be seen as one way that researchers have given otherwise subjective data an aura of objectivity, dare we say an aura of rationality, standardization and order. This is not something that we subscribe to, for using a computer software package inevitably creates distance between yourself and the data you have collected (for a sensible critique of tactile versus digital in the context of ethnographic research, see Hinchcliffe *et al.* 1997). Whether you decide to use such computer packages to analyse your data depends a lot on how close a distance you wish to have to your data and research. It also depends on the amount of time you are prepared to invest in learning to use these packages effectively. Low-tech paper-based solutions should not be discarded out of hand! One positive aspect of the digital solution, however, is the reduction of field experiences. We have all at one time or another sought to put too much material into an ethnographic account. We have all had the task of telling students that they have to be selective, that owing to dissertation word limits they must be selective in using qualitative data. Software programs can alleviate some of the pain involved in this. Another positive aspect of such programmes is that they help us avoid mining material for the most 'choice' (or 'juicy' as some would have it) quotations – a process that can introduce a dimension of bias into any analysis.

Most qualitative data analysis software packages code text by categorizing segments according to topics derived from the subjects or theoretical approach of the researcher. The main differences between software packages are in the ways they categorize material. Researchers need to be aware of these differences and of 'their specific possibilities in specific research contexts' (Crang *et al.* 1997: 772). We also need to be aware of the approach to interpretation that we wish to take (see previous section);

Crang and colleagues (1997) identify three: (1) theoretically selective analysis – illustration through 'informant quotation'; (2) analytically inductive analysis or grounded theory – involving categorizing and coding statements to demonstrate the reasoning behind actions, attitudes or beliefs; and (3) formal structural analysis – which focuses on narrative form, tropes, metaphors and metonymity in order to examine structures of a text that generate meaning.

The software packages available have been overwhelmingly focused on the written word, but this is changing fast with developments in graphics, interactivity and increased focus on the WWW. In what follows we do not offer a thorough insight into each software package; rather we outline those we have found most useful to date.

ATLAS.ti. This can be used for the visual analysis of text, audio or video qualitative data. It is especially good for content analysis. It helps the researcher to code and annotate significant elements of a text. It enables you to build networks that visually connect passages and codes to help with the exploration of the meanings embedded in your data. Here images are incorporated as primary material with codes attached to them and links to other text or images. This is an especially useful package for those researchers who have focused on visual data collection.

*QSR NUD*IST* (Non-numerical Unstructured Data Indexing Searching and Theory-building). This supports various methodologies from survey data to focus groups to Action Research. It is a good package for a research team to collate their efforts. Its main added value is that it links ideas and supposedly constructs theory (for criticism of its theory building, see Crang *et al.* 1997: 778). Basically NUD*IST allows the researcher to do the following:

- process texts such as interviews, focus groups, field notes, letter extracts, historical documents, and so on;
- code the data on screen and explore all coding and documents interactively;
- add comments or memos to documents and codes;
- search text(s) for words, phrases or patterns and code the results;
- search indexed categories so as to explore and/or identify complex relationships or patterns;
- create categories of data that are either structured (hierarchical) or unstructured;
- export results into concept mapping software, such as Inspiration or Decision Explorer.

Kitchin and Tate (2000: ch 9) provide a good step-by-step introduction to using QSR NUD*IST. They also include a practice run that can be downloaded from the Pearson Education website (http://www.awl-he.com/). Gahan and Hannibal (1997) also offer a practical guide for using QSR NUD*IST.

HyperRESEARCH enables the researcher to work with both text and multimedia data. Its features include the following:

- Autocode – assigns a code looking for multiple words or phrases.
- Hypothesis tester – presents your data in a similar way to a statistical analysis.

- Code Map – this diagram tool enables you to display your data graphically and graphically explore relationships between codes.
- The ability to use less structured notes.
- Allows the creation of diagrams and the creation of graphics.
- It is good for content analysis.

Ethnograph. This software package has not been discussed in great detail in geography methods texts. Ethnograph is a program for the analysis of textual data. Like NUD*IST, it is primarily an instrument that files information systematically, enabling the researcher to sift and manipulate without having to trawl through reams of paper. Unlike NUD*IST, which is designed for multi-researcher and larger projects, Ethnograph is designed for building complex categories and codings from a smaller set of materials. Amongst the advantages of Ethnograph, the following might be noted:

- If you type in your ethnographic field notes, Ethnograph will number every line of the transcript. You can then identify that particular lines are about a certain topic and assign code words.
- Ethnograph will allow you to search for segments of text according to whether code words overlap.
- Ethnograph will allow you to search for an interviewee by means of interviewee identifiers.
- Details about an interviewee, such as their age, gender, ethnicity, etc., can be attached to each interviewee identifier.

Read Seidel and associates (1995) for a user's guide to Ethnograph.

Textbase Alpha and *HyperQual* are slightly different software packages in that they 'have explicit functions to integrate a range of different materials; they both allow the precoding of prestructured data, such as structured interview or questionnaire responses. Hence it is possible to do preliminary sorting of answers to specific questions, as well as incorporate freeform materials such as interviews or fieldnotes' (Crang *et al.* 1997: 778).

As we have regularly stated, when undertaking ethnographic research there is a need to analyse as you gather your data, which is a practice that provides an overt integration of theory and data collection. Indeed, from our experience, from the minute a researcher collects data, analysis tends to kick into gear, even if this is unintentional. But, while this coalescing might seem to ease the task of analysis, commentators regularly make clear that the analysis of ethnographic information is not straightforward. Wolcott (1994) offers one vision of this, when holding that the real mystique of research comes not from data gathering but from its analysis. As Lee Harvey (1990) points out, one reason for this is that ethnographic research tends to yield a mammoth quantity of information that has to be sifted in analysis.

This is one reason why computer software packages are becoming more popular as data analysis aids (Box 7.9). But such aids, like the theory that has been developed or underscores a project, simply help with the process of bringing materials together (see e.g. Fielding and Lee 1991). The issue of what is selected and what is relegated during analysis and writing-up cannot be articulated in a series of self-evident steps. That analysis should not be expected to be easy is made clear by Dan Rose (1987). Here, in vivid terms, he reveals how the intensity of real-life participant observation led him to question theories and assumptions in the academic literature. In particular, reflecting on the central analytical notion of an event or situation, Dan Rose (1987: 29) found that these commonly employed units of analysis failed to reflect the reality of the lives he studied. They distorted rather than revealed the temporal grounds upon which social life took place. Such a thought-provoking critique of the basic building blocks of so much analysis draws out real difficulties in writing up research. Yet we have sought to show in this book that analysts are not restricted in the ways they examine data. For example, in Chapter 4 we examined the relative merits and shortcomings of content analysis, semiotics and discourse analysis in investigating meaning and key messages in texts. As Box 7.9 shows, these approaches have been introduced into computer software packages for analysing ethnographic data. Similarly, in Chapter 6 we explored the capacity of conversation analysis to provide insight on human beliefs and rationalities. Such procedures are the substance of ethnographic work, which, as Geertz (1973/1975) has argued, is visibly directed toward the interpretation of culture. Yet we note that ethnographic research also focuses on the materiality of social life – not so much with meaning and culture as with the raw exertion of power and processes that create social discriminations and inequalities. Here, as our passage from William Foote Whyte (1984) in Chapter 4 revealed, attention can centre more on what happened than on what it meant. Classic studies that have adopted participant observation strategies to draw out such insights include the work of Dollard (1937/1957) and Griffin (1962) on racial discrimination or Dahl (1961) on the exercise of power in US city politics.

For all these methods the reader will find a substantial number of books and papers that offer specific advice. As various commentators make clear (e.g. Plummer 1983; Barnes and Duncan 1992; Riessman 1993; Ely *et al.* 1997), a regular failing in these accounts is the scant regard they give to writing up research. Our agreement with this has been made apparent on numerous occasions, in making clear that writing creates a 'fiction' that does not exist 'in the real world'. Our writing is a distorted abbreviation of a particular interpretation of the world. Precisely this vision has underscored the social constructivist approach that has been written about in this chapter, for, as Wolcott (1994: 16) reminds us, 'data are tainted with an analytical or interpretive cast in the very process of becoming data', with the same colouring process extending into the translation of 'data' into a research report. But in our awareness of this constructivist process we are informed by a coalescing of viewpoints. This arises from one of the valued products of engaging in multi-author work, for the three authors of this book would position themselves at somewhat different points along hypothetical scales that distinguish whether their main interests are with materiality or meaning (acknowledging the falseness of implying that this is a binary divide) and that specify the research tools they are most comfortable using. This has prompted an interchange of ideas that has provoked learning and self-reflection. Fortunately in this regard we have shared a common sense that attention to questions of representation in human geography have often been troubled by a watering-down of materiality issues. In this regard it is worth indicating for those who feel more comfortable focusing on material concerns that this stance is not inconsistent with learning from the serious critiques social constructivism raises. The fact that writing-up creates a 'fiction' does not mean central elements to this fiction will not be accepted by most analysts – at least if they are viewing the social world you write about from the perspective you adopt. If the social constructivist argument is taken too literally, it can leave the impression that research accounts are unique. This impression is unfortunate. Social constructivism has much to offer those who are mainly concerned with materiality, not just in providing a powerful message that critical self-reflectivity is needed in data gathering, analysis and presentation, but in destabilizing notions of a single truth. Insight from another keyhole might produce a different emphasis and understanding. This should be a welcome addition to understanding – something to prompt reassessment of one's own work, not a threat to it.

But let us finish where we could have begun. For many who are reading this chapter the idea of undertaking participant observation research might seem a frightening prospect. Certainly we would encourage you to approach the method with caution, for it is not easy and needs to be eased into, so that you learn from experience. But if you have not undertaken ethnographic research before, how might you start to learn and then hone your skills? Our suggestion is that you choose different social settings, well before you wish to enter the field using this method. Undertake mini-projects in them. Why not start by having a coffee at Starbucks (or, if you are opposed to the McDonaldization of society, try a local coffee shop where people sit and read or talk for a long time). Perhaps compare this with a pub or bar. Seek to record your observations about what is happening in the settings you select. Do this with a view to trying to reconstruct the social fabric of the place after the event. Ask yourself what the people are doing, how the room is laid out, what signs there are of social relationships and of social differentiation. Hold back from trying to interpret what you 'see' when you are there – practise recording. Do this for a number of occasions in each of the place settings you select. Do not assume the social world that inhabits a location is uniform across the hours of the day or the days of a week (and so on). Think through what makes social situations different (and similar) from one another. A key factor in appreciating events might be consistency over time, but it is just as likely to be inconsistency. When you have completed each session, write up your notes as fully as possible but leave analysis to one side until you have completed all the sessions you want to observe (as you practice more, try to analyse as you go on). Now try to interpret what you saw. Seek to do this not from your own perspective but from that of the people observed. The key in approaching this as a learning exercise is to identify the difficulties you have in this reconstruction. Are your notes inadequate – did you forget to record information about people, so you cannot be sure whether people of one ethnic background engaged equally with others or only with those of a particular class or gender (or)? Are your difficulties linked more to a sense of theoretical shortcoming? Did you record more when social activity was high, so you missed a great deal when it seemed boring? Contrast your notes and interpretations on different settings (and at different times) so you get an idea of whether your notes are too descriptive or not focused enough. Keep asking yourself why am I finding this difficult? Go back and try to observe (and record) the issues you are having difficulty with. Make a presentation to colleagues. Do you convince them? Ask them to critique your account – get them to keep asking 'but how do you come to that conclusion'. Their prompts will probably be different from yours and should help you think in new ways. Practice will not make you perfect. But done effectively it should allow you to enter the field with more confidence and should enable you to undertake higher-quality research. Confidence is important, for first impressions amongst informants can be critical for project success. This means preparatory work is critical.

Aversion to ethnography in geography, according to Herbert (2000), comes from three major criticisms; it is unscientific, too limited to enable generalization and fails to consider its inherent representational practices. In this chapter we have exposed the limitations of such critiques. We agree with Herbert (2000: 564) that 'Geography's neglect of ethnography diminishes the discipline'. It is a different method from the surveys and interviews we outlined in previous chapters because it looks at what people *do*, as well as what people *say*. It is an approach that exposes meaning through action, not just words, and as such is associated with various methods like participant observation, Action Research and more recently performativity. It is a method that is particularly self-conscious about the process of interpretation. It pushes us as researchers to be explicit about how we interpret and the politics therein. It highlights that the dilemma of interpretation is unavoidable, for interpretative practices and subjectivity are central to all social scientific methodologies. We will leave this discussion with the words of David Ley (1988: 126):

> *The geographer's charge to interpret the complex relations of people and place requires a methodology of engagement not detachment, of informal dialogue as well as formal documentation. There is both an ontological and epistemological requirement that place as a human construction be granted more respect and complexity than the profile it displays from the pages of the census.*

Postscript: multi-layered conundrums

Additional information is worthless unless the capacity exists to analyze, criticize and reflect on it.

(Gow 1990: 145)

Throughout this book we have taken positions on issues that have no correct answers. Many of these embody ethical considerations, and ethics is a messy terrain. Much of what is considered to be ethical (or unethical) derives from your own vision of integrity, which draws on your own values, as well as the degree of awareness you have about specific places, people, processes or events. We are not priests and have no intention of preaching. Moreover, as we have noted at various points, unintentionally or not, in one guise or another, all research can be portrayed as having an unethical dimension, if only from one viewpoint. Merely taking someone's time away in an interview, asking certain kinds of questions, using published statistics with in-built biases, telling a 'true' story that unexpectedly causes grief for those portrayed, might all be unethical for some. In this context it is unlikely any book on procedures will not stand on someone's toes. This should not lead us to hide in a bunker. The research process should be a learning experience, a personal voyage of discovery. The personal dimension of this should not be underestimated. To state a point we have made often, a critical element in good research is to be self-reflective. This applies as much to your capacities to undertake a project or employ specific methods of data collection and analysis, as it does to seeking counter-evidence to 'challenge' the 'validity' of interpretations you develop.

This might appear to suggest that anything goes. This is not so. There is good research and bad research, but interpretations of what these are vary. This is not just across individuals but over time, between countries and in some measure within peer groups. In this framework there is no cook's book guide to ethical practices. All we can do is offer prompts for you to think about. That these are only prompts is signified by the 'conundrums' in the title of this postscript. That they are challenges for all research should be emphasized. As Howard (1994) honestly explains, whatever their good intentions, researchers can be overtaken by the pressures of academic expectations, along with personal costs. We need to recognize that competing demands on us as individuals, which are heightened if we see the 'field' as a distinct entity, which can be 'abstracted' once we return 'home' to write up. Recognition of the force with which such inconsistencies can impose themselves lies behind this commentary being the leaving message of the book. Ethical dimensions penetrate to the heart of research in one way or another. For those who wish to undertake research that is self-reflective as well as sensitive to wider potential gain than themselves, these are not add-ons. They are woven into all aspects of research, whether project design, data collection, data analysis or written output.

Putting these ideas in a broad framework, Marcia Taylor (1994) holds that research should be informed by three key norms. These are: *beneficence*, which means efforts should be made to maximize positive outcomes while minimizing the prospect of harm or risk to those taking part in a study (and we could add those reading it); *respect*, which relates to protecting the autonomy of participants; and *justice*, which involves ensuring non-exploitative and carefully considered research procedures are used. Taking such considerations on board starts with thinking about the design of a project, in the sense of what is looked for in data. Here Eichler (1988) is helpful. What Eichler brings out when examining non-sexist research methods are principles with broader applicability, concerning the way data can be 'silent' about particular populations. The shortcomings Eichler identified are sevenfold: (1) androcentrality or seeing the world from a male-centred angle (to which we can age being race, age, class and so on blind); (2) overgeneralization, in which one subgroup is investigated but interpretation and result presentation extend to others, such as referring to parents when what is meant is mothers; (3) insensitivity to the social importance of variables, as with showing that migration is related to social mobility without exploring whether this fits both men and women – which it does differentially (e.g. Halfacree 1995); (4) using double standards, such as measuring variables unequally for population groups, as with allocating married women a social 'class' based on their husband's employment; (5) inappropriate role ascription, such as ascribing domestic or unpaid work 'naturally' to women; (6) inappropriate analytical units, such as investigating behaviour as family actions when individuals engage in them; and (7) dichotomization, in which population groups are taken to be mutually exclusive when their actions or attributes overlap. Each of these points relates to all elements in a research process, whether design, implementation or presentation.

In processes of design, implementation and presentation, presentation has received least attention in the literature, and indeed in deliberations of learned societies about the ethics of research. Thus, while review boards or ethics committees for research proposals, alongside learned society or governmental guidance for the treatment of 'subjects' and gaining 'informed consent', are not unusual, you will look hard for commentary on the ethics of writing up results. As various commentators have made clear, the literature is strangely silent about almost all aspects of writing up (e.g. Plummer 1983; Barnes and Duncan 1992). This is unfortunate. As various authors have revealed, in writing up what they saw as a fair and honest account of a social situation, they have at times caused great offence amongst those they investigated (e.g. Vidich and Bensman 1968; Morin 1970). Indeed, any elucidation of social processes and values in organizations, social groups or communities is likely to reveal elements some people do not want 'outsiders' to believe. Making a pointed comment in this regard, Duncan and Barnes (1992: 252) remind us that 'to speak for another [when writing up, in this case] is not a politically neutral act. We have appropriated their voice – colonized their perspective'. In this context, written reports, most obviously when about wealthy or powerful groups, can result in threats of legal action (Wallis 1977). Research notes can also be called on by the state, with no recourse in most countries to protecting the confidentiality of informants if (say) the police request your data (Mitchell and Draper 1982; Lee 1993). Moreover, despite efforts by many researchers to conceal the identity of informants, perhaps by using fictitious names for places, organizations and people, the literature is replete with exposures of informants, even 50 years after a study (Plummer 1983). Having put informants in the way of such 'threats', ethical questions are even posed by acknowledging your indebtedness to them (Howard 1994; Robson 1994).

But the ethics of research is not just about informants, but also about result presentation. One problem with mentally distinguishing 'home' from 'field' when data are collected is that it encloses writing up in a kind of synthetic cotton wool. This can diminish the capacity of 'field' experiences to penetrate the cocoon of the writing process. When you are sitting in the confines of 'home', theoretical nicety, scoring in examinations and ascending career ladders can take precedence over the uncomfortable roughness of 'the field'.

Two aspects of this raise little-acknowledged ethical questions. On the one hand, there is a tendency to neglect the past, to exaggerate its imperfections and to embellish the novelty of what is proclaimed as new. This tendency is not new, as Plummer (1983) notes. Fox Keller (1985: 5) captures the nature of such exaggerations in a specific context: 'The proposition that science is subject to the influence of special interests has been transformed, in some quarters, to relativism – to the view that science is nothing but the expression of special interests'. This tendency has existed in human geography at least since the quantitative revolution of the 1960s (Bunge 1966; D. W. Harvey 1969). Think of the structuralists who suddenly 'found' gender and race after more than a decade of acknowledging little except class conflict. Alternatively, we note with bemusement, as one of us has sought to achieve this for some time, that some cultural geographers are recognizing that they overstated their case in the past and have begun to call for a rematerialization of theorizations (Jackson 2000; Philo 2000: 30–6).

A second tendency is convenient neglect of 'uncomfortable facts'. We find this in many guises. We restrict ourselves to two. One was brought to mind in exploring the outrage that accompanied Fogel and Engerman's work (1974/1989) on slavery (see Box 3.5). As these researchers have since made clear, while contradicting certain inaccurate images about the slave economy, they should have acknowledged more forcefully the immorality of slavery. Our concern is not with this shortcoming but with reactions to their conclusions. These carried an implication that casting any aspect of slavery in a 'more favourable' light cannot be accepted, even if evidence supports it. A similar point has been made by investigators like Sender and Smith (1986) over inaccurate claims by commentators like Rodney (1976) that slavery did not exist in Africa before white traders came. Uncomfortable as some will find it, those who visit former slave-trading forts in countries like Ghana soon find that the first white engagement with the African slave trade was sales to local chiefs. Ignoring 'uncomfortable evidence' can only weaken theorization, as it changes the weight given to different 'causal forces'. Although the point has broader applicability, being theoretically blinkered is most noticeable when arguments carry an ideological tone, including wanting to make 'amends' for past misdeeds (even if only by being sympathetic). One example is found in the image that European colonialism was built on a vision that subjugated people were inferior. Contrast this with Trautman's work (1997), which shows that in the early years of British rule in India common ground was sought with Indian peoples, based on the vision that the British and Indian populations were both Aryans with a shared ethnic position. Only in Victorian times was this reading changed, in part as a consequence of the Indian Rebellion – what UK history calls the Indian Mutiny – after which the British withdrew from Indian society, in part for fear of treading on cultural distinctions. As the 'divide-and-rule' policy that followed showed, this change marked a more aggressive colonialism, which many contemporary commentaries seem to take as always present (e.g. Tammita-Delgoda 1997; Grover and Grover 2001). Recognizing periodicity in British stances should strengthen theory. Assuming Victorian themes fit all colonial times diminishes it. We provide the above examples deliberately, to indicate that, even for practices widely regarded as heinous (most certainly by us), there is a more complex 'reality' than contemporary accounts are prone to paint. The ethics of ignoring 'the uncomfortable' goes beyond intellectual advancement, as such imageries too readily find voice in the public psyche.

Of course we have to recognize the power politics of dominant themes (and omissions) in academic accounts. Offering a particular slant on Kuhn's vision (1970) on how 'normal science' constrains insight, Lee (1993: 34) reminds us of the practice of 'chilling' – in which researchers are deterred from disseminating research results because they anticipate a hostile reaction from colleagues. In a sense this is a form of intellectual dishonesty. Yet, as Longino (1990) reminds us, social interests bear heavily on what are considered acceptable beliefs, hypotheses or conclusions, for research is penetrated by social mores and power relations in wider society (R. J. Johnston 1991; Demeritt 2000). These forces cannot be set aside easily. But for us researchers have a responsibility not to follow the mass unthinkingly. After all, according to popular legend, lemmings come to a sticky end.

A sense of questioning is imperative in deciding on research methods. One of the less than praiseworthy aspects of the history of human geography has been periods of methodological dogma. Assertions made in the 1970s that qualitative research was 'un-scientific' were no more credible or creditworthy than assertions made in the 1980s and into the 1990s that quantitative research has nothing to offer. At least some contemporary commentators are recognizing that both qualitative and quantitative methods have much to offer (e.g. Philo *et al.* 1998; Hodson 1999). We have sympathy with his sentiment. In this book we show that all research methods have weaknesses as well as potentialities. They need to be treated with care. They are like pet ferrets: handle them well and they bring joy; mess around with them casually and disorder accompanies personal disfigurement.

Glossary

Anti-foundationalism Foundationalist claims to knowledge are based on systematic claims about how knowledge is made possible, which are used to assess competing claims. Most commonly associated with logical positivism and critical realism, these claims have been challenged by ideas from postmodernism and post-structuralism (amongst others), which reject claims of 'grand theory'. Anti-foundationalist philosophies challenge the idea of a (singular) dominant philosophy.

Bias Bias is multidimensional and can occur at every phase of the research process. It is used to refer to a prejudice, or any special influence that sways the mind.

Causality As John Marshall (1985: 121) notes, causality is 'an explanatory frame imposed upon the data by the observer'. The implication is that statements about causality are inevitably theory laden. When referring to causality, researchers are concerned with what causes (or creates) a process, event or phenomenon. This does not mean what occurs regularly alongside it, as statistical correlation shows (minus theoretical explanation of how the two are linked). Moreover, for causation to be assumed, the implication is that a change in condition X (that which is being explained) is caused by a change in condition Y. In other words, causality implies process over time, not coexistence at one point in time.

Contingency Contingency is dependency on something else; or an event that is liable but not certain to happen: 'Contingency is the quality or state of being contingent' (*Chambers Dictionary*, 1989).

Critical realism Developed in particular by Bhasker (1978, 1986), critical realism is sometimes thought of as the 'middle way' between positivism/empiricism and the relativism of postmodern positions. It suggests that, while there is an 'out-there' reality to be studied, this reality is inevitably constructed through our interpretations, leading to the existence of multiple understandings of those 'realities'.

Deconstruction Promoted in particular by Derrida, deconstruction is an approach that undermines claims to authority by exposing the rhetorical strategies used by authors. Deconstruction is the attempt 'to undo claims to truth and coherence by uncovering the incoherences within texts and tracking down the traces of oppositional elements, each in the other' (R. J. Johnston *et al.* 1994: 468). This process often occurs by first reversing oppositions such as male/female and culture/nature and then displacing them.

Deduction Deduction is an approach to research that proceeds from theory formulation to testing empirically.

Discourse Under the influence of Foucault, discourse refers to systems of knowledge and their associated practices. However, the concept of discourse is used in manifold ways. Ron Johnston and associates (1994: 136) define it as 'the ensemble of social practices through which the world is made meaningful and intelligible to oneself and others: frameworks that embrace particular combinations of narratives, concepts, ideologies and signifying practices, each relevant to a particular realm of social action'. Discourses are heterogeneous, regulated, embedded, situated and performative (R. J. Johnston *et al.* 2000: 180).

Ecological fallacy Ecological fallacy refers to the error of assuming that relationships between variables at one geographical scale also occur at another scale. So, if more ice cream is eaten in towns with a high proportion of elderly people, this does not mean that it is elderly people who are eating the ice cream. It might simply be that high proportions of elderly people are often found in resorts, where others come for visits during which they eat ice cream.

Emancipation Emancipation is the process through which people can be set free from restraint.

Empirical The term 'empirical' is usually applied to practical primary research that is conducted in the 'field'. It is often used to refer to work resting on trial or experiment.

Empiricism Crude empiricism sees the scientist as a spectator, with reality external to the scientist. Empiricism is a doctrine proclaiming that all knowledge ultimately originates in experience. Crude empiricism dictates that an investigation is begun without anticipating research results (Doyal and Harris 1986).

Empowerment In the research context, empowerment is used to refer to mechanisms that can be used to facilitate participants to take control of their lives in some way; to feel they have more power and efficacy over their everyday life.

Epistemology An epistemology is a conception of what constitutes valid knowledge (or a particular kind of knowledge, such as scientific knowledge). An epistemology offers solutions to three key problems: (1) what is the origin of knowledge; (2) how is the merit of knowledge established once it is created; and (3) what is meant by 'scientific' progress and how can this be identified (Doyal and Harris 1986).

Essentialism Essentialism is often used to describe negatively the practice of reducing human experience to universal essences, in opposition to the view that human nature is a social construction.

Ethnography 'In its most characteristic form it involves the ethnographer in participating overtly or covertly in people's daily lives for an extended period of time, watching what happens, listening to what is said, asking questions – in fact collecting whatever data are available to throw light on the issues that are the focus of the research' (Hammersley and Atkinson 1995: 1). As such, while ethnography is commonly associated with active participation in 'community' life, it can also involve in-depth interviews, documentary analysis, social surveys and whatever other data collection devices add insight on the problem at hand (see e.g. Vidich and Bensman 1968).

Ethnomethodology Ethnomethodology involves examining the ways people produce orderly social interaction on a routine, everyday basis. It is concerned with the study of the 'common-sense reasoning skills and abilities through which ordinary members of a culture produce and recognize intelligible courses of action' (Heritage 1989: 21). It provides the theoretical underpinning for conversation analysis in looking at these abilities in conversational encounters.

Fordism Fordism refers to the 'set of industrial and broader societal practices associated with the work-place innovations pioneered by Henry Ford in Detroit, in the second decade of the 20th Century' (R. J. Johnston *et al.* 1994: 203). In general, the term is used to describe the processes and practices of mass production.

Foundationalism Foundationalist claims to knowledge are based on systematic claims about how knowledge is made possible, which are used to assess competing claims. Most commonly associated with logical positivism and critical realism, these claims have been challenged by ideas from postmodernism and post-structuralism (amongst others), which reject claims of 'grand theory'.

Grounded theory Grounded theory refers to a strategy of theory development that seeks to discover theory from data, using the method of comparative analysis. Its underlying philosophy holds that society and people are changing too rapidly for deductive approaches to theory verification to work. Rather than developing 'grand theory', all we can hope for is short-term theory, with the most important criterion for evaluating theory being that 'it works'. As a consequence this approach is associated with theoretical sampling rather than statistical sampling techniques (Glaser and Strauss 1967; Strauss and Corbin 1990).

Hermeneutics Hermeneutics is the study of interpretation and meaning. It is based on the ontological position that the world is subjective, with the epistemological project being to make interpretations of this subjective world.

Humanism Humanistic geography accords an active role to human agency focusing on understanding the meaning, value and human significance of life events; basically what a person is and what he or she can do. Humanism in this sense is often associated with the French School of human geography (Vidal de la Blache), but more recently it has influenced the development of the 'new' cultural geography.

Hypothesis Ron Johnston and associates (1994: 267) define hypothesis as 'a provisional statement which guides empirical work in several scientific epistemologies. Within positivism, a hypothesis is an empirical statement not yet accepted as true – they structure empirical research programmes ... in Critical rationalism hypotheses are designed not to be validated, but to be refuted – or falsified.'

Induction Induction defines the process of research that proceeds from research investigations to theory (in contrast to deduction, which moves the other way).

Method Whereas methodology is a coherent set of rules and procedures that function within epistemological and ontological frameworks, method refers to the 'process' or technical means of collecting data.

Methodology Methodology embraces issues of methods of data collection and analysis when these are grounded in the bedrock of a specific view on the nature of 'reality' (ontology) and the basis on which knowledge claims are made (epistemology).

Mimesis Mimesis is the belief that researchers should seek to produce as accurate a representation of the world as they possibly can.

Objectivity Something that is 'objective' relates to that which is external to the consciousness (as in contrast to subjective, which relates to human consciousness). In research methodology, 'objectivity' is often used to describe something as 'value free' or uncoloured by one's own sensations or emotions.

Ontology Ontology refers to a set of assumptions that underpin a theory or system of ideas – that which can be known.

Paradigm Based on a notion developed by Kuhn, paradigm refers to the 'working assumptions, procedures and findings routinely accepted by a group of scholars, which together define a stable pattern of scientific activity; this in turn defines the community which shares it' (R. J. Johnston *et al.* 1994: 432).

Phemonenology Phemonenology describes a philosophy that is 'founded on the importance of reflecting on the ways in which the world is made available for intellectual inquiry: this means that it pays particular attention to the active, creative function of language and discourse in making the world intelligible' (R. J. Johnston *et al.* 1994: 438)

Positionality Positionality refers to historically generated circumstances that create the 'position' of the researcher, such as age, gender, ethnic heritage, education and life experiences. These define what Denzin and Lincoln (1998: 23) term the 'biographically situated researcher'. These conditions nurture experiences, and therefore understandings, of the world and situate the researcher with regard to research subjects.

Positivism Loosely, positivism is an approach that emphasizes the discovery of laws of society, often involving an empiricist commitment to quantitative methods. Logical positivism is based on the ontological argument that the world is objectively given. The epistemological effort therefore is to apply objective methods to acquire the truth.

Post-Fordism In contrast to Fordism, post-Fordism refers to a 'a collection of work-place practices, modes of industrial organization and institutional forms identified with the period since the mid-1970s, characterized by the application of production methods considered to be more flexible than those of the Fordist era' (R. J. Johnston *et al.* 1994: 459).

Postmodernism Postmodernism is a complex and highly contested concept. In general, postmodernists argue that modernist meta-narratives, which seek absolute and universal truths in their modes of enquiry, fail to account for the inherent differences in society. Postmodernists emphasize multiple readings of texts and the existence of multiple understandings of reality.

Rapport Rapport is the development of a positive, empathetic relationship with research participants.

Realism Realists argue that a reality exists independently of our thoughts and beliefs. Realism (in contrast to positivism) makes a fundamental distinction between the identification of causal mechanisms and the identification of empirical regularities, because 'what causes something to happen has nothing to do with the number of times it happens' (R. J. Johnston *et al.* 2000: 673).

Reflexivity According to England (1994: 82), reflexivity is 'self-critical sympathetic introspection and the self-conscious analytical scrutiny of the self as researcher'.

Relativism Related to postmodernism, relativism is a doctrine that suggests that accepted standards of right or wrong vary across social environments and from person to person.

Reliability Reliability refers to repeatability or consistency in research findings. It is based on the idea that research should not be biased by an investigator, so results should be 'repeatable' to check on this. The core idea that researchers should seek to avoid imposing their own biases such that these overwhelm contradictory evidence has much to commend it. The idea that data collected by one researcher are capable of 'exact' repetition is less credible, given (amongst other things) the capacity of data providers to learn from previous rounds of information provision. Expectations of reliability conditions being sustained are commonly associated with positivism.

Semiotics Semiotics is preoccupied with the process of interpretation. Elements of a text are seen to derive their meaning from their interrelation with a code.

Social constructionism Social constructionism is the idea that society constructs the world as people experience it. Extreme social constructionists do not believe that an objective world exists 'out there', only our interpretations of it.

Structuralism Structuralism is the view that deep structures exist behind the social and cultural realities that we perceive. In geography, French structuralism, structural Marxism and post-structuralism exhibit some similarities but many differences.

Subjectivity Subjectivity is defined by Ron Johnston and associates (1994: 604) as 'a concept that grounds our understanding of who we are, as well as our knowledge claims'. Within different epistemological and methodological traditions the notion of subjectivity is conceptualized in different ways (but shares a common opposition to objectivity).

Text Text is 'a set of signifying practices commonly associated with the written page but over the past several decades increasingly broadened to include other types of cultural production such as landscapes, maps, paintings as well as economic, political and social institutions' (R. J. Johnston *et al.* 2000: 824–5).

Theory A theory is frequently conceived of as a set of connected statements used in the process of explanation (R. J. Johnston *et al.* 1994). The nature and status of theories can be seen to differ among the various philosophies of social science.

Thick description Thick description is used by Geertz (1973/1975) to describe the process whereby research is carried our through intensive interrogation of participants actions and understandings to uncover 'structures of significance'. It is usually associated with the ethnographic method and hermeneutics.

Triangulation Triangulation involves employing complementary methods or data sources to circumvent the potential inadequacies of single data sources. A key issue is that, if the conclusions reached using different sources coalesce, the researcher should be able to put forward research conclusions with more confidence.

Validity Analysts have recognized various forms of validity, with a basic distinction being between internal and external validity. Most commonly, when analysts refer to 'validity' they mean internal validity. External validity refers to the extent to which research conclusions or theories can be extended beyond specific contexts; to the generalizability of findings (Hedrick *et al.* 1993). As Bernard (1994) noted, it is not worth debating this if internal validity is absent. What is meant by internal validity is also open to different interpretation, with Vaus (1991) distinguishing between criterion, content and construct validity. However, the first two of these are specific to answers to questions, whereas construct validity is concerned with the specifics of data collection as well as the conceptualization of research problems and research design. Construct validity is what most commentators mean when they use the word validity on its own. In a nutshell, construct validity occurs to the extent that a concept is represented/measured in a manner that is consistent with the underlying theoretical perspective of the study. In reality, although often proclaimed as a primary criterion for good research, evaluation of validity is not easy (and is rare).

References

Ackers, L. 1985: Racism and political marginalization in the city: the relationship between black people and the Labour Party in London. Ph. D. thesis, London School of Economics.

Addison, P. 1994: *The road to 1945: British politics and the Second World War*. Revised edition. London: Pimlico.

Agar, M. and MacDonald, J. 1995: Focus groups and ethnography. *Human Organization* **54**, 78–86.

Ahmad, W. 1999: Ethnic statistics: better than nothing or worse than nothing? In D. F. L. Dorling and S. Simpson (eds), *Statistics in society*. London: Arnold, 124–31.

Aiken, C. S. 1987: Race as a factor in municipal underbounding. *Annals of the Association of American Geographers* **77**, 564–79.

Aitken, S. C. 1997: Analysis of texts: armchair theory and couch-potato geography. In R. Flowerdew and D. Martin (eds), *Methods in human geography*. Harlow: Longman, 197–212.

Alford, B. W. E., Lowe, R. and Rollings, N. 1992: *Economic planning 1943–1951: a guide to documents in the Public Record Office*. London: PRO Publications.

Allen, C. 1975: *Plain tales from the Raj: images of British India in the twentieth century*. London: André Deutsch.

—— 1983: *Tales from the South China Seas: images of the British in South-East Asia in the twentieth century*. London: André Deutsch.

Allen, J. and McDowell, L. 1989: *Landlords and property: social relations in the private rented sector*. Cambridge: Cambridge University Press.

Alonso, W. and Starr, P. (eds) 1987: *The politics of numbers*. New York: Russell Sage Foundation.

Ambrose, P. J. and Colenutt, R. J. 1975: *The property machine*. Harmondsworth: Penguin.

Amedeo, D. and Golledge, R. G. 1975: *An introduction to scientific reasoning in geography*. New York: Wiley.

Anderson, E. 1978: *A place on the corner*. Chicago: University of Chicago Press.

Anderson, K. and Gale, F. (eds) 1991: *Inventing places: studies in cultural geography*. Melbourne: Longman Cheshire.

—— —— (eds) 1999: *Cultural geographies*. South Melbourne: Addison-Wesley.

Anderson, M. J. 1982: *The American census: a social history*. New Haven: Yale University Press.

Anderson, N. 1923: *The hobo*. Chicago: University of Chicago Press.

Andranovich, G. D. and Riposa, G. 1993: *Doing urban research*. London: Sage.

Anzaldua, G. 1987: *Borderlands/La Frontera: the new Mestiza*. San Francisco: Spinsters/ Aunt Lute Foundation Books.

Anzaldua, G. (ed.) 1990: *Making face, making soul = Haciendo Caras: creative and critical perspectives by women of color*. San Francisco: Aunt Lute Foundation Books.

Armstrong, G. 1993: Like that Desmond Morris. In R. Hobbs and T. May (eds) *Interpreting the field*. Oxford: Oxford University Press, 3–43.

Armstrong, J. S. 1967: Derivation of theory by means of factor analysis or Tom Swift and his electronic factor analysis machine. *American Statistician* **21**(5), 17–21.

Asbury, J. E. 1995: Overview of focus group research. *Qualitative Health Research* **5**, 414–20.

Atherton, L. 1994: *Never complain, new explain: records of the Foreign Office and State Paper Office 1500–c.1960*. London: PRO Publications.

Atkinson, J. M. 1978: *Discovering suicide*. London: Macmillan.

Atkinson, P. 1990: *The ethnographic imagination*. London: Routledge.

—— and Coffey, A., 1997: Analysing documentary realities. In D. Silverman (ed.), *Qualitative research: theory, method and practice*. London: Sage, 45–62.

—— —— Delamont, S., Lofland, J. and Lofland, L. H. 2001: *Handbook of ethnography*. London: Sage.

Baktin, M. 1981: *The dialogic imagination*. Austin: University of Texas Press.

Balchin, P. N. 1979: *Housing improvement and social inequality: case study of an inner city*. Farnborough: Saxon House.

Ball, S. J. 1984: Beachside reconsidered: reflections on a methodological apprenticeship. In R. G. Burgess (ed.), *The research process in educational settings*. London: Falmer, 69–96.

—— 1994: Political interviews and the politics of interviewing. In G. Walford (ed.), *Researching the powerful in education*. London: UCL Press, 96–115.

Banks, M. 1995: Visual research methods. *Social Research Update* **11**.

Banthiya, R. 1994: *From historicity to postmodernity: a case of South Asia*. Jaipur: Rawat.

Barbour, R. and Kitzinger, J. 1999: *Developing focus group research*. London: Sage.

Barke, M. and France, L. A. 1988: Second homes in the Balaeric Islands. *Geography* **73**, 143–5.

Barley, N. 1983: *The innocent anthropologist: notes from a mud hut*. Harmondsworth: Penguin.

—— 1988: *Not a hazardous sport*. Harmondsworth: Penguin.

Barn, R. 1994: Race and ethnicity in social work: some issues for anti-discriminatory research. In B. Humphries and C. Truman (eds), *Re-thinking social research*. Aldershot: Avebury, 37–58.

Barnes, T. J. 1996: *Logics of dislocation: models, metaphors, and meanings of economic space*. New York: Guilford.

—— and Duncan, J. S. 1992: Introduction: writing worlds. In T. J. Barnes and J. S. Duncan (eds), *Writing worlds*. London: Routledge, 1–17.

—— and Hannah, M. 2001: Theme issue: quantitative geography. *Environment and Planning D: Society and Space* **19**.

Barnett, C. 1998: The cultural turn: fashion or progress in human geography? *Antipode* **30**, 379–94.

—— 1999: Deconstructing context: exposing Derrida. *Transactions of the Institute of British Geographers* **24**, 277–93.

—— and Low, M. 1996: Speculating on theory: towards a political economy of academic publishing. *Area* **28**, 13–24.

Bassett, K. A. 1995: On reflexivity: further comments on Barnes and the sociology of science. *Environment and Planning* **A27**, 1527–33.

Batterbury, S. 1994: Alternative affiliations and the personal politics of overseas research. In E. Robson and K. Willis (eds), *Postgraduate fieldwork in developing areas*. London: Institute of British Geographers Developing Areas Research Group Monograph No. 8, 60–89.

Baty, P. 1999a: A quality game where cheating is allowed? *Times Higher Education Supplement* 12 Mar., 4.

—— 1999b: Teach-only contracts threat. *Times Higher Education Supplement* 2 Apr., 40.

Baudrillard, J. 1983: *Simulations*. New York: Semiotext(e).

—— 1986: *America*. London: Verso.

Bauer, M. and Gaskell, G. 2000: *Qualitative researching with text, image and sound: a practical handbook*. London: Sage.

Bax, M. 1976: *Harpstrings and confessions: machine-style politics in the Irish Republic*. Assen: Van Gorcum.

Baxter, J. and Eyles, J. D. 1997: Evaluating qualitative research in social geography: establishing 'rigour' in interview analysis. *Transactions of the Institute of British Geographers* **22**, 505–25.

Beck, U. 1992: *Risk society: towards a new modernity*. London: Sage. Translation of *Risikogeselleschaft: auf dem weg in eine andere moderne*. Frankfurt am Main: Suhrkamp Verla, 1986.

Becker, H. S. and Geer, B. 1957: Participant observation and interviewing: a comparison. *Human Organization* **16**(3), 28–32.

Beevor, A. 1998: *Stalingrad*. Harmondsworth: Penguin.

Bell, C. 1977: Reflections on the Banbury restudy. In C. Bell and H. E. Newby (eds), *Doing sociological research*. London: Allen & Unwin, 47–62.

—— 1984: The SSRC: restructured and defended. In C. Bell and H. Roberts (eds), *Social researching*, London: Routledge & Kegan Paul, 14–31.

—— and Newby, H. E. 1971: *Community studies*. London: Allen & Unwin.

—— —— (eds) 1977: *Doing sociological research*. London: Allen & Unwin.

Bell, J. 1993: *Doing your research project*. 2nd edition. Buckingham: Open University Press.

Bell, M. M. 1994: *Childerley: nature and morality in a country village*. Chicago: University of Chicago Press.

Belson, W. A. 1981: *The design and understanding of survey questions*. Aldershot: Gower.

Benjamin, B. 1970: *The population census*. London: Heinemann.

Bennett, R. J. 1985: The impact on city finance of false registrations in second homes. *Tijdschrift voor Economische en Sociale Geografie* **76**, 298–308.

Berg, L. 1994: Masculinity, place and a binary discourse of 'theory' and 'empirical investigation' in the human geography of Aotearoa/New Zealand. *Gender, Place and Culture* **1**, 245–60.

Berger, P. and Luckmann, T. 1967: *The social construction of reality*. London: Doubleday.

Bernard, H. R. 1994: *Research methods in anthropology*. 2nd edition. Thousand Oaks, CA: Sage.

Bernstein, C. and Woodward, B. 1974: *All the President's men*. New York: Warner Books.

Bernstein, R. 1983: *Beyond objectivism and relativism: science, hermeneutics, and praxis*. Philadelphia: University of Pennsylvania Press.

Berry, B. J. L. 1971: DIDO data analysis: GIGO or pattern analysis. In H. McConnell and

D. W. Yaseen (eds), *Perspectives in geography one: models of spatial variation*. DeKalb: Northern Illinois University Press, 105–31.

Bettie, J. 1995: Roseanne and the changing face of working-class iconography. *Social Text* **14**(4), 125–49.

Bhaskar, R. 1978: *A realist theory of science*. Brighton: Harvester.

—— 1986: *Scientific realism and human emancipation*. London: Verso.

Billinge, M., Gregory, D. J. and Martin, R. L. (eds) 1984: *Recollections of a revolution*. London: Macmillan.

Blair, E. H. and Robertson, J. A. 1903–19: *The Philippine Islands 1493–1898*. 55 vols. Cleveland: A. H. Clark Co.

Block, F. 1977: *The origins of international economic disorder*. Berkeley and Los Angeles: University of California Press.

Blomley, N. 1994: Activism and the academy. *Environment and Planning D: Society and Space* **12**, 380–5.

Bloor, D. 1976: *Knowledge and social imagery*. London: Routledge.

Blowers, A. 1980: *The limits of power: the politics of local planning policy*. Oxford: Pergamon.

Blunt, A. 1999: The flight from Lucknow: British women travelling and writing home, 1857–8. In J. S. Duncan and D. J. Gregory (eds), *Writes of passage*. London: Routledge, 92–113.

—— and Rose, G. (eds) 1994: *Writing women and space: colonial and postcolonial geographies*. New York: Guildford.

Bodman, A. R. 1991: Weavers of influence: the structure of contemporary geographic research. *Transactions of the Institute of British Geographers* **16**, 21–37.

—— 1992: Holes in the fabric: more on the master weavers in human geography. *Transactions of the Institute of British Geographers* **17**, 108–9.

Bolotin, F. N. and Cingranelli, D. L. 1983: Equity and urban policy. *Journal of Politics* **45**, 209–19.

Bondi, L. 1990: Progress in geography and gender. *Progress in Human Geography* **14**, 438–45.

Booth, C. 1891: *London labour and the London poor: first series – poverty*. London: Methuen.

Boots, B. 1996: Referees as gatekeepers: some evidence from geographical journals. *Area* **28**, 177–85.

Bordo, S. 1986: The cartesian masculinization of thought. *Signs* **11**, 439–56.

Boruch, R. F. and Pearson, R. W. 1988: Assessing the quality of longitudinal surveys. *Evaluation Review* **12**, 3–58.

Bouma, G. D. 1993: *The research process*. Revised edition. Melbourne: Oxford University Press Australia.

Bourdieu, P. 1990: *In other words: essays towards a reflexive sociology*. Cambridge: Polity.

Bourque, L. and Clarke, V. 1992: *Processing data: the survey example*. Newbury Park, CA: Sage.

Box, S. 1983: *Power, crime and mystification*. London: Tavistock.

Boyer, R. and Savageau, D. 1989: *Places rated almanac: your guide to finding the best places to live in America*. New York: Prentice Hall.

Boyne, G. A. and Powell, M. 1991: Territorial justice. *Political Geography Quarterly* **10**, 263–81.

—— —— 1993: Territorial justice and Thatcherism. *Environment and Planning C: Government and Policy* **11**, 35–53.
Bradley, J. E., Kirby, A. M. and Taylor, P. J. 1978: Distance decay and dental decay. *Regional Studies* **12**, 529–40.
Brenner, M., Brown, J. and Canter, D. (eds) 1985: *The research interview*. London: Academic.
Bridgen, P. and Lowe, R. 1998: *Welfare policy under the Conservatives 1951–1964*. London: PRO Publications.
Briggs, C. L. 1986: *Learning how to ask: a socio-linguistic appraisal of the role of the interview in social science research*. Cambridge: Cambridge University Press.
Brindle, D. 1991: Poverty data ground rules to be revised. *Guardian* 14 Dec., 3.
Broadfoot, B. 1973: *Ten lost years, 1929–1939: memories of Canadians who survived the Depression*. Toronto: Doubleday Canada.
—— 1976: *The pioneer years, 1895–1914: memories of settlers who opened the West*. Toronto: Doubleday Canada.
—— 1977: *Years of sorrow, years of shame: the story of Japanese Canadians in World War Two*. Toronto: Doubleday Canada.
Brody, H. 1971: *Indians on skid row*. Ottawa: Department of Indian Affairs and Northern Development Northern Science Research Group Report No. 70–2.
Brown, J. R. 1994: *Smoke and mirrors: how science reflects reality*. New York: Routledge.
Brown, K. M. 1991: *Mama Lola: a vodou priestess in Brooklyn*. Berkeley and Los Angeles: University of California Press.
Brownill, S. 1990: *Developing London's Docklands: another great planning disaster?* London: Paul Chapman.
Brunsden, D. and Thornes, J. B. 1979: Landscape sensitivity and change. *Transactions of the Institute of British Geographers* **4**, 463–84.
Bryman, A. 1988: *Quantity and quality in social research*. London: Routledge.
—— 1994: *Quantity and quality in social research*. 3rd edition. London: Routledge.
Buller, H. and Hoggart, K. 1986: Nondecision-making and community power: the case of residential development control in rural areas. *Progress in Planning* **25**, 133–203.
—— —— 1994: *International counterurbanization: British migrants in rural France*. Aldershot: Avebury.
Bulmer, M. 1983: Sampling. In M. Bulmer and D. P. Warwick (eds), *Social research in developing countries*. Chichester: Wiley, 91–9.
—— 1984: Why don't sociologists make more use of official statistics? In M. Bulmer (ed.), *Sociological research methods*. Basingstoke: Macmillan, 131–52.
—— and Warwick, D. P. 1983: Data collection. In M. Bulmer and D. Warwick (eds), *Social research in developing countries*. Chichester: Wiley, 145–60.
Bunge, W. 1966: *Theoretical geography*. 2nd edition. Lund: Lund Studies in Geography Series C No. 1.
—— 1971: *Fitzgerald: geography of a revolution*. Cambridge, MS: Schenkman.
Buraway, M. 1992: *Ethnography unbound*. Berkeley and Los Angeles: University of California Press.
Burgess, J. 1996: Focusing on fear: the use of focus groups in a project for the Community Forest Unit, Countryside Commission. *Area* **28**, 130–5.
—— and Wood, P. A. 1988: Decoding Docklands. In J. D. Eyles and D. M. Smith (eds), *Qualitative methods in human geography*. Cambridge: Polity, 94–117.

—— Limb, M. and Harrison, C. M. 1988a: Exploring environmental values through the medium of small groups: 1. theory and practice. *Environment and Planning* **A20**, 309–26.
—— —— —— 1988b: Exploring environmental values through the medium of small groups: 2. illustrations of a group at work. *Environment and Planning* **A20**, 457–76.
Burgess, R. G. 1984: *In the field*. London: Routledge.
Burke's peerage and baronetage (irregular). London: Fitzroy Dearborn.
Burnett, J. 1982: *Destiny obscure: autobiographies of childhood education and family from the 1820s to the 1920s*. London: Allen Lane.
Burningham, K. and Cooper, G. 1999: Being constructive: social constructionism and the environment. *Sociology* **33**, 297–316.
Burton, R. F. and Wilson, G. A. 1999: The yellow pages business directory as a sampling frame for farm surveys. *Journal of Rural Studies* **15**, 91–102.
Butler, J. 1990: *Gender trouble, feminism and the subversion of identity*. New York: Routledge.
—— 1993: Critically queer. *GLQ* **1**, 17–32.
Butler, T. 1997: *Gentrification and the middle classes*. Aldershot: Ashgate.
Butterworth, D. 1992: *Waiting for rain: a farmer's story*. Chapel Hill, NC: Algonquin Books.
Buzzard, S. 1990: Surveys. In K. Finsterbusch, J. Ingersoll and L. Llewellyn (eds), *Methods for social analysis in developing countries*, Boulder, CO: Westview, 71–87.
Byron, M. 1993: Audio-visual aids and geography research. *Area* **25**, 379–85.
—— 1994: *The unfinished cycle: post war migration from the Caribbean to Britain*. Aldershot: Avebury.
Campbell, D. T. and Stanley, J. C. 1963: *Experimental and quasi-experimental designs for research*. Chicago: Rand McNally.
Cannell, C. F. and Fowler, F. J. 1968: Comparison of a self-enumerative procedure and a personal interview. *Public Opinion Quarterly* **32**, 250–64.
Cantley, C. 1992: Negotiating research. In J. Vincent and S. Brown (eds), *Critics and customers: the control of social policy research*. Aldershot: Avebury, 27–43.
Cantwell, J. D. 1998: *The Second World War: a guide to documents in the Public Record Office*. London: PRO Publications.
Cappelletto, F. 1998: Memories of Nazi-Fascist massacres in two central Italian villages. *Sociologia Ruralis* **38**, 69–85.
Carley, J. 1983: Tracing your matrilineal ancestry. *History Workshop* **16**, 137–42.
Casley, D. and Lury, D. A. 1987: *Data collection in developing countries*. 2nd edition. Oxford: Clarendon Press.
Castells, M. 1977: *The urban question*. London: Edward Arnold. Translation of *La question urbaine*. Paris: Maspero, 1972.
—— 1983: *The city and the grassroots*. London: Edward Arnold.
—— 1989: *The informational city*. Oxford: Blackwell.
—— 2000: *The rise of network society*. 2nd edition. Oxford: Blackwell.
Castree, N. 1999: 'Out there?' 'in here?' domesticating critical geography. *Area* **31**, 81–6.
Caunce, S. 1994: *Oral history and the local historian*. Harlow: Longman.
CEC 1999: *What do Europeans think about the environment?* Luxembourg: Office for Official Publications of the European Communities.
—— annual: *Eurobarometer/Eurobarometre: public opinion in the European Community*. Luxembourg: Office for Official Publications of the European Communities.

CEC–London 1995: *Do you still believe all you read in the newspapers?* London: Commission of the European Communities.
Chakravarty, S. 1989: *The Raj syndrome: a study in imperial perceptions*. New Delhi: Penguin India.
Chambers, A. F. 1982: *What is this thing called science?* 2nd edition. Milton Keynes: Open University Press.
Chambers, R. 1994a: The origins and practice of participatory rural appraisal. *World Development* **22**, 953–69.
—— 1994b: Participatory Rural Appraisal (PRA): analysis of experience. *World Development* **22**, 1253–68.
—— 1994c: Participatory Rural Appraisal (PRA): challenges, potentials and paradigm. *World Development* **22**, 1437–54.
Chapman, C. R. 1998: *Pre-1941 censuses and population listings*. 3rd edition. London: Society of Genealogists.
Chaudhary, V. 1998: School league tables blamed for boom in exclusions. *Guardian* 31 Mar., 7.
Checkland, S. J. 1995: Infra-red scrutiny of the Arnolfini Marriage leaves an art-historical industry short of a good theory. *Guardian* 13 Feb., 20.
Cherry, N. and Rodgers, B. 1979: Using a longitudinal study to assess the quality of retrospective data. In L. Moss and H. Goldstein (eds), *The recall method in social surveys*. London: University of London Institute of Education, 31–47.
Cheshire, P. C., Furtado, A. and Magrini, S. 1996: Quantitative comparisons of European regions and cities. In L. Hantrais and S. Mangen (eds), *Cross-national research methods in the social sciences*. London: Pinter, 39–50.
Chester, L., Fay, S. and Young, H. 1967: *The Zinoviev letter*. London: Heinemann.
Chouinard, V. 1997: Making space for disability differences. *Environment and Planning D: Society and Space* **15**, 2–6.
CIPFA annual: *Housing revenue account statistics*. London: Chartered Institute of Public Finance and Accountancy.
Clark, T. N. and Hoffmann-Martinot, V. (eds) 1998: *The new political culture*. Boulder, CO: Westview.
Clifford, J. 1986: Introduction: partial truths. In J. Clifford and G. Marcus (eds), *Writing culture*. Berkeley and Los Angeles: University of California Press, 1–26.
—— 1988: *The predicament of culture*. Cambridge, MA: Harvard University Press.
—— 1990: Notes on (field)notes. In R. Sanjek (ed.), *Fieldnotes: the making of anthropology*. Ithaca, NY: Cornell University Press, 47–70.
—— and Marcus, G. (eds) 1986: *Writing culture*. Berkeley and Los Angeles: University of California Press.
Cloke, P. J., Philo, C. and Sadler, D. 1991: *Approaching human geography*. London: Paul Chapman.
—— Goodwin, M. and Milbourne, P. 1997: *Rural Wales*. Cardiff: University of Wales Press.
Coffey, A. 1999: *The ethnographic self*. London: Sage.
Cole, K. 1994: Data modifications, data suppression, small populations and other features of the 1991 Small Area Statistics. *Area* **26**, 69–78.
Collins, H. M. 1985: *Changing order: replication and induction in scientific practice*. London: Sage.

—— and Pinch T. 1993: *The golem: what everybody should know about science.* Cambridge: Cambridge University Press.

Comte, A. 1903: *A discourse on the positive spirit.* London: Reeves.

Congressional Information Service, monthly: *American statistical index: monthly supplements.* Bethesda, MY [annual compilations also].

Connell, J. 1973: The geography of development or the development of geography. *Antipode* **5**(2), 27–39.

Converse, J. M. and Presser, S. 1986: *Survey questions.* Beverly Hills, CA: Sage University Paper No. 63.

Cook, I. 1997: Participant observation. In R. Flowerdew and D. Martin (eds), *Methods in human geography.* Harlow: Longman, 127–49.

—— and Crang, M. 1995: *Doing ethnographies.* Norwich: Environmental Publications, CATMOG No. 58.

—— Crouch, D., Naylor, S. and Ryan, J. (eds) 2000: *Cultural turns/geographical turns.* London: Prentice Hall.

Cook, T. D., Cooper, H., Cordray, D. S., Hartmann, L., Hedges, L. V., Light, R. V., Louis, T. A. and Mosteller, F. 1992: *Meta-analysis for explanation.* New York: Russell Sage Foundation.

Cook-Lynn, E. 1996: *Why I can't read Wallace Stegner and other essays.* Madison: University of Wisconsin Press.

Cooper, D. 1967: *Psychiatry and anti-psychiatry.* London: Tavistock.

Cornwell, T. 2000: The work that has rocked anthropology. *Times Higher Education Supplement* 27 Oct., 20–1.

Cotterill, P. 1992: Interviewing women: issues of friendship, vulnerability and power. *Women's Studies International Forum* **15**, 593–606.

Coulson, I. and Crawford, A. 1995: *Archives in education.* London: PRO Publications.

Cox, K. R. and Golledge, R. G. (eds) 1969: *Behavioral problems in geography.* Evanston, ILL: Northwestern University Studies in Geography No. 17.

Cox, R. W. 1987: *Production, power and world order: social forces in the making of history.* New York: Columbia University Press.

Craig, J. 1996: *The consistency of statements of age in censuses and at death registration.* London: City University Social Statistics Research Unit LS Working Paper No. 76.

Crang, M. 1997: Picturing practices: research through the tourist gaze. *Progress in Human Geography* **24**, 359–73.

—— Hudson, A., Reimer, S. and Hinchcliffe, S. 1997: Software for qualitative research: 1. prospectus and overview. *Environment and Planning* **A29**, 771–87.

Creaton, H. 1998: *Sources for the history of London 1939–45.* London: British Records Association Archives and the User No. 9.

Crenson, M. A. 1971: *The un-politics of air pollution.* Baltimore: Johns Hopkins University Press.

Cribier, F. and Kych, A. 1993: A comparison of retirement migration from Paris and London. *Environment and Planning* **A25**, 1399–420.

Crompton, R. 1991: Three varieties of class analysis. *International Journal of Urban and Regional Research* **15**, 108–13.

Cronon, W. 1992: A place for stories: nature, history, and narrative. *Journal of American History* **78**, 1347–76.

Crossman, R. 1975: *The diaries of a Cabinet Minister: volume one – Minister of Housing 1964–66*. London: Hamish Hamilton and Jonathan Cape.
Croucher, S. 1997: Constructing the image of ethnic harmony in Toronto, Canada. *Urban Affairs Review* **32**, 319–47.
Cryer, P. 1996: *The research student's guide to success*. Buckingham: Open University Press.
Cullen, M. J. 1975: *The statistical movement in early Victorian Britain*. Hassocks: Harvester.
Cunningham, F. 1973: *Objectivity in social science*. Toronto: University of Toronto Press.
Curran, J. and Seaton, J. 1988: *Power without responsibility: the press and broadcasting in Britain*. 3rd edition. London: Routledge.
Dahl, R. A. 1961: *Who governs?* New Haven, CT: Yale University Press.
Dale, A. and Marsh, C. 1993: *The 1991 census user's guide*. London: HMSO.
Dalyell, T. 1983: *A science policy for Britain*. London: Longman.
Daniels, A. K. 1983: Self-deception and self-discovery in fieldwork. *Qualitative Sociology* **6**(2), 195–214.
Daniels, S. 1992: The implication of industry: Turner and Leeds. In T. J. Barnes and J. S. Duncan (eds), *Writing worlds*. London: Routledge, 38–49.
—— and Cosgrove, D. 1988: Introduction: iconography and landscape. In D. Cosgrove and S. Daniels (eds), *The iconography of landscape*. Cambridge: Cambridge University Press, 1–10.
Darga, K. 1999: *Sampling and the census: the case against the proposed adjustment for undercount*. Washington: AEI Press.
Darwin, C. 1861: *On the origin of species by means of natural selection*. London: John Murray.
Davies, A. K. 1988: *Invisible careers: women civic leaders from the volunteer world*. Chicago: University of Chicago Press.
Davies, A. R. 1999: Where do we go from here? Environmental focus groups and planning policy. *Local Environment* **4**, 295–316.
—— and Gathorne-Hardy, F. 1996: *Making connections: community involvement in environmental initiatives*. Cambridge: University of Cambridge Committee for Interdisciplinary Environmental Studies.
Davies, C. 1980: Making sense of the census in Britain and the USA. *Sociological Review* **28**, 581–609.
Davies, N. 2000a: Wrong turn: the trouble with special measures. *Guardian* 11 July 2000, 7.
—— 2000b: Fiddling the figures to get the right results. *Guardian*, 11 July, 1, 6–7.
Davies, R. B. and Dale, A. 1994: Introduction. In A. Dale and R. B. Davies (eds), *Analysing social and political change*. London: Sage, 1–19.
Davies, R. L. 1968: Effects of consumer income differences on the business provision of small shopping centres. *Urban Studies* **5**, 144–64.
Davis, J. A. and Smith, T. W 1992: *The NORC General Social Survey*. Newbury Park, CA: Sage.
de Saussure, F. 1974: *Course in general linguistics*. London: Fontana.
Dean, L. and Whyte, W. F. 1970: How do you know if the informant is telling the truth? In L. A. Dexter (ed.), *Elite and specialized interviewing*. Evanston, ILL: Northwestern University Press, 119–30. Reprinted from *Human Organization* **17**(2), 1958, 34–8.
Dean, M. 1995: Social Trends chief censors its history. *Guardian* 21 Jan., 5.

Dear, M. J. 1986: Postmodernism and planning. *Environment and Planning D: Society and Space* **4**, 367–84.
—— 1988: The postmodern challenge: reconstructing human geography. *Transactions of the Institute of British Geographers* **13**, 262–74.
—— and Flusty, S. 1998: Postmodern urbanism. *Annals of the Association of American Geographers* **88**, 50–72.
Debrett's peerage, baronetage, knightage and companionage (irregular). London: Debrett's Peerage.
Deleria, V. 1997: *Red earth, white lies: native Americans and the myth of scientific fact.* Golden, CO: Fulcrum.
Demeritt, D. B. 1994: The nature of metaphors in cultural geography and environmental history. *Progress in Human Geography* **18**, 163–85.
—— 1996: Social theory and the reconstruction of science and geography. *Transactions of the Institute of British Geographers* **21**, 484–503.
—— 1998: Science, social constructivism and nature. In B. Braun and N. Castree (eds), *Remaking reality*. London: Routledge, 173–93.
—— 2000: The new social contract for science: accountability, relevance and value in US and UK science and research policy. *Antipode* **32**, 308–29.
—— 2001a: Scientific forest conservation and the statistical picturing of nature's limits in the Progressive-era United States. *Environment and Planning D: Society and Space* **19**.
—— 2001b: The construction of global warming and the politics of science. *Annals of the Association of American Geographers*.
—— 2001c: Being constructive about nature. In N. Castree (ed.), *Social nature*. Oxford: Blackwell.
Denny, C. 1998: Labour's statistical u-turn. *Guardian* 25 Feb., 19.
Denzin, N. 1997: *Interpretative ethnography*. Thousand Oaks, CA: Sage.
—— and Lincoln, Y. 1994: *Handbook of qualitative research*. London: Sage.
—— —— (eds) 1998: *The landscape of qualitative research*. London: Sage.
Derman, W. 1990: Informant interviews in international social impact assessment. In K. Finsterbusch, J. Ingersoll and L. Llewellyn (eds), *Methods for social analysis in developing countries*. Boulder, CO: Westview, 107–24.
Desrosières, A. 1996: Statistical traditions: an obstacle to international comparisons? In L. Hantrais and S. Mangen (eds), *Cross-national research methods in the social sciences*. London: Pinter, 17–27.
Devereaux, S. and Hoddinott, J. (eds) 1992: *Fieldwork in developing countries*. Hemel Hempstead: Harvester Wheatsheaf.
Dey, I. 1993: *Qualitative data analysis*. London: Routledge.
Dickenson, J. P. and Clarke, C. G. 1972: Relevance and the newest geography. *Area* **3**, 25–7.
Dijkstra, W. and van der Zouwen, J. (eds) 1982: *Response behaviour in the survey interview*. London: Academic.
Dilthey, W. 1923/1988: *Introduction to human science: an attempt to lay a foundation for the study of society and history*. London: Harvester Wheatsheaf. Translation of *Einleitung in die geiteswissenschaften*. Volume One. Berlin: BG Teubner, 1923.
Directory of directors (annual). East Grinstead: Thomas Skinner Directories.
Dixon, C. J. and Leach, B. 1977: *Sampling methods for geographical research*. Norwich: Geo Books CATMOG No. 17.

—— —— 1984: *Survey research in underdeveloped countries*. Norwich: Geo Books CATMOG No. 39.
Dodd, D. 1993: *Historical statistics of the states of the United States*. Westport, CT: Greenwood.
Dodd, V. 2000: How the web of lies was unravelled. *Guardian* 12 Apr., 4.
Doel, M. 1999: *Poststructuralist geographies*. Lanham, MD: Rowman & Littlefield.
Dollard, J. 1937/1957: *Caste and class in a southern town*. New York: Doubleday Anchor.
Domhoff, G. W. 1978: *Who really rules? New Haven and community power reexamined*. New Brunswick, NJ: Transaction.
Domschke, E. and Goyer, D. S. 1986: *The handbook of national population censuses: Africa and Asia*. Westport, CT: Greenwood.
Dorling, D. F. L. 1994: The negative equity map of Britain. *Area* 26, 327–342.
—— and Simpson, S. 1999: Introduction to statistics in society. In D. F. L. Dorling and S. Simpson (eds), *Statistics in society: the arithmetic of politics*. London: Arnold, 1–5.
—— Pattie, C. J., Rossiter, D. J. and Johnston, R. J. 1996: Missing voters in Britain 1992–96. In F. M. Farrell, D. Broughton, D. Denver and J. Fisher (eds), *British elections and parties yearbook 1996*. London: Frank Cass, 37–49.
Dorst, J. D. 1989: *The written suburb: an American site, an ethnographic dilemma*. Philadelphia: University of Pennsylvania Press.
Douglass, W. A. 1984: *Emigration in a south Italian town*. New Brunswick, NJ: Rutgers University Press.
Doyal, L. and Harris, R. 1986: *Empiricism, explanation and rationality*. London: Routledge & Kegan Paul.
Doyle, L. 1999: *The Big Issue*: empowering homeless women through academic research. *Area* **31**, 239–46.
Drake, M. and Finnegan, R. (eds) 1994: *Sources and methods for family and community historians – 19th and 20th centuries*. Cambridge: Cambridge University Press.
Driver, F. 2001: *Geography militant: cultures of exploration and empire*. Oxford: Blackwell.
Duncan, J. S. 1993: Sites of representation: place, time and the discourse of the 'other'. In J. S. Duncan and D. Ley (eds), *Place/culture/representation*. London: Routledge, 35–56.
—— 1999: Dis-orientation: on the shock of the familiar in a far-away place. In J. S. Duncan and D. J. Gregory (eds), *Writes of passage*. London: Routledge, 151–63.
—— and Barnes, T. J. 1992: Afterword. In T. J. Barnes and J. S. Duncan (eds), *Writing worlds*. London: Routledge, 248–53.
—— and Duncan, N. 1988: (Re)reading the landscape. *Environment and Planning D: Society and Space* **6**, 117–26.
—— and Gregory, D. J. (eds) 1999: *Writes of passage: reading travel writing*. London: Routledge.
—— and Ley, D. 1993: Introduction: representing the place of culture. In J. S. Duncan and D. Ley (eds), *Place/culture/representation*. London: Routledge, 1–21.
Duncan, S. S. 1991: Gender divisions of labour. In K. Hoggart and D. R. Green (eds), *London*. London: Edward Arnold, 95–122.
—— and Goodwin, M. 1988: *The local state and uneven development*. Cambridge: Polity.
Dunleavy, P. J. 1977: Protest and acquiescence in urban politics. *International Journal of Urban and Regional Research* **1**, 193–218.
—— and O'Leary, B. 1987: *Theories of the state*. London: Macmillan.

Durkheim, E. 1952: *Suicide*. London: Routledge & Kegan Paul. Translation of *Le Suicide: Étude de sociologie*. Paris: F. Alcan, 1897.
Dyck, I., Lynam, J. and Anderson, J. 1995: Women talking: creating knowledge through difference in cross-cultural research. *Women's Studies International Forum* **18**(5–6), 611–26.
Eagleton, T. 1983: *Literary theory*. Oxford: Blackwell.
—— 1990: *The ideology of the aesthetic*. Oxford: Blackwell.
Ebdon, D. 1985: *Statistics in geography*. 2nd edition. Oxford: Blackwell.
Eco, U. 1989: *Foucault's pendulum*. London: Secker & Warburg.
Edgell, S. 1993: *Class*. London: Routledge.
Edmondson, B. 1997: The wired bunch: on-line surveys and focus groups might solve the toughest problems in market research. But can internet users really speak for everyone? *American Demographics* **19**(6), 10–15.
Edwards, D., Ashmore, M. and Potter, J. 1995: Death and furniture: the rhetoric, politics and theology of bottom line arguments against relativism. *History of the Human Sciences* **8**, 25–49.
Edwards, P. 1993: *Rural life: guide to local records*. London: B. T. Batsford.
Eichler, M. 1988: *Nonsexist research methods*. London: Allen & Unwin.
Elliott, L. 1995: CBI at loggerheads with Government over output figures. *Guardian* 26 May, 21.
—— and Denny, C. 1998: Re-write for the Tory years: Brown backs Treasury plan to include increasing poverty gap in figures. *Guardian* 11 Nov., 21.
—— and Thomas, R. 1997: Dole figures 'are flawed'. *Guardian* 15 Apr., 21.
Ely, M., Anzul, M., Freidman, T. and Garner, D. 1996: *Doing qualitative research: circles within circles*. London: Falmer.
—— Vinz, R., Downing, M. and Anzul, M. 1997: *On writing qualitative research: living by words*. London: Falmer.
Emmison, M. and Smith, P. 2000: *Researching the visual*. London: Sage.
England, K. V. L. 1994: Getting personal: reflexivity, positionality and feminist research. *Professional Geographer* **46**, 80–9.
Erickson, K. and Stull, D. 1997: *Doing team ethnography*. London: Sage.
Erikson, K. T. 1966: *The wayward puritans*. New York: Wiley.
Errington, A. J. 1985: Sampling frames for farm surveys in the UK. *Journal of Agricultural Economics* **36**, 251–8.
European Commission 1995a: *Europe's environment: the Dobris assessment*. Luxembourg: Office for Official Publications of the European Communities.
—— 1995b: *Europe's environment: statistical compendium for the Dobris assessment*. Luxembourg: Office for Official Publications of the European Communities.
—— 1999: *Sixth periodic report on the social and economic situation and development of the regions of the European Union*. Luxembourg: Office for Official Publications of the European Communities.
Eurostat 1999: *Regions: nomenclature of territorial units for statistics*. Luxembourg: Office for Official Publications of the European Communities.
Evans, H. 1983: *Good times, bad times*. London: Weidenfeld & Nicolson.
Evans, M. 1988: Participant observation. In J. D. Eyles and D. M. Smith (eds), *Qualitative methods in human geography*. Cambridge: Polity, 197–218.

Evans-Pritchard, E. E. 1937: *Witchcraft, oracles and magic among the Azande*. Oxford: Clarendon Press.

—— 1940: *The Nuer: a description of the modes of livelihood and political institutions of a Nilotic people*. Oxford: Clarendon Press.

Eyles, J. D. 1968: *The inhabitants' images of Highgate Village (London)*. London: London School of Economics Graduate School of Geography Discussion Paper No. 15.

—— and Smith, D. M. (eds) 1988: *Qualitative methods in human geography*. Cambridge: Polity.

Fainstein, S. S. 1994: *The city builders: property, politics and planning in London and New York*. Oxford: Blackwell.

FAO, annual: *FAO yearbook – production*. Rome: Food and Agricultural Organization.

—— annual: *FAO yearbook – trade*. Rome: Food and Agricultural Organization.

—— annual: *Fisheries statistics yearbook*. Rome: Food and Agricultural Organization.

—— annual: *Forest products yearbook*. Rome: Food and Agricultural Organization.

FBI, annual: *Uniform crime reports for the United States*. Washington: US Government Printing Office.

Feldman, M. S. 1995: *Strategies for interpreting qualitative data*. Thousand Oaks, CA: Sage.

Ferber, R. and Wells, H. G. 1952: Detection and correction of interviewer bias. *Public Opinion Quarterly* **16**, 107–27.

Feyerabend, P. 1975: *Against method*. London: New Left Books.

Fieldhouse, E. A. and Tye, R. 1996: Deprived people or deprived places? Exploring the ecological fallacy in studies of deprivation with Samples of Anonymized Records. *Environment and Planning* **A28**, 237–59.

Fielding, A. J. 1982: Counterurbanization in Western Europe. *Progress in Planning* **17**, 1–52.

—— 1992: Migration and social mobility: South East England as an escalator region. *Regional Studies* **26**, 1–15.

—— and Halford, S. 1999: A longitudinal and regional analysis of gender-specific social and spatial mobilities in England and Wales 1981–91. In P. Boyle and K. H. Halfacree (eds), *Migration and gender in the developed world*. London: Routledge, 30–53.

Fielding, J. 1993: Coding and managing data. In N. Gilbert (ed.), *Researching social life*, London: Sage, 218–38.

Fielding, N. G. and Lee, R. M. (eds) 1991: *Using computers in qualitative research*. London: Sage.

Finch, J. 1986: *Research and policy*. London: Falmer.

Fincher, R. and Jacobs, J. (eds) 1998: *Cities of difference*. New York: Guilford.

Fine, M. 1998: Working the hyphens: reinventing self and other in qualitative research. In N. Denzin and Y. Lincoln (eds), *The landscape of qualitative research*. London: Sage, 131–52.

—— and Vanderslice, V. 1992: Reflections on qualitative research. In E. Posavac (ed.), *Methodological issues in applied social psychology*. New York: Plenium, 271–87.

Fink, A. (ed.) 1995: *The survey kit*. 9 vols. London: Sage.

—— and Kosecoff, J. 1985: *How to conduct surveys*. Beverly Hills, CA: Sage.

—— —— 1998: *How to conduct surveys*. 2nd edition. Thousand Oaks, CA: Sage.

Finlay, R. 1981: *Parish registers*. London: Institute of British Geographers Historical Geography Research Series No. 7.

Finsterbusch, K. Ingersoll, J. and Llewellyn, L. (eds) 1990: *Methods for social analysis in developing countries*. Boulder, CO: Westview.
Firey, W. 1947: *Land use in central Boston*. Cambridge, MA: Harvard University Press.
Fiske, J. 1987: *Television culture*. London: Methuen.
—— and Hartley, J. 1978: *Reading television*. 1998 edition. London: Routledge.
Floud, R., Wachter, K. and Gregory, A. 1990: *Height, health and history*. Cambridge: Cambridge University Press.
Flowerdew, R. and Martin, D. (eds) 1997: *Methods in human geography*. Harlow: Longman.
Foddy, W. 1993: *Constructing questions for interviews and questionnaires*. Cambridge: Cambridge University Press.
Fogel, R. W. and Engerman, S. L. 1974/1989: *Time on the cross: the economics of American Negro slavery*. New York: W. W. Norton.
Forrest, R. and Murie, A. 1990: Moving strategies among home owners. In J. H. Johnson and J. Salt (eds), *Labour migration*, London: David Fulton, 191–209.
—— —— 1992: Change on a rural council estate: an analysis of dwelling histories. *Journal of Rural Studies* **8**, 53–65.
—— —— 1994: The dynamics of the owner-occupied housing market in southern England in the late 1980s. *Regional Studies* **28**, 275–89.
Foster, J. and Sheppard, J. 1995: *British archives: a guide to archive sources in the United Kingdom*. 3rd edition. London: Macmillan.
Foucault, M. 1982: The subject and power. In *Michel Foucault*, ed. H. Dreyfus and P. Rabinow. Brighton: Harvester, 208–26.
Foulkes, S. H. 1948/1983: *Introduction to group-analytic psychology*. London: Maresfield Reprints.
Fowler, F. J. 1984: *Survey research methods*. Beverly Hills, CA: Sage.
—— and Mangione, T. 1990: *Standardized survey interviewing*. Newbury Park, CA: Sage.
Fowler, S. 1995: *Sources of labour history*. London: PRO Publications.
Fox Keller, E. F. 1985: *Reflections on gender and science*. New Haven, CT: Yale University Press.
Frank, A. 1947/1954: *The diary of Anne Frank*. London: Pan.
Frankfort-Nachmias, C. and Nachmias, D. 1992: *Research methods in the social sciences*. 4th edition. London: Edward Arnold.
Freud, S. 1976: *The interpretation of dreams*. Harmondsworth: Penguin.
Fried, M. 1966: Grieving for a lost home. In J. Q. Wilson (ed.), *Urban renewal*. Cambridge, MA: MIT Press, 359–79.
Friedland, W. H. and Nelkin, D. 1971: *Migrant: agricultural workers in America's North East*. New York: Holt, Rinehart & Winston.
Friedman, J. J. 1977: Community action on water pollution. *Human Ecology* **5**, 329–53.
Fuller, D. 1999: Part of the action, or 'going native'? *Area* **31**, 221–7.
Fuller, L. 1992: *The Cosby Show*. Westport, CT : Greenwood.
Fyfe, N. 1992: Observations on observations. *Journal of Geography in Higher Education* **6**, 127–33.
Gahan, C. and Hannibal, M. (eds) 1997: *Doing qualitative research using QSR NUD*IST*. London: Sage.
Gallent, N. and Tewdwr-Jones, M. 2000: *Rural second homes in Europe*. Aldershot: Ashgate.

Gardiner, J. 1968: Police enforcement of traffic laws. In J. Q. Wilson (ed.), *City politics and public policy*. New York: Wiley, 151–72.
Geertz, C. 1963: *Agricultural involution: the process of ecological change in Indonesia*. Berkeley and Los Angeles: University of California Press.
—— 1973/1975: *The interpretation of cultures*. London: Hutchinson.
—— 1983: *Local knowledge*. London: Fontana.
—— 1988: *Works and lives: the anthropologist as author*. Cambridge: Polity.
Gerwirtz, S. and Ozga, J. 1994: Interviewing the educational policy elite. In G. Walford (ed.), *Researching the powerful in education*. London: UCL Press, 186–203.
Giddens, A. 1987: *Social theory and modern society*. Cambridge: Polity.
—— 1991: *Modernity and self-identity*. Cambridge: Polity.
Gilg, A. W. and Kelly, M. 1997: The delivery of planning policy in Great Britain. *Environment and Planning C: Government and Policy* **15**, 19–36.
Gillham, B. 2000: *Developing a questionnaire*. London: Continuum.
Gittins, D. 1979: Oral history, reliability and recollection. In L. Moss and H. Goldstein (eds), *The recall method in social surveys*. London: University of London Institute of Education, 82–97.
Glaser, B. S. and Strauss, A. L. 1967: *The discovery of grounded theory*. Chicago: Aldine.
Glasgow University Media Group 1976: *Bad news*. London: Routledge & Kegan Paul.
—— 1980: *More bad news*. London: Routledge & Kegan Paul.
—— 1982: *Really bad news*. London: Writers & Readers.
Glasscock, R. E. (ed.) 1975: *The lay subsidy of 1334*. Oxford: Oxford University Press.
Gleeson, R. 2000: Enabling geography: exploring a new political ethical ideal. *Ethics, Place and Environment* **3**(1), 65–70.
Glover, J. 1996: Epistemological and methodological considerations in secondary analysis. In L. Hantrais and S. Mangen (eds), *Cross-national research methods in the social sciences*. London: Pinter, 28–38.
Goddard, A. 2000: Exeter targets its staff in RAE move. *Times Higher Education Supplement* 19 May, 4.
Goheen, P. G. 1970: *Victorian Toronto 1850 to 1900*. Chicago: University of Chicago, Department of Geography Research Paper 127.
Goldthorpe, J. H. 1987: *Social mobility and class structure in modern Britain*. 2nd edition. Oxford: Clarendon Press.
Goss, J. D. 1996: Introduction to focus groups. *Area* **28**, 113–14.
—— 1999: Once-upon-a-time in the commodity world: an unofficial guide to Mall of America. *Annals of the Association of American Geographers* **89**, 45–75.
Gottdiener, M. 1995: *Postmodern semiotics*. Oxford: Blackwell.
Gould, P. R. 1970: Is statistix inferens the geographical name for a wild goose? *Economic Geography* 46, 439–48.
Gow, D. 1990: Rapid rural appraisal: social science as investigative journalism. In K. Finsterbusch, J. Ingersoll and L. Llewellyn (eds), *Methods for social analysis in developing countries*, Boulder, CO: Westview, 143–63.
Goyder, J. 1987: *The silent minority: nonrespondents in sample surveys*. Cambridge: Polity.
Goyer, D. S. and Domschke, E. 1983: *The handbook of national population censuses: Latin America, the Caribbean, North America and Oceania*. Westport, CT: Greenwood.
Grant, U. R. and Purcell, L. E. (eds), 1995: *The farm diary of Elmer G. Powers*. DeKalb: Northern Illinois University Press.

Grant, W. 1995: Is agricultural policy still exceptional? *Political Quarterly* **66**, 156–69.
Gray, G. and Guppy, N. 1994: *Successful surveys*. Toronto: Harcourt Brace.
Gray, P. G. 1955: The memory factor in social surveys. *Journal of the American Statistical Association* **50**, 344–63.
—— and Gee, F. A. 1967: *Electoral registration for parliamentary elections*. London: HMSO, Government Social Survey SS391.
Greele, R. J. 1991: *Envelopes of sound: the art of oral history*. 2nd edition. New York: Praeger.
Green, M. and MacColl, G. 1987: *Reagan's reign of error*. Revised edition. New York: Pantheon.
Greenbaum, T. 1998: *The handbook for focus group research*. 2nd edition. Thousand Oaks, CA: Sage.
Greenwood, D. and Levin, M. 1998: *Introduction to action research*. Thousand Oaks, CA: Sage.
Greer, A. L. and Greer, S. 1976: Suburban political behavior. In B. Schwartz (ed.), *The changing face of the suburbs*. Chicago: University of Chicago Press, 203–19.
Gregory, D. J. 1978: *Ideology, science and human geography*. London: Hutchinson.
—— 1994: *Geographical imaginations*. Oxford: Blackwell.
—— 1999: Scripting Egypt: orientalism and the cultures of travel. In J. S. Duncan and D. J. Gregory (eds), *Writes of passage*. London: Routledge, 114–50.
Gregson, N. 1995: And now it's all consumption. *Progress in Human Geography* **19**, 135–41.
Griffin, J. H. 1962: *Black like me*. London: Collins.
Gross, P. R. and Levitt, N. 1994: *Higher superstition: the academic left and its quarrels with science*. Baltimore: Johns Hopkins University Press.
Grover, B. L. and Grover, S. 2001: *A new look at modern Indian history*. 18th edition. New Delhi: S. Chand & Co.
Groves, R. M., Biemer, P. P., Lyberg, L. E., Massey, J. T., Nicholls, W. L. and Waksberg, J. (eds) 1988: *Telephone survey methodology*. New York: Wiley.
Gupta, A. and Ferguson, J. (eds) 1997: *Anthropological locations*. Berkeley and Los Angeles: University of California Press.
Gutting, D. 1996: Narrative identity and residential history. *Area* **28**, 482–90.
Habermas, J. 1984: *The theory of communicative action*. Cambridge: Cambridge University Press. Translation of *Theorie des kommunikativen handeln*. Frankfurt am Main: Suhrkamp, 1982.
Hacking, I. 1990: *The taming of chance*. Cambridge: Cambridge University Press.
Haines-Young, R. and Petch, J. 1986: *Physical geography: its nature and methods*. London: Harper & Row.
Hakim, C. 1982: *Secondary analysis in social research*. London: Allen & Unwin.
—— 1987: *Research design*. London: Unwin Hyman.
Halfacree, K. H. 1995: Household migration and the structuration of patriarchy: evidence from the USA. *Progress in Human Geography* **19**, 159–82.
Hall, C. 1982: Private archives as sources for historical geography. In A. R. H. Baker and M. Billinge (eds), *Period and place*. Cambridge: Cambridge University Press, 274–80.
Hall, R. and Hall, J. 1995: Missing in the 1991 census. *Area* **27**, 53–61.
—— Ogden, P. E. and Hill, C. 1997: The pattern and structure of one-person

households in England and Wales and France. *International Journal of Population Geography* **3**, 161–81.
Hall, S. 1995: New cultures for old. In D. B. Massey and P. Jess (eds), *A place in the world?* Oxford: Oxford University Press, 175–213.
Hambro Company guide, quarterly. London: Hemmington Scott Publishing Ltd.
Hamilton, C. 1991: *The Hitler diaries: fakes that fooled the world*. Lexington, KY: University of Kentucky Press.
Hamm, M. and Ferrell, J. 1998: Confessions of danger and humanity. In J. Ferrell and M. Hamm (eds), *Ethnography at the edge*. Boston: Northeastern University Press, 254–72.
Hammersley, M. 1995: *The politics of social research*. London: Sage.
—— and Atkinson, P. 1995: *Ethnography*. 2nd edition. London: Routledge.
Hamnett, C. R. 1973: Home improvement grants as an indicator of gentrification in inner London. *Area* **5**, 252–61.
——: The spatial impact of the British home ownership market slump, 1989–91. *Area* **25**, 217–27.
—— 1994: Social polarization in global cities. *Urban Studies* **31**, 401–24.
—— 1997: A stroke of the Chancellor's pen: the social and regional impact of the Conservative 1988 higher rate taxation cut. *Environment and Planning* **A29**, 129–47.
—— and Randolph, B. 1988: Labour and housing market change in London. *Urban Studies* **25**, 380–98.
Hamshere, J. D. 1987: Data sources in historical geography. In M. Pacione (ed.), *Historical geography: progress and prospect*. London: Croom Helm, 46–69.
Handcock, W. D. 1977: *English historical documents: volume XII(2) 1874–1914*. London: Eyre and Spottiswoode.
Hanson, S. and Pratt, G. 1995: *Gender, work and space*. London: Routledge.
Haraway, D. J. 1988: Situated knowledges. *Feminist Studies* **14**, 575–99.
—— 1989: *Primate visions: gender, race, and nature in the world of modern science*. New York: Routledge.
—— 1991: *Simians, cyborgs, and women: the reinvention of nature*. London: Free Association.
—— 1992: The promises of monsters: a regenerative politics for inappropriate/d others. In L. Grossberg, C. Nelson, and P. A. Treichler (eds), *Cultural studies*. New York: Routledge, 295–337.
Harber, J. 1978: Labour archives in the Calder Valley. *History Workshop* **6**, 147–54.
Harding, S. 1991: *Whose science? Whose knowledge? Thinking from women's lives*. Milton Keynes: Open University Press.
Harley, J. B. 1988: Maps, knowledge and power. In D. Cosgrove and S. Daniels (eds), *The iconography of landscape*. Cambridge: Cambridge University Press, 277–312.
—— 1989: Deconstructing the map. *Cartographica* **26**, 1–20. Reprinted in T. J. Barnes and J. S. Duncan (eds) 1992: *Writing worlds*. London: Routledge, 231–47.
Harper, L. 1990: Industry output figures dropped. *Guardian* 16 Apr., 3.
Harris, R. C. and Phillips, E. 1984: *Letters from Windermere 1912–1914*. Vancouver: University of British Columbia Press.
Harvey, D. W. 1969: *Explanation in geography*. London: Edward Arnold.
—— 1973: *Social justice and the city*. London: Edward Arnold.
—— 1989: *The condition of postmodernity*. Oxford: Blackwell.
—— 1992: Local militancy and the politics of a research project. *Social Text* **42**, 69–98.

—— 1993: From space to place and back again: reflections on the condition of postmodernity. In J. Bird, B. Curtis, T. Putnam, G. Robertson and L. Tickner (eds), *Mapping the futures*. London: Routledge, 3–29.

—— 2000: *Spaces of hope*. Edinburgh: Edinburgh University Press.

Harvey, L. 1990: *Critical social theory*. London: Unwin Hyman.

Hattersley, L. and Cresser, R. 1995: *Longitudinal study 1971–1991: history, organization and quality of data*. London: HMSO.

Hawkes, T. 1992: *Structuralism and semiotics*. London: Routledge.

Hawkings, D. T. 1992: *Criminal ancestors*. Stroud: Alan Sutton.

Hawtin, M., Hughes, G., Percy-Smith, J. and Foreman, A. 1994: *Community profiling*. Buckingham: Open University Press.

Healey, M. J. and Rawlinson, M. B. 1993: Interviewing business owners and managers. *Geoforum* **24**, 339–55.

Healey, P. 1997: *Collaborative planning: shaping places in fragmented societies*. Basingstoke: Macmillan.

Heath, A. and McMahon, D. 1997: Education and occupational attainments. In V. Karn (ed.), *Ethnicity in the 1991 census: volume four – employment, education and housing among the ethnic minority populations of Britain*. London: HMSO, 91–113.

Hebbert, M. 1991: The borough effect in London's geography. In K. Hoggart and D. R. Green (eds), *London*. London: Edward Arnold, 191–206.

Hedges, A. 1985: Group interviewing. In R. Walker (ed.), *Applied qualitative research*. Aldershot: Gower, 71–91.

Hedrick, T. E., Bickman, L. and Rog, D. J. 1993: *Applied research design*. Thousand Oaks, CA: Sage.

Hellevik, O. 1984: *Introduction to causal analysis*. London: Allen & Unwin.

Henerson, M. E., Morris, L. L. and Fitz-Gibbon, C. T. 1978: *How to measure attitudes*. Beverly Hills, CA: Sage.

Henley, P. 1998: Film-making and ethnographic research. In J. Prosser (ed.), *Image-based research*. London: Falmer, 42–59.

Herbert, S. 2000: For ethnography. *Progress in Human Geography* **24**: 550–68.

Heritage, J. 1989: Current developments in conversation analysis. In D. Roger and P. Bull (eds), *Conversation*. Avon: Multilingual Matters, 21–47.

—— 1997: Conversation analysis and institutional talk: analyzing data. In D. Silverman (ed.), *Qualitative research*. London: Sage, 161–82.

Herod, A. 1993: Gender issues in the use of interviewing as a research method. *Professional Geographer* **45**, 305–17.

Higgs, E. 1989: *Making sense of the census: the manuscript returns for England and Wales 1801–1901*. London: HMSO.

—— 1996: *A clearer sense of the census: the Victorian censuses and historical research*. London: HMSO.

—— (ed.) 1998: *History and Electronic Artefacts*. Oxford: Clarendon Press.

Hill, B. E. 1999: Farm household income: perceptions and statistics. *Journal of Rural Studies* **15**, 345–58.

Hill, M. R. 1993: *Archival strategies and techniques*. London: Sage.

Hillery, G. A. 1955: Definitions of community: areas of agreement. *Rural Sociology* **20**, 111–23.

Hinchcliffe, S., Crang, M., Reimer, S. and Hudson, A. 1997: Software for qualitative research: 2. some thoughts on aiding analysis. *Environment and Planning* **A29**, 1109–124.

Hindess, B. 1973: *The use of official statistics in sociology*. Basingstoke: Macmillan.

Hine, C. 2000: *Virtual ethnography*. Thousand Oaks, CA: Sage.

Hoare, A. G. 1983: Pork-barrelling in Britain. *Environment and Planning C: Government and Policy* **1**, 413–38.

Hodson, R. 1999: *Analyzing documentary accounts*. Thousand Oaks, CA: Sage.

Hoggart, K. 1978: Consumer shopping strategies and purchasing activity. *Geoforum* **9**, 415–23.

—— 1990: Riots and urban public expenditure: New Jersey and Pennsylvania 1962–74. *Urban Geography* **11**, 347–72.

—— 1997: The middle classes in rural England 1971–1991. *Journal of Rural Studies* **13**, 253–73.

Holbrook, B. and Jackson, P. A. 1996: Shopping around: focus group research in north London. *Area* **28**, 136–42.

Hollway, W. and Jefferson, T. 2000: *Doing qualitative research differently: free association, narrative and the interview method*. London: Sage.

Holmes, C. 1981: Government files and privileged access. *Social History* **7**, 333–50.

Holmes, S. A. 1998: US fight over census heats up. *International Herald Tribune* 14 Apr., 7.

Honadle, G. and Cooper, L. 1990: Closing the loops: workshop approaches to evaluating development projects. In K. Finsterbusch, J. Ingersoll and L. Llewellyn (eds), *Methods for social analysis in developing countries*. Boulder, CO: Westview, 185–202.

hooks, B. 1991: *Yearning: race, gender and cultural politics*. London: Turnaround.

Howard, S. 1994: Methodological issues in overseas fieldwork: experiences from Nicaragua's northern Atlantic coast. In E. Robson and K. Willis (eds), *Postgraduate fieldwork in developing areas*. London: Institute of British Geographers Developing Areas Research Group Monograph No. 8, 19–35.

Howell, N. 1990: *Surviving fieldwork*. Washington: American Anthropological Association.

Hoy, D. 1985/1990: Jacques Derrida. In Q. Skinner (ed.), *The return of grand theory in the human sciences*. Canto edition. Cambridge: Cambridge University Press, 41–64.

Hubbard, P. 1999: Researching female sexwork: reflections on geographical exclusion, critical methodologies and 'useful' knowledge. *Area* **31**, 229–37.

Huff, C. 1985: *British women's diaries: a descriptive bibliography of selected nineteenth century women's manuscript diaries*. New York: AMS Press.

Huff, D. 1954: *How to lie with statistics*. New York: W. W. Norton.

Hughes, A., Morris, C. and Seymour, S. (eds) 2000: *Ethnography and rural research*. Cheltenham: Countryside and Community Press.

Hughes, B. 2000: History unsealed. *Guardian (Saturday Review)* 12 Feb., 3.

Hughes, J. 1980: *The philosophy of social research*. London: Longman.

Humphrey, J. G. and Winegarten, R. 1996: *Black Texas women: a sourcebook – documents, biographies, timeline*. Austin, TX: University of Texas Press.

Humphries, S. 1984: *The handbook of oral history*. London: Inter-Action Inprint.

Hunt, J. 1984: The development of rapport through the negotiation of gender in fieldwork among police. *Human Organization* **43**, 283–96.

Hunter, F. 1953: *Community power structure*. Chapel Hill: University of North Carolina Press.

Hussey, S. 1997: Low pay, underemployment and multiple occupations: men's work in the inter-war countryside. *Rural History* **8**, 217–35.
Hutton, W. 1992: How Whitehall cut the dole queues. *Guardian* 11 Nov., 13.
Huxley, A. 1929: *Do what you will*. London: Allen Unwin.
Hyman, H. 1944: Do they tell the truth? *Public Opinion Quarterly* **8**, 557–9.
IMF, annual: *Directional trade statistics yearbook*. Washington: International Monetary Fund.
Inglehart, R. 1977: *The silent revolution: changing values and political styles among Western publics*. Princeton: Princeton University Press.
—— 1990: *Culture shift in advanced industrial society*. Princeton: Princeton University Press.
—— 1997: *Modernization and postmodernization: cultural, economic and political change in 43 societies*. Princeton: Princeton University Press.
Institute for Social Research Survey Research Center 1976: *Interviewer's manual*. Revised edition. Ann Arbor: University of Michigan Press.
Isaac, L., Christiansen, L., Miller, J. and Nickel, T. 1998: Temporally recursive regression and social historical inquiry. *International Review of Social History* **43**, 9–32.
Jackson, P. A. 1983: Principles and problems of participant observation. *Geografiska Annaler* **65B**, 39–46.
—— 1984: Social disorganization and moral order in the city. *Transactions of the Institute of British Geographers* **9**, 168–80.
—— 1989: *Maps of meaning*. London: Unwin Hyman.
—— 1991: The cultural politics of masculinity. *Transactions of the Institute of British Geographers* **16**, 199–213.
—— 1994: Black male: advertising and the cultural politics of masculinity. *Gender, Place and Culture* **1**, 49–59.
—— 1999: Postmodern urbanism and the ethnographic void. *Urban Geography* **20**, 400–2.
—— 2000: Rematerializing social and cultural geography. *Social and Cultural Geography* **1**, 9–14.
—— and Smith, S. J. 1984: *Exploring social geography*. London: Allen & Unwin.
—— and Thrift, N. J. 1995: Geographies of consumption. In D. Miller (ed.), *Acknowledging consumption*. London: Routledge, 203–37.
James, S. 1984: *The content of social explanation*. Cambridge: Cambridge University Press.
Jencks, C. 1991: *The language of post-modern architecture*. New York: Rizzoli.
Johnson, A. K. 1996: *Urban ghetto riots 1965–1968: a comparison of Soviet and American press coverage*. New York: Columbia University Press.
Johnson, J. C. 1990: *Selecting ethnographic informants*. Newbury Park, CA: Sage.
Johnson, J. B. and Joslyn, R. A. 1995: *Political science research methods*. 3rd edition. Washington: Congressional Quarterly Inc.
Johnston, L. and Valentine, G. 1995: Wherever I lay my girlfriend, that's my home. In D. Bell and G. Valentine (eds), *Mapping desire*. London: Routledge, 99–113.
Johnston, R. J. 1979: The spatial impact of fiscal changes in Britain: regional policy in reverse. *Environment and Planning* **A11**, 1439–44.
—— 1986: *Philosophy and human geography*. 2nd edition. London: Edward Arnold.
—— 1991: *Geography and geographers*. 4th edition. London: Edward Arnold.
—— 2000: Intellectual respectability and disciplinary transformation? Radical geography

and the institutionalization of geography in the USA since 1945. *Environment and Planning* **A32**, 971–90.
—— Gregory, D. J. and Smith, D. M. (eds) 1994: *Dictionary of human geography*. 3rd edition. Oxford: Blackwell.
—— —— Pratt, G. and Watts, M. J. (eds) 2000: *Dictionary of human geography*. 4th edition. Oxford: Blackwell.
Jones, J. P. III, Nast, H. and Roberts, S. (eds) 1997: *Thresholds in feminist geography*. New York: Rowman & Littlefield.
—— Natter, W. and Schatski, T. R. (eds) 1993: *Postmodern contentions*. New York: Guildford.
Jones, B. D., Greenberg, S. R., Kaufman, C. and Drew, J. 1977: Bureaucratic response to citizen-initiated contacts: environmental enforcement in Detroit. *American Political Science Review* **71**, 148–65.
Jones, E. L., Porter, S. and Turner, M. 1984: *A gazetteer of English urban fire disasters 1500–1900*. London: Institute of British Geographers Historical Geography Research Series No. 13.
Jones, H., Caird, J. B., Berry, W. and Dewhurst, J. 1986: Peripheral counter-urbanization: findings from an integration of census and survey data in northern Scotland. *Regional Studies* **20**, 15–26.
Junker, B. 1960: *Fieldwork*. Chicago: University of Chicago Press.
Kain, R. J. P. and Prince, H. C. 1985: *The tithe surveys of England and Wales*. Cambridge: Cambridge University Press.
Katz, C. 1992: All the world is staged: intellectuals and the projects of ethnography. *Environment and Planning D: Society and Space* **10**, 495–510.
—— 1994: Playing the field: questions of fieldwork in geography. *Professional Geographer* **46**, 67–72.
—— and Monk, J. (eds) 1993: *Full circles: geographies of women over the life course*. London: Routledge.
Katzman, D. M. and Tuttle, D. M. 1983: *Plain folk: the life stories of undistinguished Americans*. Urbana, IL: University of Illinois Press.
Kearns, G. 1985: *Urban epidemics and historical geography: cholera in London 1848–49*. London: Institute of British Geographers Historical Geography Research Series No. 15.
Keith, M. 1992: Angry writing: (re)presenting the unethical world of the ethnographer. *Environment and Planning D: Society and Space* **10**, 551–68.
—— and Peach, C. 1983: The conditions in England's inner cities on the eve of the 1981 riots. *Area* **15**, 316–19.
Kellner, D. 1992: Popular culture and the construction of postmodern identities. In S. Lash and J. Friedman (eds), *Modernity and identity*, Oxford: Blackwell, 141–77.
Kent, R. 1981: *A history of British empirical sociology*. Aldershot: Gower.
Kershaw, R. and Pearsall, M. 2000: *Immigrants and aliens: a guide to sources on UK immigration and citizenship*. London: PRO Publications.
Kettle, M. 1998: Sacred ideal founded on Jefferson's fudge. *Guardian* 8 June, 20.
King, R. L., Connell, J. and White, P. E. (eds) 1995: *Writing across worlds: literature and migration*. London: Routledge.
Kinsman, P. 1995: Landscape, race and national identity: the photography of Ingrid Pollard. *Area* **27**, 300–10.

Kirby, A. M. 1985: Voluntarism and state funding of housing. *Tijdschrift voor Economische en Sociale Geografie* **76**, 53–62.

Kitchin, R. M. and Hubbard, P. J. 1999: Research, action and 'critical' geographies. *Area* **31**, 195–8.

—— and Tate, N. J. 2000: *Conducting research into human geography*. London: Prentice Hall.

—— and Wilton, R. 2000: Disability, geography and ethics. *Ethics, Place and Environment* **3**(1), 61–4.

Kitzinger, J. 1994: The methodology of focus groups. *Sociology of Health and Illness* **16**, 103–26.

Kneale, P. E. 1999: *Study skills for geography students*. London: Arnold.

Knightsbridge, A. A. H. 1983: National archives policy. *Journal of the Society of Archivists* **8**, 213–23.

Knox, P. L. 1978: The intraurban ecology of primary medical care. *Environment and Planning* **A10**, 415–36.

Kogevinas, M., Marmott, M. G., Fox, A. J. and Goldblatt, P. O. 1991: Socio-economic differences in cancer survival. *Journal of Epidemiology and Community Health* **45**, 216–19.

Kolakowski, L. 1972: *Positivistic philosophy*. Harmondsworth: Penguin.

Kollock, P. and Smith, M. 1999: Communities in cyberspace. In M. Smith and P. Kollock (eds), *Communities in cyberspace*, London: Routledge, 3–25.

Krueger, R. 1994: *Focus groups*. London: Sage.

—— 1997a: *Developing questions for focus groups*. London: Sage.

—— 1997b: *Moderating focus groups*. London: Sage.

—— 1997c: *Analysing and reporting focus groups*. London: Sage.

—— 1998: *Focus groups: a practical guide for applied research*. London: Sage.

—— and King, J. 1997: *Involving community members in focus groups*. London: Sage.

Kuhn, T. S. 1970: *The structure of scientific revolutions*. 2nd edition. Chicago: University of Chicago Press.

Kuper, A. 1983: *Anthropology and anthropologists: the modern British school*. London: Routledge & Kegan Paul.

La Fleur, J. D. 2000: *Pieter Van der Broecke's journal of voyages to Cape Verde, Guinea and Angola (1605–1612)*. London: The Hakluyt Society.

Lainhart, A. S. 1992: *State census records*. Baltimore: Genealogical Publishing Co.

Lakatos, I. 1978: *Imre Lakatos: philosophical papers – volume one: the methodology of scientific research programmes*. Cambridge: Cambridge University Press.

—— and Musgrave, A. (eds) 1974: *Criticism and the growth of knowledge*. Cambridge: Cambridge University Press.

Lakhani, S. 2000: Integration/exclusion? young British Asians and the politics of ethnicity. Ph.D. thesis, King's College London.

Land, A., Lowe, R. and Whiteside, N. 1992: *The development of the welfare state 1939–1951*. London: PRO Publications.

Langevin, B., Begeot, F. and Pearce, D. 1992: Censuses in the European Community. *Population Trends* **68**, 33–6.

Lasswell, H. D. 1941: The world attention survey. *Public Opinion Quarterly* **5**, 456–62.

Lather, P. 1991: *Getting smart: feminist research and pedagogy with/in the postmodern*. New York: Routledge.

Latour, B. 1987: *Science in action*. Milton Keynes: Open University Press.
—— and Woolgar, S. 1979: *Laboratory life: the construction of scientific facts*. Beverly Hills, CA: Sage.
Lawrence, M. 1998: Miles from home in the field of dreams: rurality and the social at the end of history. *Environment and Planning D: Society and Space* **16**, 705–32.
Lawton, R. 1978: Introduction. In R. Lawton (ed.), *The census and social structure*. London: Frank Cass, 1–27.
Le Roy Ladurie, E. 1978: *Montaillou*. Harmondsworth: Penguin.
Leal, D. L. and Hess, F. M. 1999: Survey bias on the front porch: are all subjects interviewed equally? *American Politics Quarterly* **27**, 468–87.
LeCompte, M. and Schensul, J. 1999: *Designing and conducting ethnographic research*. Walnut Creek, CA: Altamira.
Lee, R. M. 1993: *Doing research on sensitive topics*. London: Sage.
—— 1995: *Dangerous fieldwork*. Thousand Oaks, CA: Sage.
Lees, L. 1994: Rethinking gentrification: beyond the positions of economics and culture. *Progress in Human Geography* **18**, 137–50.
—— 1996: In the pursuit of difference: representations of gentrification. *Environment and Planning* **A28**, 453–70.
—— 1997: Ageographia, heterotopia, and Vancouver's new public library. *Environment and Planning D: Society and Space* **15**, 321–47.
—— 1998: Urban renaissance and the street: spaces of control *and* contestation. In N. Fyfe (ed.), *Images of the street*. London: Routledge, 236–53.
—— 1999: Critical geography and the opening up of the academy: lessons from 'real life' attempts. *Area* **31**, 377–83.
—— 2001: Towards a critical architectural geography: the case of an ersatz Colosseum. *Ecumene* **8**, 51–86.
—— and Longhurst, R. 1995: Feminist geography in Aotearoa/New Zealand: a workshop. *Gender, Place and Culture* **2**, 219–24.
Lefkowitz, M. 1996: *Not out of Africa: how Afrocentrism became an excuse to teach myth as history*. New York: Basic.
Lewin, E. and Leap, W. L. 1996: Introduction. In E. Lewin and W. L. Leap (eds), *Out in the field: reflections of lesbian and gay anthropologists*. Urbana, IL: University of Illinois Press, 1–28.
Lewin, H. S. 1947: Hitler youth and the Boy Scouts of America. *Human Relations* **1**, 206–27.
Lewis, O. 1951: *Life in a Mexican village: Tepoztlan revisited*. Urbana, IL: University of Illinois Press.
Ley, D. 1974: *The black inner city as frontier outpost*. Washington: Association of American Geographers Monograph No. 7.
—— 1987: Styles of the times: liberal and neo-conservative landscapes in inner Vancouver. *Journal of Historical Geography* **13**, 40–56.
—— 1988: Interpretative social research in the inner city. In J. D. Eyles (ed.), *Research in human geography*. Oxford: Blackwell, 121–38.
—— 2000: *Seeking homo economicus: the strange story of Canada's Business Immigration Program*. Vancouver: Simon Fraser University, Vancouver Centre of Excellence Working Paper No. 00–02. [htpp://www. riim. metropolis. net]
—— and Cybriwsky, R. 1974: Graffiti as territorial markers. *Annals of the Association of American Geographers* **64**, 491–505.

—— and Samuels, M. (eds) 1978: *Humanistic geography: prospects and problems*. London: Croom Helm.

Lieberson, S. 1985: *Making it count*. Berkeley and Los Angeles: University of California Press.

Lineberry, R. L. 1975: Equality, public policy and public services: the underclass hypothesis and the limits of equality. *Policy and Politics* **4**(2), 67–84.

Lipman, V. D. 1949: *Local government areas 1834–1945*. Oxford: Blackwell.

Lipsky, M. 1980: *Street-level bureaucracy*. New York: Russell Sage Foundation.

Little, C. E. 1975: Preservation policy and personal perception: a 200 million acre misunderstanding. In R. Bruch, J. G. Fabos and E. H. Zube (eds), *Landscape assessment*. Stroudsburg, NJ: Dowden, Hutchinson and Ross, 46–58.

Littlejohn, J. 1963: *Westrigg: the sociology of a Cheviot parish*. London: Routledge & Kegan Paul.

Livingstone, D. 1984: Natural theology and Neo-Lamarckism: the changing context of nineteenth-century geography in the United States and Great Britain. *Annals of the Association of American Geographers* **74**, 9–28.

—— 1985: Evolution, science and society: historical reflections on the geographical experiment. *Geoforum* **16**, 119–30.

London Federation of Housing Associations 1996: *LFHA directory 1996*. London: National Federation of Housing Associations.

London Research Centre, annual: *London housing statistics*. London.

Longhurst, R. 1996: Refocusing groups: pregnant women's geographical experiences of Hamilton, New Zealand/Aotearoa. *Area* **28**, 143–9.

Longino, H. E. 1990: *Science as social knowledge*. Princeton: Princeton University Press.

Lowe, P. D. and Ward, N. 1997: Field level bureaucrats and the making of new moral discourse in agri-environmental discourses. In D. E. Goodman and M. J. Watts (eds), *Globalizing food*, London: Routledge, 256–72.

Lukes, S. 1974: *Power: a radical view*. London: Macmillan.

Lumas, S. 1997: *Making use of the census*. 3rd edition. London: HMSO.

Lynch, K. 1960: *The image of the city*. Cambridge, MA: MIT Press.

Lynd, R. S. and Lynd, H. M. 1929/1956: *Middletown: a study in American culture*. New York: Harcourt, Brace & World.

Lyons, M. and Simister, J. 2000: From rags to riches? Migration and intergenerational change in London's housing market, 1971–1991. *Area* **32**, 271–85.

McCracken, G. 1988: *The long interview*. Newbury Park, CA: Sage.

Macdonald, K. and Tipton, C. 1993: Using documents. In N. Gilbert (ed.), *Researching social life*. London: Sage, 187–200.

Macdonald, S. 1997: The museum as mirror: ethnographic reflections. In W. James, J. Hochey and A. Dawson (eds), *After writing culture*. London: Routledge, 161–77.

McDowell, L. 1992a: Valid games? A response to Erica Schoenberger. *Professional Geographer* **44**, 212–15.

—— 1992b: Doing gender: feminism, feminists and research methods in human geography. *Transactions of the Institute of British Geographers* **17**, 399–416.

—— and Court, G. 1994: Performing work: bodily representations in merchant banks. *Environment and Planning: Society and Space* **A12**, 725–50.

—— and Sharp, J. P. (eds) 1999: *A feminist glossary of human geography*. London: Arnold.

MacFarlane, A. 1977: *Reconstructing historical communities*. Cambridge: Cambridge University Press.

McGillivray, A. V. and Scammon, R. M. 1994: *America at the polls: a handbook of American presidential election statistics*. Two vols. Washington: Congressional Quarterly.

McGranahan, D. and Wayne, I. 1948: German and American traits reflected in popular drama. *Human Relations* **1**, 429–55.

Mackay, H. (ed.) 1997: *Consumption and everyday life*. London: Sage.

McKendrick, J. H. (ed.) 1995: *Multi-method research in population geography*. London: Institute of British Geographers Population Geography Research Group.

MacKenzie, D. 1999: Eugenics and the rise of mathematical statistics in Britain. In D. F. L. Dorling and S. Simpson (eds), *Statistics in society*, London: Arnold, 55–61.

Macnaghten, P. 1995: Public attitudes to countryside leisure. *Journal of Rural Studies* **11**, 135–47.

—— Grove-White, R., Jacobs, M. and Wynne, B. 1995: *Public perceptions and sustainability in Lancashire: indicators, institutions and participation*. Preston: Lancashire County Council and Lancaster University.

Madge, C. 1993: Boundary disputes. *Area* **25**, 294–9.

—— 1994: The ethics of research in the 'Third World'. In E. Robson and K. Willis (eds), *Postgraduate fieldwork in developing areas*. London: Institute of British Geographers Developing Areas Study Group Monograph No. 8, 91–102.

—— Raghuran, P., Skelton, T., Willis, K. and Williams, J. 1997: Methods and methodologies in feminist geographies. In Women and Geography Study Group, *Feminist geographies*. Harlow: Longman, 86–111.

Maguire, P. 1987: *Doing participatory research: a feminist approach*. Amherst, MA: University of Massachusetts.

Maier, M. H. 1995: *The data game: controversies in social science statistics*. Armonk, NY: M. E. Sharpe.

Malinowski, B. 1922: *Argonauts of the western Pacific: an account of native enterprise and adventure in the archipelagos of Melanesian New Guinea*. London: Routledge & Kegan Paul.

—— 1935/1978: *Coral gardens and their magic: a study of the methods of tilling the soil and of agricultural rites in the Trobriand Islands*. New York: Dover.

—— 1967: *A diary in the strict sense of the term*. New York: Harcourt, Brace & World.

Malthus, T. R. 1798/1998: *An essay on the principle of population*. Amherst, NY: Promotheus.

Mann, C. and Stewart, F. 2000: *Internet communication and qualitative research*. London: Sage.

Mann, P. H. 1985: *Methods in social investigation*. 2nd edition. Oxford: Blackwell.

Maquis, K. H. 1970: Effects of social reinforcement on health reporting in the household interview. *Sociometry* **33**, 203–15.

Marchand, B. 1974: Quantitative geography: revolution or counter revolution? *Geoforum* **17**, 15–23.

Marcus, G. 1997: Critical cultural studies as one power/knowledge like, among and in engagement with others. In E. Long (ed.), *From sociology to cultural studies*. Malden, MA: Blackwell, 399–425.

—— and Fisher, M. 1986: *Anthropology as cultural critique*. Chicago: University of Chicago Press.

Marsden, T. K. 1999: Rural futures: the consumption countryside and its regulation. *Sociologia Ruralis* **39**, 501–20.

Marshall, G., Newby, H. E., Rose, D. and Vogler, C. 1988: *Social class in modern Britain.* London: Hutchinson.

Marshall, J. U. 1985: Geography as a scientific enterprise. In R. J. Johnston (ed.), *The future of geography.* London: Methuen, 123–38.

Marshall, M. 1987: *Long waves of regional development.* Basingstoke: Macmillan.

Martin, D. 2000: Towards the geographies of the 2001 UK Census of Population. *Transactions of the Institute of British Geographers* **25**, 321–32.

Mascia-Lees, F., Sharpe, P. and Ballerino Cohen, C. 1989: The postmodern turn in anthropology: cautions from a feminist perspective. *Signs* **15**, 7–33.

Massey, D. B. 1995: *Spatial divisions of labour.* 2nd edition. Basingstoke: Macmillan.

—— and Wield, D. 1992: Science parks. *Environment and Planning D: Society and Space* **10**, 411–22.

Matthews, T. S. 1957: *The sugar pill: an essay on newspapers.* London: Victor Gollancz.

Mead, M. 1977: *Letters from the field 1925–1975.* New York: Harper & Row.

Medlicott, W. N., Dakin, D. and Lambert, M. E. 1969: *Documents on British foreign policy 1919–1939: volume X – Far Eastern affairs.* 2nd series. London: HMSO.

Mercer, D. 1984: Unmasking technocratic geography. In M. Billinge, D. J. Gregory and R. L. Martin (eds), *Recollections on a revolution.* London: Macmillan, 153–99.

Merrett, S. 1979: *State housing in Britain.* London: Routledge & Kegan Paul.

Merrifield, A. K. 1995: Situated knowledge through exploration: reflections on Bunge's geographical expeditions. *Antipode* **27**, 49–70.

Merton, R. K. 1938/1970: *Science, technology and society in seventeenth-century England.* New York: Howard Fertig.

Midanik, L. 1982: The validity of self-regulated alcohol consumption and alcohol problems. *British Journal of Addiction* **77**, 357–82.

Middleton, E. 1995: Samples of anonymized records. In S. Openshaw (ed.), *Census users' handbook.* Cambridge: GeoInformation International, 337–62.

Mikkelsen, B. 1995: *Methods for development work and research.* New Delhi: Sage.

Miles, I. and Irvine, J. 1979: The critique of official statistics. In J. Irvine and J. Evans (eds), *Demystifying social statistics.* London: Pluto, 113–29.

Miles, M. and Huberman, M. 1984: *Qualitative data analysis.* London: Sage.

Miller, J. C. 1988: Municipal annexation and boundary change. In *The municipal yearbook 1988.* Washington: International City Management Association, 59–67.

Millett, K. 1971: *The prostitution papers: a candid dialogue.* New York: Avon.

Million dollar directory (annual). New York: Dun and Bradstreet.

Ministerio de Cultura, Fundación Ramón Areces and IBM España 1990: *Computerization project for the Archivo General de Indias.* Madrid.

Minogue, M. 1977: *Documents on contemporary British government: volume two – local government in Britain.* Cambridge: Cambridge University Press.

Mishler, E. G. 1986: *Research interviewing.* Cambridge, MA: Harvard University Press.

Mitchell, B. and Draper, D. 1982: *Relevance and ethics in geography.* London: Longman.

Mitchell, B. R. 1988a: *British historical statistics.* Cambridge: Cambridge University Press.

—— 1998b: *International historical statistics: Africa, Asia and Oceania 1750–1993.* 3rd edition. Basingstoke: Macmillan.

—— 1998c: *International historical statistics: the Americas 1750–1993*. 4th edition. New York: Stockton.
—— 1998d: *International historical statistics: Europe 1750–1993*. 4th edition. Basingstoke: Macmillan.
Modood, T. 1997: Employment. In T. Modood and R. Bethoud (eds), *Ethnic minorities in Britain: diversity and disadvantage*, London: Policy Studies Institute, 83–149.
Mohammad, R. 2001: 'Insiders' and/or 'outsiders': positionality, theory and praxis. In M. Limb and C. Dwyer (eds), *Qualitative methodologies for geographers*. London: Arnold.
Molotch, H. 1976: The city as a growth machine. *American Journal of Sociology* **82**, 309–32.
Moodie, D. W. and Catchpole, A. J. W. 1975: *Environmental data from historical documents by content analysis*. Winnipeg: University of Manitoba Geographical Study No. 5.
Morgan, B. S. 1979: Residential segregation, marriage and the evolution of the stratification system. *Environment and Planning* **A11**, 209–17.
Morgan, D. 1988: *Focus groups as qualitative research*. London: Sage
—— 1997a: *Focus groups as qualitative research*. 2nd edition. London: Sage
—— 1997b: *The focus group guidebook*. London: Sage.
—— 1997c: *Planning focus groups*. London: Sage.
Morin, E. 1970: *The red and the white: report from a French village*. New York: Random House.
Morris, C. and Potter, C. 1995: Recruiting the new conservationists: farmers' adoption of agri-environmental schemes in the UK. *Journal of Rural Studies* **11**, 51–63.
Mortimer, I. 1999: *Record repositories in Great Britain*. 11th edition. London: HMSO.
Morton, A. 1997: *Education and the state from 1833*. London: PRO Publications.
Moser, C. A. and Kalton, G. 1971: *Survey methods in social investigation*. London: Heinemann.
Moss, L. 1991: *The government social survey: a history*. London: HMSO.
Moss, S. 2000: History's verdict on Holocaust upheld. *Guardian* 12 Apr., 5.
Mostyn, B. 1979: *Personal benefits and satisfactions derived from participation in urban wildlife projects*. Shrewsbury: Nature Conservancy Council.
Mueller, C. 1997: International press coverage of East German protest events 1989. *American Sociological Review* **62**, 820–32.
Mullins, D. 1993: *An evaluation of the Housing Corporation rural programme*. Salisbury: Rural Development Commission.
Murphy, M., Goldblatt, P. O., Thornton-Jones, H. and Silcocks, P. 1990: Survival amongst women with cancer of the uterine cervix. *Journal of Epidemiology and Community Health* **44**, 291–6.
Myers, G. and Macnaghten, P. 1999: Can focus groups be analysed as talk? In R. Barbour and J. Kitzinger (eds), *Developing focus group research*. London: Sage, 173–85.
Myrdal, G. 1969: *Objectivity in social research*. New York: Pantheon.
Nast, H. 1994: Opening remarks on 'Women in the field'. *Professional Geographer* **46**, 54–66.
Nath, J. 1991: A Bengali woman in three different field situations. In M. N. Panini (ed.), *From the female eye: accounts of women fieldworkers studying their own communities*. Delhi: Hindustan Publishing, 59–65.
National Assembly for Wales, annual: *Welsh housing statistics*. Cardiff.

National Federation of Housing Associations, annual: *CORE annual statistics*. London.

Natter, W. and Jones, J. P. III 1993: Pets or meat: class, ideology, and space in Roger and Me. *Antipode* **25**, 140–58.

Nelson, C., Treicher, P. and Grossberg, L. 1992: Cultural studies. In L. Grossberg, C. Nelson and P. Treicher (eds), *Cultural studies*. New York: Routledge, 1–16.

Newby, H. E. 1977: In the field: reflections on the study of Suffolk farm workers. In C. Bell and H. E. Newby (eds), *Doing sociological research*. London: Allen & Unwin, 108–29.

—— 1986: Locality and rurality: the restructuring of rural social relations. *Regional Studies* **20**, 209–15.

Newson, L. A. 1995: *Life and death in early colonial Ecuador*. Norman: University of Oklahoma Press.

Nichols, P. 1991: *Social survey methods: a fieldguide for development workers*. Oxford: Oxfam.

Norris, C. 1995: Truth, science and the growth of knowledge. *New Left Review* **210**, 105–23.

Norton-Taylor, R. 1998: MI5 rushes to shred files on subversives. *Guardian* 12 Jan., 7.

Nove, A. 1975: Is there a ruling class in the USSR?. *Soviet Studies* **27**, 615–35.

—— 1977: Can Eastern Europe feed itself? *World Development* **5**, 417–24.

Novick, P, 1988: *That noble dream: the objectivity question and the American historical profession*. Cambridge: Cambridge University Press.

Oakley, A. 1981: Interviewing women: a contradiction in terms. In H. Roberts (ed.), *Doing feminist research*. London: Routledge, 30–61.

—— and Oakley, R. 1979: Sexism in official statistics. In J. Irvine and J. Evans (eds), *Demystifying social statistics*. London: Pluto, 172–89.

O'Brien, L. 1992: *Introductory quantitative geography*. London: Routledge.

OECD (SOPEMI), annual: *Trends in international migration*. Paris: Organization for Economic Cooperation and Development.

—— annual: *OECD in figures*. Paris: Organization for Economic Cooperation and Development.

Okely, J. and Callaway, H. (eds) 1992: *Anthropology and autobiography*. London: Routledge.

Oldenburg, V. T. 1989: *The making of colonial Lucknow 1856–1877*. Delhi: Oxford University Press.

Olsson, G. 1980: *Birds in egg: eggs in bird*. London: Pion.

Openshaw, S. 1995: A quick introduction to most of what you need to know about the 1991 census. In S. Openshaw (ed.), *Census users' handbook*. Cambridge: GeoInformation International, 1–26.

—— 1998: Towards a more computationally minded scientific human geography. *Environment and Planning* **A30**, 317–32.

—— and Turton, I. 1996: New opportunities for geographical census analysis using individual level data. *Area* **28**, 167–76.

Oppenheim, A. N. 1986: *Questionnaire design and attitude measurement*. Aldershot: Gower.

Owen, C. 1999: Government household surveys: using government household surveys in social research. In D. F. L. Dorling and S. Simpson (eds), *Statistics in society*. London: Arnold, 19–28.

Pahl, R. E. 1966: The rural–urban continuum. *Sociologia Ruralis* **6**, 299–329.
Palriwala, R. 1991: Researcher and women: dilemmas of a fieldworker in a Rajasthan village. In M. N. Panini (ed.), *From the female eye: accounts of fieldworkers studying their own communities*. Delhi: Hindustan Publishing, 27–35.
Panini, M. N. (ed.) 1991: *From the female eye: accounts of women fieldworkers studying their own communities*. Delhi: Hindustan Publishing.
Panofsky, E. 1962: *Studies in iconology*. New York: Harper & Row.
Papke, D. R. 1999: *The Pullman case*. Lawrence: University Press of Kansas.
Parenti, M. 1986: *Inventing reality: the politics of the mass media*. New York: St Martin's.
Park, R. E. 1925: The city: suggestions for the investigation of human behaviour in the urban environment. In R. E. Park and E. W. Burgess (eds), *The city*. Chicago: University of Chicago Press, 1–46.
—— 1950: *Race and culture*, ed. E. C. Hughes, C. S. Johnson, J. Masouka, R. Redfield and L. Wirth. Glencoe, ILL: Free Press.
—— 1968: The urban community as a spatial pattern and a moral order. In C. Peach (ed.), *Urban social segregation*. London: Longman, 21–31. Originally part of E. W. Burgess (ed.), *The urban community*. Chicago: University of Chicago Press, 1926.
Parr, H. 1998: Mental health, ethnography and the body. *Area* **30**, 28–37.
Parry, B. 1998: Hunting the gene-hunters: the role of hybrid networks, status, and chance in conceptualizing and accessing 'corporate elites'. *Environment and Planning* **A30**, 2147–62.
Parsons, A. J. and Knight, P. G. 1995: *How to do your dissertation in geography and related disciplines*. London: Chapman & Hall.
Pateman, C. 1970: *Participation and democratic theory*. Cambridge: Cambridge University Press.
Pattie, C. J. 1986: Positive discrimination in the provision of primary education in Sheffield. *Environment and Planning* **A18**, 1249–57.
Payne, S. L. 1951: *The art of asking questions*. Princeton: Princeton University Press.
Peake, L. 1986: A conceptual enquiry into urban politics and gender. In K. Hoggart and E. Kofman (eds), *Politics, geography and social stratification*, London: Croom Helm, 62–85.
Peet, R. 1998: *Modern geographic thought*. Oxford: Blackwell.
Penhale, B. 1990: *Households, families and fertility*. London: City University Social Statistics Research Unit LS User Guide No. 1.
Péroz-Mallaína, P. E. 1998: *Spain's men of the sea*. Baltimore: Johns Hopkins University Press.
Perks, R. and Thomson, A. (eds) 1998: *The oral history reader*. London: Routledge.
Perroux, F. 1983: *A new concept of development*. London: Croom Helm.
Personal Narratives Group 1989: *Interpreting women's lives: feminist theory and personal narratives*. Bloomington: Indiana University Press.
Pfeffer, M. J. 1983: Social origins of three systems of farm production in the United States. *Rural Sociology* **48**, 540–62.
Philip, L. S. 1998: Combining quantitative and qualitative approaches to social research in human geography. *Environment and Planning* **A30**, 261–76.
Phillips, E. M. and Pugh, D. S. 1993: *How to get a Ph.D.* 2nd edition. Milton Keynes: Open University Press.
Phillips, M., Huhne, C. and Fairhall, D. 1989: How the cards are stacked. *Guardian* 15 Mar., 21.

Philo, C. 1992: Neglected rural geographies. *Journal of Rural Studies* **8**, 193–207.
—— (ed.) 1998: Theme issue: reconsidering quantitative geography. *Environment and Planning* **A30**(2).
—— 2000: More words, more worlds: reflections on the 'cultural turn' and human geography. In I. Cook, D. Crouch, S. Naylor and J. Ryan (eds), *Cultural turns/geographical turns*. London: Prentice Hall, 26–53.
—— Mitchell, R. and More, A. 1998: Reconsidering quantitative geography: the things that count. *Environment and Planning* **A30**, 191–201.
Pile, S. J. 1991: Practising interpretative geography. *Transactions of the Institute of British Geographers* **16**, 458–69.
—— 1993: Human agency and human geography revisited: a critique of 'new models' of the self. *Transactions of the Institution of British Geographers* **18**, 122–39.
—— 1996: *The body and the city: psychoanalysis, space and subjectivity*. London: Routledge.
—— and Thrift, N. J. (eds) 1995: *Mapping the subject*. London: Routledge.
Platt, J. 1981: Evidence and proof in documentary research. *Sociological Review* **29**, 31–52, 53–66.
Plummer, K. 1983: *Documents of life*. London: Unwin Hyman.
Pollitt, C., Harrison, S., Hunter, D. J. and Marnoch, G. 1992: No hiding place: on the discomforts of researching the contemporary policy process. In J. Vincent and S. Brown (eds), *Critics and customers: the control of social policy research*. Aldershot: Avebury, 54–72.
Polsby, N. W. 1980: *Community power and political theory*. 2nd edition. New Haven, CT: Yale University Press.
Pooley, C. G. and Turnbull, J. 1998: *Migration and mobility in Britain since the eighteenth century*. London: UCL Press.
Poovey, M. 1998: *A history of the modern fact: problems of knowledge in the sciences of wealth and society*. Chicago: University of Chicago Press.
Popper, K. R. 1963: *Conjectures and refutations: the growth of scientific knowledge*. London: Routledge & Kegan Paul.
—— 1965: Unity of method in the natural and social sciences. In D. Braybrooke (ed.), *Philosophical problems of the social sciences*. New York: Macmillan, 32–41.
Porter, S. 1990: *Exploring urban history*. London: Batsford.
Poster, M. 1990: *The mode of information*. Cambridge: Polity.
—— 1995: *The second media age*. Cambridge: Polity.
Potter, J. 1996: *Representing reality*. London: Sage.
—— and Wetherell, M 1994: Analyzing discourse. In A. Bryman and R. G. Burgess (eds), *Analyzing qualitative data*. London: Routledge, 47–66.
Pratt, A. C. 1994: *Uneven re-production*. Oxford: Pergamon.
—— 1995: Putting critical realism to work. *Progress in Human Geography* **19**, 61–74.
Pratt, M. L. 1986: Fieldwork in common places. In J. Clifford and G. Marcus (eds), *Writing culture*. Berkeley and Los Angeles: University of California Press, 27–50.
Prince, H. 1980: Review of 'Humanistic geography'. *Annals of the Association of American Geographers* **70**, 294–6.
Probyn, E. 1993: *Sexing the self: gendered positions in cultural studies*. London: Routledge.
Professional Geographer 1994: Women in the field: critical feminist methodologies and theoretical perspectives. *Professional Geographer* **46**, 54–102.

Prosser, J. (ed.) 1998: *Image-based research*. London: Falmer.
—— and Schwartz, D. 1998: Photographs within the sociological research process. In J. Prosser (ed.), *Image-based research*. London: Falmer, 115–30.
Punch, M. 1986: *The politics and ethics of fieldwork*. Beverly Hills, CA: Sage.
Rabow, J. and Neuman, C. E. 1984: Garbaeology as a method of cross-validating interviewer data on sensitive topics. *Sociology and Social Research* **68**, 480–97.
Ragin, C. C. 1994: *Constructing social research*. Thousand Oaks, CA: Pine Forge.
—— and Becker, H. S. (eds) 1992: *What is a case?* Cambridge: Cambridge University Press.
Rallings, C. and Thrasher, M. 2000: *British electoral facts 1832–1999*. Abingdon: Ashgate.
—— —— annual: *Local elections handbook*. Plymouth: Local Government Chronicle Elections Centre.
Raper, J. F., Rhind, D. W. and Shepherd, J. W. 1992: *Postcodes*. Harlow: Longman.
Rapport, N. 1993: *Diverse world-views in an English village*. Edinburgh: Edinburgh University Press.
Real, M. 1977: *Mass mediated culture*. Englewood Cliffs, NJ: Prentice Hall.
Reason, P. and Bradbury, H. (eds) 2000: *Handbook of action research*. London: Sage.
Reay, D. 1995: The fallacy of easy access. *Women's Studies International Forum* **18**, 205–13.
Redfield, R. 1930/1973: *Tepoztlan, a Mexican village*.Chicago: University of Chicago Press.
Rees, A. D. 1950: *Life in a Welsh countryside: a social study of Llanfihangel yng Ngwynfa*. Cardiff: University of Wales Press.
Rees, P. H. 1995: Putting the census on the researcher's desk. In S. Openshaw (ed.), *Census users' handbook*. Cambridge: GeoInformation International, 27–81.
Reinharz, S. 1992: *Feminist methods in social science*. New York: Oxford University Press.
Rhind, D. W. (ed.) 1983: *A census user's handbook*. London: Macmillan.
Rice, S. 1931: *Methods in social science*. Chicago: University of Chicago Press.
Richards, K. S. 1990: 'Real' geomorphology. *Earth Surface Processes and Landforms* **15**, 195–7.
Richardson, R. C. and James, T. B. 1983: *The urban experience: a sourcebook – English, Scottish and Welsh towns, 1450–1700*. Manchester: Manchester University Press.
Richmond, P. 1985: The state and the role of the housing association sector in rural areas: a case study in Devon. Ph.D. thesis, University of Exeter.
Ricoeur, P. 1971: The model of the text: meaningful action considered as text. *Social Research* **38**, 529–62.
Riessman, C. K. 1993: *Narrative analysis*. Newbury Park, CA: Sage.
Riis, J. 1971: *How the other half lives: studies among the tenements of New York*. New York: Dover.
Ritchie, D. A. 1995. *Doing oral history*. New York: Twayne.
Robinson, G. 1998: *Methods and techniques in human geography*. Chichester: Wiley.
Robinson, W. S. 1950: Ecological correlations and the behavior of individuals. *American Sociological Review* **15**, 351–7.
Robson, E. 1994: From teacher to taxi driver: reflections on research roles in developing areas. In E. Robson and K. Willis (eds), *Postgraduate fieldwork in developing areas*. London: Institute of British Geographers Developing Areas Research Group Monograph No. 8, 36–59.
—— and Willis, K. (eds) 1994: *Postgraduate fieldwork in developing areas*. London: Institute of British Geographers Developing Areas Research Group Monograph No. 8.

Rocheleau, D. 1995: Maps, numbers, text and context: mixing methods in feminist political ecology. *Professional Geographer* **47**, 458–66.
Rodney, W. 1976: *How Europe underdeveloped Africa*. London: Bogle-L'Ouverture.
Roethlisberger, F. J. and Dickson, M. J. 1939: *Management and the worker*. Cambridge, MA: Harvard University Press.
Rorty, R. 1980: *Philosophy and the mirror of nature*. Princeton: Princeton University Press.
—— 1991: *Objectivity, relativism, and truth: philosophical papers – volume one*. Cambridge: Cambridge University Press.
Rose, D. 1987: *Black American street life: south Philadelphia, 1969–1971*, Philadelphia: University of Pennsylvania Press.
—— 1990: *Living the ethnographic life*. Newbury Park, CA: Sage.
—— Buck, N. and Johnston, R. J. 1994: The British Household Panel Study: a valuable new resource for geographical research. *Area* **26**, 368–76.
Rose, G. 1993: *Feminism and geography*. Cambridge: Polity.
—— 1994: The cultural politics of place: local representation and oppositional discourse in two films. *Transactions of the Institute of British Geographers* **19**, 46–60.
—— 2000: Psychoanalytic theory, geography and. In R. J. Johnston, D. J. Gregory, G. Pratt and M. J. Watts (eds), *The dictionary of human geography*. Oxford: Blackwell, 653–5.
—— 2001: *Visual methodologies*. London: Sage.
Rouse, J. 1987: *Knowledge and power: toward a political philosophy of science*. Ithaca, NY: Cornell University Press.
Routledge, P. 1996: The third space as critical engagement. *Antipode* **28**, 399–419.
Rubin, H. and Rubin, I. 1995: *Qualitative interviewing*. London: Sage.
Runyon, R. P. 1981: *How numbers lie*. Lexington, MA: Lewis Publishing.
Ryan, M. 1988: Postmodern politics. *Theory, Culture and Society* **5**, 559–76.
Rycroft, S. 1996: Changing lanes: textuality off and on the road. *Transactions of the Institute of British Geographers* **21**, 425–8.
Sacks, H. 1984: On doing 'being ordinary'. In J. Atkinson and J. Heritage (eds), *Structures of social action*. Cambridge: Cambridge University Press, 413–29.
Sacks, H. 1992: *Lectures on conversations*. Two vols. Oxford: Blackwell.
Salamon, L. M. and van Evera, S. 1973: Fear, apathy and discrimination: a test of three explanations of political participation. *American Political Science Review* **67**, 1288–306.
Sanjek, R. 1990: On ethnographic validity. In R. Sanjek (ed.), *Fieldnotes: the makings of anthropology*. Ithaca, NY: Cornell University Press, 385–418.
Sarup, M. 1988: *An introductory guide to post-structuralism and postmodernism*. London: Harvester Wheatsheaf.
—— 1993: *An introductory guide to post-structuralism and postmodernism*. 2nd edition. Hemel Hempstead: Harvester Wheatsheaf.
Sassen, S. 1991: *The global city*. Princeton: Princeton University Press.
Sayer, A. 1984: *Method in social science: a realist approach*. London: Hutchinson. (2nd edition, London: Routledge, 1992.)
—— 1989: Dualistic thinking and rhetoric in geography. *Area* **21**, 301–5.
—— 1991: Behind the locality debate: deconstructing geography's dualisms. *Environment and Planning* **A23**, 283–308.
—— 1997: Essentialism, social constructionism, and beyond. *Sociological Review* **45**, 453–87.
Scharfstein, B.-A. 1989: *The dilemma of context*. New York: New York University Press.

Schartz, D. 1989: Visual ethnography. *Qualitative Sociology* **12**, 119–54.
Schensul, S. L., Schensul, J. J. and LeCompte, M. D. 1999: *Essential ethnographic methods: observations, interviews and questionnaires*. London: Sage.
Schoenberger, E. 1991: The corporate interview as a research method in economic geography. *Professional Geographer* **43**, 180–9.
—— 1992: Self criticism and self awareness in research: a reply to Linda McDowell. *Professional Geographer* **44**, 215–18.
Schuman, H. 1992: Context effects. In N. Schwarz and S. Sudman (eds), *Context effects in social and psychological research*. New York: Springer-Verlag, 5–20.
—— and Converse, J. 1971: The effects of black and white interviewers on black responses in 1968. *Public Opinion Quarterly* **35**, 44–68.
Schutz, A. 1960: The social world and theory of social action. *Social Research* **27**, 203–21.
Scott, J. 1990: *A matter of record*. Cambridge: Polity.
Scott, J. C. 1985: *Weapons of the weak: everyday forms of peasant resistance*. New Haven, CT: Yale University Press.
Seale, C. (ed.) 1998: *Researching society and culture*. London: Sage.
Sebald, H. 1962: Studying national character through comparative content analysis. *Social Forces* **40**, 318–22.
Seidel, J., Friese, S. and Leonard, D. 1995: *The ETHNOGRAPH Version 4. 0: a users' guide*. Amherst, MA: Qualis Research Associates.
Sellers, J. 2000: Translocal orders and urban environmentalism. In K. Hoggart and T. N. Clark (eds), *Citizen responsive government*. Greenwich, CT: JAI Press, 117–47.
Sender, J. and Smith, J. 1986: *The development of capitalism in Africa*. London: Methuen.
Sharpe, L. J. 1978: The social scientist and policy-making in Britain and America. In M. Bulmer (ed.), *Social policy research*. Basingstoke: Macmillan, 302–12.
Shaw, G. 1982: *British directories as sources of historical geography*. London: Institute of British Geographers Historical Geography Research Series No. 8.
Shaw, J. and Matthews, J. 1998: Communicating academic geography – the continuing challenge. *Area* **30**, 367–72.
Sheskin, I. M. 1985: *Survey research for geographers*. Washington: Association of American Geographers.
Shorney, D. 1996: *Protestant nonconformity and Roman Catholics: a guide to sources in the Public Record Office*. London: PRO Publications.
Shurmer-Smith, P. 2000: Hélène Cixous. In M. Crang and N. J. Thrift (eds), *Thinking space*. London: Routledge, 154–66.
Sibley, D. 1995: *Geographies of exclusion*. London: Routledge.
Sidaway, J. D. 1992: In other worlds: on the politics of research by 'First World' geographers in the 'Third World'. *Area* **24**, 403–8.
Siegel, S. 1956: *Nonparametric statistics for the behaviural sciences*. New York: McGraw-Hill.
Silverman, D. 1993: *Interpreting qualitative data*. London: Sage.
—— 1997: *Qualitative research*. London: Sage.
—— 1998: Analysing conversation. In C. Seale (ed.), *Researching society and culture*. London: Sage, 261–74.
Simmons, P. A. (ed.) 1999: *Housing statistics of the United States*. 2nd edition. Washington: Berman.
Simpson, S. 1996: Non-response to the 1991 census: the effect of ethnic group

enumeration. In D. Coleman and J. Salt (eds), *Ethnicity in the 1991 census*. London: HMSO, 63–79.

Singleton, A. 1999: Measuring international migration. In D. Dorling and S. Simpson (eds), *Statistics in society*, London: Arnold, 148–58.

Skeels, A. 1993: A passage to postmodernity: Carl Sauer repositioned in the field. MA thesis, University of British Columbia.

Slater, D. 1995: Photography and modern vision: the spectacle of 'natural' magic. In C. Jencks (ed.), *Visual culture*. London: Routledge, 218–37.

—— 1998: Content analysis and semiotics. In C. Seale (ed.), *Researching society and culture*. London: Sage, 233–44.

Slattery, M. 1986: *Official statistics*. London: Tavistock.

Smart, C. 1984: *The ties that bind: law, marriage and the reproduction of patriarchal relations*. London: Routledge.

Smith, A. 1776/1991: *The wealth of nations*. London: Everyman's Library.

Smith, F. M. 1996: Problematizing language: limitations and possibilities in 'foreign language' research. *Area* **28**, 160–6.

Smith, J. C. and Horton, C. P. (eds) 1995: *Historical statistics of black America*. New York: Gale Research.

Smith, N. 1984: *Uneven development*. Oxford: Blackwell.

Smith, S. J. 1981: Humanistic method in contemporary social geography. *Area* **13**, 193–8.

—— 1984: Practising humanistic geography. *Annals of the Association of American Geographers* **74**, 353–74.

—— 1988: Constructing local knowledge: the analysis of self in everyday life. In J. D. Eyles and D. M. Smith (eds), *Qualitative methods in human geography*. Cambridge: Polity, 17–38.

—— 1997: Beyond geography's visible worlds: a cultural politics of music. *Progress in Human Geography* **21**, 502–29.

—— 2000: Performing the (sound)world. *Environment and Planning D: Society and Space* **18**, 615–38.

Smith, V. 1991: Split affinities. In M. Hirsch and E. Keller (eds), *Conflicts in feminism*. New York: Routledge, 271–87.

Sokal, A. 1996: A physicist experiments with cultural studies. *Lingua Franca* **6** (May–June), 62–4.

Southall, H. and Gilbert, D. 1996: A good time to wed? Marriage and economic distress in England and Wales 1839–1914. *Economic History Review* **49**, 35–57.

—— —— and Bryce, C. 1994: *Nineteenth century trade union records*. London: Institute of British Geographers Historical Geography Research Series No. 27.

Sparke, M. 1996: Displacing the field in fieldwork: masculinity, metaphor and space. In N. Duncan (ed.), *Bodyspace*. London: Routledge, 212–33.

Spector, M. and Kitsue, J. 1987: *Constructing social problems*. New York: Aldine de Gruyter.

Spector, P. E. 1993: Research designs. In M. S. Lewis-Beck (ed.), *Experimental design and methods*. Thousand Oaks, CA: Sage, 1–74.

Social register (irregular). New Orleans: Social Register Association.

Spradley, J. P. 1979: *The ethnographic interview*. Orlando, FL: Harcourt, Brace, Jovanovich.

Stanley, L. and Wise, S. 1993: *Breaking out again: feminist ontology and epistemology*. London: Routledge.

Steckel, R. H. 1991: The quality of census data for historical enquiry: a research agenda. *Social Science History* **15**, 579–99.
Stein, L. and Fleischmann, A. 1987: Newspaper and business endorsements in municipal elections. *Journal of Urban Affairs* **9**, 325–36.
Stewart, D. W. and Kamins, M. A. 1993: *Secondary research*. 2nd edition. Newbury Park, CA: Sage.
—— and Shamdasani, P. N. 1990: *Focus groups*. Newbury Park, CA: Sage.
Stewart, E. P. 1913/1938: *Letters of a woman homesteader*. New York: Houghton Mifflin.
Stocking, G. W. 1987: *Victorian anthropology*. New York: Free Press.
Strathern, M. 1987: An awkward relationship: the case of feminism and anthropology. *Signs* **12**, 276–92.
Strauss, A. and Corbin, J. 1990: *Basics of qualitative research: grounded theory procedures and techniques*. Newbury Park, CA: Sage.
Stringer, E. 1999: *Action research*. London: Sage.
Stryker, R. 1996: Beyond history versus theory. *Sociological Methods and Research* **23**, 304–52.
Suttles, G. D. 1968: *The social order of the slum*. Chicago: University of Chicago Press.
—— 1972: *The social construction of communities*. Chicago: University of Chicago Press.
Szucs, L. B. 1986: *Chicago and Cook County sources: genealogy and historical guide*. Salt Lake City: Ancestry Publishing.
Talbot, C. 1992: On-line data sources. In A. Rogers, H. Viles and A. Goudie (eds), *The student's companion to geography*, Oxford: Blackwell, 311–14.
Tammita-Delgoda, S. 1997: *A traveller's history of India*. 2nd edition. Moreton-in Marsh: Windrush.
Taylor, A. J. P. 1963: *The First World War*. Harmondsworth: Penguin.
Taylor, M. F. 1994: Ethical considerations in European cross-national research. *International Social Science Journal* **142**, 523–32.
The Times, annual: *The Times 1000*. London: Times Books.
Thomas, J. 1993: *Doing critical ethnography*. London: Sage.
Thomas, R. 1995: Changes to jobless figures planned to defuse criticism. *Guardian* 24 July, 14.
Thomas, W. I. and Znaniecki, F. 1918: *The Polish peasant in Europe and America: monograph of an immigrant group – volume one*. Chicago: University of Chicago Press.
Thrift, N. J. 1986: Little games and big stories: accounting for the practice of personality and politics in the 1945 General Election. In K. Hoggart and E. Kofman (eds), *Politics, geography and social stratification*. London: Croom Helm, 86–143.
—— 1996: *Spatial formations*. London: Sage.
—— 1997a: Us and them: re-imagining places, reimagining identities. In H. Mackay (ed.), *Consumption and everyday life*. London: Sage, 159–212.
—— 1997b: The still point: resistance, expressive embodiment and dance. In S. J. Pile and M. Keith (eds), *Geographies of resistance*. London: Routledge, 124–53.
—— 2000a: Entanglement of power: shadows? In J. Sharp, P. Routledge, C. Philo and R. Paddison (eds), *Entanglements of power*. London: Routledge, 262–78.
—— 2000b: Non-representational theory. In R. J. Johnston, D. J. Gregory, G. Pratt and M. J. Watts (eds) *The dictionary of human geography*. 4th edition. Oxford: Blackwell, 556.
Tilly, C. 1981: *As sociology meets history*. New York: Academic.

—— 1984: *Big structures, large processes, huge comparisons*. New York: Russell Sage Foundation.
Todd, J. and Eldridge, J. 1987: *Electoral registration in inner city areas, 1983–1984*. London: Office of Population Censuses and Survey Social Survey Division.
Toffler, A. 1970: *Future shock*. London: Bodley Head.
Tonkiss, F. 1998a: The history of the social survey. In C. Seale (ed.), *Researching society and culture*. London: Sage, 58–71.
—— Tonkiss, F. 1998b: Analysing discourse. In C. Seale (ed.), *Researching society and culture*. London: Sage, 245–60.
Tooze, J. A. 1998: Imagining national economies, in G. Cubbitt (ed.), *Imagining nations*. Manchester: Manchester University Press, 212–28.
Tortora, R. D. 1994: Sampling frames for rural household surveys. In J. A. Christenson, R. C. Maurer and N. L. Strang (eds), *Rural data, people and policy*. Boulder, CO: Westview, 113–25.
Townsend, J. G. 1995: *Women's voices from the rainforest*. London: Routledge.
Tran, M. 1990: NY up in arms at 'wrong' census. *Guardian* 31 Aug., 8.
Trautmann, T. R. 1997: *Aryans and British India*. Berkeley and Los Angeles: University of California Press.
Travers, T. 1989: Community charge and other financial changes. In J. Stewart and G. Stoker (eds), *The future of local government*. Basingstoke: Macmillan, 9–29.
Travis, A. 1994: Study undermined popular conceptions of refugees. *Guardian* 4 July, 2.
Trist, E. and Bamforth, K. 1951: Some social and psychological consequences of the longwall method of coal getting. *Human Relations* **4**, 3–38.
Truman, C. and Humphries, B. 1994: Re-thinking social research: research in an unequal world. In B. Humphries and C. Truman (eds), *Re-thinking social research*. Aldershot: Avebury, 1–20.
Tuma, N. 1994: Event history analysis. In A. Dale and R. B. Davies (eds), *Analyzing social and political change*. London: Sage, 138–66.
Turkel, S. 1967/1993: *Division Street*. New York: New Press.
—— 1970/1986: *Hard times: an oral history of the Great Depression*. New York: Pantheon.
—— 1992: *Race: how blacks and whites think and feel about the American obsession*. New York: Anchor.
Tuson, P. 1979: *The records of the British residency and agencies in the Persian Gulf*. London: India Office Library and Records.
Tyler, S. 1986: Post-modern ethnography: from document of the occult to occult document. In J. Clifford and G. Marcus (eds), *Writing culture*. Berkeley and Los Angeles: University of California Press, 12–140.
UK Department of the Environment, Transport and the Regions, annual: *Housing and construction statistics*. London: The Stationery Office.
—— irregular: *English housing condition survey*. London: The Stationery Office.
—— quarterly: *Local housing statistics*, London: The Stationery Office.
—— quarterly: *Households found accommodation under the homeless provisions of the 1985 Housing Act, England*. London: The Stationery Office.
UK Ministry of Health, annual: *Local taxation returns*. London: HMSO. [available for 1920–39, except for 1921/2]
UK Office of National Statistics, annual (a): *Annual abstract of statistics*. London: The Stationery Office.

—— annual (b): *Family spending*. London: The Stationery Office.
—— annual (c): *New earnings survey*. London: The Stationery Office.
UK Office of National Statistics Social Survey Division, annual: *Housing in England*. London: The Stationery Office.
UK Office of National Statistics Social Statistics Division, annual: *Living in Britain*. London: The Stationery Office.
UK Office of Population Censuses and Surveys 1975: *Reorganization of local government areas: correlation of new and old areas*. London: HMSO.
UK Public Record Office (irregular): *Guide to the contents of the Public Record Office*. London: The Stationery Office.
UK Royal Commission on Historical Manuscripts 1995: *Principal family and estate collections: family names A–K*. London: HMSO.
UN, annual: *Industrial commodity statistics yearbook – production and consumption*. New York: United Nations.
—— annual: *International trade statistics yearbook*. New York: United Nations Department of Economics and Social Affairs Statistics.
—— annual: *Statistical yearbook*. New York: United Nations.
—— monthly: *Bulletin of statistics*. New York: United Nations.
UNESCO, annual: *Statistical yearbook*. Paris: United Nations Educational, Scientific and Cultural Organization.
US Bureau of Census, irregular: *County and city databook*. Washington: US Government Printing Office.
—— irregular: *State and metropolitan databook*. Washington: US Government Printing Office.
—— annual: *Statistical abstract of the United States*. Washington: US Government Printing Office.
US Department of Commerce, International Trade Administration 1999: *Metropolitan area exports 1993–98*. Washington: US Government Printing Office.
US Department of Justice, Office of Justice Programs, Bureau of Justice Statistics 1999: *Sourcebook of criminal justice statistics 1998*. Washington: US Government Printing Office.
Valentine, G. 1997: Tell me about ...: using interviews as a research methodology. In R. Flowerdew and D. Martin (eds), *Methods in human geography*. Harlow: Longman, 110–26.
—— 1999: Doing household research: interviewing couples together and apart. *Area* **31**, 67–74.
Valuation Office, biannual: *Property market report*. London.
van de Berg, L. and Wenner, L. (eds) 1991: *Television criticism: approaches and applications*. London: Longman.
van Maanen, J. 1988: *Tales from the field: on writing ethnography*. Chicago: University of Chicago Press.
—— 1995: An end to innocence: the ethnography of ethnography. In J. van Maanen (ed.), *Representation in ethnography*, Thousand Oaks, CA: Sage, 1–35.
van Meter, K. M. 1994: Sociological methodology. *International Social Science Journal* 139, 15–25.
Vaus, D. A. de 1991: *Surveys in social research*. 3rd edition. London: UCL Press.
Vidich, A. J. and Bensman, J. 1968: *Small town in mass society*. 2nd edition. Princeton: Princeton University Press.

—— and Shapiro, G. 1955: A comparison of participant observation and survey data. *American Sociological Review* **20**, 28–33.

Wagstaffe, M. and Moyser, G. 1987: The threatened elite: studying leaders in an urban community. In G. Moyser and M. Wagstaffe (eds), *Research methods for elite studies*. London: Allen & Unwin, 183–201.

Walford, G. 1994a: A new focus on the powerful. In G. Walford (ed.), *Researching the powerful in education*. London: UCL Press, 2–11.

—— (ed.) 1994b: *Researching the powerful in education*. London: UCL Press.

Walford, N. 1995: *Geographical data analysis*. Chichester: Wiley.

Walker, M. 1990a: The census inquisition. *Guardian* 7 Apr., 23.

—— 1990b: US historians say records falsified. *Guardian* 17 Apr., 6.

—— Walker, M. 1990c: US cities refuse to go down for count. *Guardian* 26 Sept., 9.

Walker, R. and Lewis, R. 1998: Media convergence and social research: the Hathaway project. In J. Prosser (ed.), *Image-based research*. London: Falmer, 162–75.

Walker, R. A. 1989: Geography from the left. In G. L. Gaile and C. J. Willmott (eds), *Geography in America*. Columbus, OH: Merrill, 619–50.

Wallace, M., Charlton, J. and Denham, C. 1995: The new OPCS area classifications. *Population Trends* **79**, 15–30.

Wallis, R. 1977: The moral career of a research project. In C. Bell and H. E. Newby (eds), *Doing sociological research*. London: Allen & Unwin, 149–69.

Walsh, D. 1998: Doing ethnography. In C. Seale (ed.), *Researching society and culture*. London: Sage, 217–32.

Walton, J. 1992: Making the theoretical case. In C. C. Ragin and H. S. Becker (eds), *What is a case?* Cambridge: Cambridge University Press, 121–37.

Ward, A. 1996: *Our bodies are scattered: the Cawnpore massacres and the Indian Mutiny of 1857*. London: John Murray.

Ward, L. 1998: Quality of life gets a higher profile. *Guardian* 24 Nov., 2.

Ward, S. V. 1988: *The geography of interwar Britain*. London: Routledge.

Warren, J. F. 1986: *Rickshaw coolie: a people's history of Singapore 1880–1940*. Oxford: Oxford University Press.

Wasserman, S. and Faust, K. 1994: *Social network analysis*. Cambridge: Cambridge University Press.

Waterton, C. and Wynne, B. 1999: Can focus groups access community views? In R. Barbour and J. Kitzinger (eds), *Developing focus group research*. London: Sage, 127–43.

Webb, E. J., Campbell, D. T., Schwartz, R. D. and Sechrest, L. 1966/2000: *Unobtrusive measures*. Sage 'classic' edition. Chicago: Rand McNally.

Webber, R. J. 1978: Which residential neighbourhoods are alike? *Population Trends* **11**, 21–6.

Weber, M. 1949: *The methodology of the social sciences*. Glencoe, ILL: Free Press.

Weber, R. 1985: *Basic content analysis*. Beverly Hills, CA: Sage.

Webster, B. and Stewart, J. 1974: The area analysis of resources. *Policy and Politics* **3**(1), 5–16.

Weneras, C. and Wold, A. 1997: Nepotism and sexism in peer review. *Nature* **387**, 341–43.

West, J. 1983: *Town records*. Chichester: Phillimore.

Western, J. 1986: Places, authorship, authority. In L. Guelke (ed.), *Geography and humanistic knowledge*, Waterloo: University of Waterloo Department of Geography Publication No. 25, 23–37.

—— 1992: *A passage to England: Barbadian Londoners speak of home*. Minneapolis: University of Minnesota Press.

Westwood, S. and Williams, J. (eds) 1997: *Imagining cities*. London: Routledge.

White, P. E. 1981: Migration at the micro-scale: intra-parochial movement in rural Normandy, 1946–1954. *Transactions of the Institute of British Geographers* **6**, 451–70.

White, R. K. 1949: Hitler, Roosevelt and the nature of war propaganda. *Journal of Abnormal and Social Psychology* **44**, 157–74.

Whitehead, P. 1983: Intra-urban spatial variations in local government service provision. *Environment and Planning C: Government and Policy* **1**, 229–47.

WHO, annual: *World health statistics annual*. Geneva: World Health Organization.

Whyte, I. D. and Whyte, K. A. 1981: *Sources for Scottish historical geography*. London: Institute of British Geographers Historical Geography Research Series No. 6.

Whyte, W. F. 1943: *Street corner society: the social order of an Italian slum*. Chicago: University of Chicago Press.

—— 1984: *Learning from the field*. Beverly Hills, CA: Sage.

—— (ed.) 1991: *Participatory action research*. Newbury Park, CA: Sage.

—— 1994: *Participant observer: an autobiography*. Ithaca, NY: ILR Press.

Widdowfield, R. 1999: The limitations of official homelessness statistics. In D. F. L. Dorling and S. Simpson (eds), *Statistics in society*. London: Arnold, 181–8.

Wieviorka, M. 1992: Case studies: history or sociology? In C. C. Ragin and H. S. Becker (eds), *What is a case?* Cambridge: Cambridge University Press, 159–72.

Wilkinson, T. 1981: *Down and out*. London: Quartet.

Williams, A. M., Shaw, G. and Greenwood, J. 1989: From tourist to tourism entrepreneur, from consumption to production: evidence from Cornwall, England. *Environment and Planning* **A21**, 1639–53.

Williams, R. 1966: *Communications*. Harmondsworth: Penguin.

Williams, W. M. 1956: *The sociology of an English village: Gosforth*. London: Routledge & Kegan Paul.

Willis, P. E. 1977: *Learning to labour: how working class kids get working class jobs*. Farnborough: Saxon House.

Who's who (annual). London: Adam & Charles Black.

Who's who in America (annual). Chicago: Marquis.

Wilson, K. 1992: Thinking about the ethics of fieldwork. In S. Deveraux and J. Hoddinott (eds), *Fieldwork in developing countries*. Hemel Hempstead: Harvester Wheatsheaf, 179–99.

Winchester, H. P. M. 1992: The construction and deconstruction of women's roles in the urban landscape. In K. Anderson and F. Gale (eds), *Inventing places*. Melbourne: Longman Cheshire, 139–56.

—— 1996: Ethical issues in interviewing as a research method in human geography. *Australian Geographer* **2**, 117–31.

Wolcott, H. F. 1994: *Transforming qualitative data*. Thousand Oaks, CA: Sage.

Women and Geography Study Group 1997: *Feminist geographies*. Harlow: Longman.

Woodeson, A. 1993: 'Going back to the land': rhetoric and reality in Women's Land Army memories. *Oral History* **21** (2), 65–71.

Woods, M. 1997: Discourses of power and rurality: local politics in Somerset in the 20th century. *Political Geography* **16**, 453–78.

Woolf, S. J. 1984: Towards the history of the origins of statistics. In J. C. Perrot and S. J. Woolf, *State and statistics in France 1789–1815*. London: Harwood, 81–194.

Woolgar, S. 1988: *Science, the very idea*. Chichester: Tavistock.

World Bank, annual: *World development report*. Washington.

World Resources Institute, biannual: *World resources*. London: World Resources Institute [with UN Development Programme, UN Environment Programme and the World Bank].

Worsley, P. 1980: One world of three? A critique of world systems theory. In R. Miliband and J. Saville (eds), *The socialist register 1980*. London: Merlin, 298–338.

Wright, P. 1987: *Spy catcher*. New York: Viking Penguin.

WTO, annual: *Yearbook of tourism statistics*. Madrid: World Tourism Organization.

Yarwood, R. and Gardner, G. 2000: Fear of crime, cultural threat and the countryside. *Area* **32**, 403–11.

Yeatman, A. 1994: *Postmodern revisionings of the political*. London: Routledge.

Yeung, H. W.-C. 1997: Critical realism and realist research in human geography. *Progress in Human Geography* **21**, 51–74.

Yin, R. K. 1984: *Case study research*. Beverly Hills, CA: Sage.

Young, I. M. 1990: *Justice and the politics of difference*. Princeton: Princeton University Press.

Young, K. and Kramer, J. 1978: Local exclusionary policies in Britain. In K. R. Cox (ed.), *Urbanization and conflict in market societies*. London: Methuen, 229–51.

Young, M. and Willmott, P. 1957: *Family and kinship in East London*. London: Routledge & Kegan Paul.

Zarkovich, S. S. 1965: *Sampling methods and censuses*. Rome: Food and Agriculture Organization of the United Nations.

Zeigler, D. O., Brunn, S. D. and Johnson, J. H. 1996: Focusing on Hurricane Andrew through the eyes of the victims. *Area* **28**, 124–9.

Zuckerman, H. 1972: Interviewing an ultra-elite. *Public Opinion Quarterly* **36**, 159–75.

Index